FOUNDATION AND GROWTH OF MACROMOLECULAR SCIENCE

Advances in Research for Sustainable Development

FOUNDATION AND GROWTH OF MACROMOLECULAR SCIENCE

Advances in Research for Sustainable Development

Edited by

Meegle S. Mathew, PhD
Józef T. Haponiuk, PhD
Sabu Thomas, PhD, DSc

AAP | APPLE ACADEMIC PRESS

First edition published 2024

Apple Academic Press Inc.
1265 Goldenrod Circle, NE,
Palm Bay, FL 32905 USA

760 Laurentian Drive, Unit 19,
Burlington, ON L7N 0A4, CANADA

CRC Press
2385 NW Executive Center Drive,
Suite 320, Boca Raton FL 33431

4 Park Square, Milton Park,
Abingdon, Oxon, OX14 4RN UK

Library and Archives Canada Cataloguing in Publication

Title: Foundation and growth of macromolecular science : advances in research for sustainable development / edited by Meegle S. Mathew, PhD, Józef T. Haponiuk, PhD, Sabu Thomas, PhD, DSc.
Names: Mathew, Meegle S., editor. | Haponiuk, Józef T., editor. | Thomas, Sabu, editor.
Description: First edition. | Includes bibliographical references and index.
Identifiers: Canadiana (print) 20230506461 | Canadiana (ebook) 2023050647X | ISBN 9781774913673 (hardcover) | ISBN 9781774913680 (softcover) | ISBN 9781003370505 (ebook)
Subjects: LCSH: Polymers. | LCSH: Polymeric composites.
Classification: LCC TA455.P58 F68 2024 | DDC 620.1/92—dc23

Library of Congress Cataloging-in-Publication Data

Names: Mathew, Meegle S., editor. | Haponiuk, Józef T., editor. | Thomas, Sabu, editor.

Title: Foundation and growth of macromolecular science : advances in research for sustainable development / edited by Meegle S. Mathew, Józef T. Haponiuk, Sabu Thomas.

Description: First edition. | Palm Bay, FL : Apple Academic Press, 2024. | Includes bibliographical references and index. | Summary: "This new volume explores the foundation of macromolecular science along with the field's current challenges, broad applications in many areas, and solutions to challenges. Foundation and Growth of Macromolecular Science: Advances in Research for Sustainable Development discusses many new applications that have emerged recently based on different polymers and their composites, some on the application of polymer composite in electromagnetic interference (EMI) shielding, sensing, and wound healing. This book discusses innovations in the synthesis and characterization of many different classes of polymers and provides in-depth explanations of the behavior of polymers in solutions and in bulk. It considers the synthesis, characterization properties, and applications of many polymer composites and nanomaterials. The diverse topics cover zein-based composites, nanocellulose extracted from nutshells as a potential filler for polymers, polymers to strengthen concrete, polymer/metal nanocomposite magnetic materials, synthetic polymers in aquatic environments, pharmaceutical 3D-printing, alginate-based wound healing dressings, and more. Many chapters in the volume place emphasis on sustainable development and green eco-friendly approaches. This volume will help researchers to understand the significance of macromolecular science and research advances for sustainable development for the benefit of humanity. It emphasizes the many roles of macromolecular science in the important fields of science, such as chemistry, physics, engineering, biology, pharmaceutical science, and more"-- Provided by publisher.

Identifiers: LCCN 2023035722 (print) | LCCN 2023035723 (ebook) | ISBN 9781774913673 (hardback) | ISBN 9781774913680 (paperback) | ISBN 9781003370505 (ebook)

Subjects: MESH: Polymers | Biocompatible Materials | Sustainable Development

Classification: LCC QP801.P64 (print) | LCC QP801.P64 (ebook) | NLM QT 37.5.P7 | DDC 572/.33--dc23 eng/20230922

LC record available at https://lccn.loc.gov/2023035722

LC ebook record available at https://lccn.loc.gov/2023035723

ISBN: 978-1-77491-367-3 (hbk)
ISBN: 978-1-77491-368-0 (pbk)
ISBN: 978-1-00337-050-5 (ebk)

About the Editors

Meegle S. Mathew, PhD
Assistant Professor, Postgraduate and Research Department of Chemistry, Mar Athanasius College, Kothamangalam, India

Meegle S. Mathew, PhD, is currently working as Assistant Professor in the Postgraduate and Research Department of Chemistry at Mar Athanasius College, Kothamangalam, India. Dr. Mathew received her PhD from the Indian Institute of Space Science and Technology, Thiruvananthapuram, in 2019, in the area of Nanomaterials for Biomedical applications. Following her PhD program, she was awarded a UGC-DR DS Kothari Postdoctoral Fellowship for the topic of photocatalytic activity of metal nanoclusters grafted TiO2. Dr. Mathew worked as a postdoctoral fellow for three years at the School of Energy Materials, Mahatma Gandhi University, Kottayam. During her research career, she has published several research articles in high-impact journals, been granted an Indian patent, edited two books, and published several book chapters. She has been working in a multidisciplinary area that combines nanoscience with biomedicine and energy conversion systems. Her research interest areas are nano-biophotonics, optical imaging, the development of biosensors and biomedical devices using fluorescent nanoparticles, polymer nanocomposites for biomedical applications, photocatalysis, and photoreduction.

Józef T. Haponiuk, PhD
Head, Polymers Technology Department, Faculty of Chemistry, Gdańsk University of Technology, Poland

Professor Józef T. Haponiuk, PhD, is the Head of the Polymers Technology Department at the Faculty of Chemistry of the Gdańsk University of Technology, Poland, since 2006. His special scientific interests include polyurethane chemistry and processing, polymer and rubber recycling, biopolymers, the use of biomass-derived raw materials in polymer technology, polymer composites and nanocomposites. Professor Haponiuk is the author or coauthor of over 300 original scientific papers, including 90 published in periodicals from the JCR list, editor and coauthor of monographs in international publications (Elsevier, Apple Academic Press, and Royal Society of Chemistry), and creator of 11 granted patents and 25 patent applications. Professor Haponiuk is a valued academic teacher in the field of polymers chemistry and engineering. He has been a supervisor for 12 PhD theses.

Sabu Thomas, PhD, DSc
Chairman, Trivandrum Engineering Science and Technology Research Park (TrEST Research Park); Former Vice Chancellor, Mahatma Gandhi University, Kottayam; Director, School of Energy Materials; Founder, Director and Professor, International and Inter University Centre for Nanoscience and Nanotechnology, Kottayam, Kerala, India

Sabu Thomas, PhD, is currently the Chairman of Trivandrum Engineering Science and Technology Research Park (TrEST Research Park), Kerala, India. He is also the former Vice Chancellor at Mahatma Gandhi University, Kottayam, as well as Director of the School of Energy Materials; and Founder, Director, and Professor at the International and Inter University Centre for Nanoscience and Nanotechnology, Kottayam. He is also a full-time Professor of Polymer Science and Engineering at the School of Chemical Sciences, Mahatma Gandhi University, Kottayam, Kerala, India. Prof. Thomas is an outstanding leader with sustained international acclaim for his work in nanoscience, polymer science and engineering, polymer nanocomposites, elastomers, polymer blends, interpenetrating polymer networks, polymer membranes, green composites, and nanocomposites, nanomedicine and green nanotechnology. Prof. Thomas has been conferred with an Honoris Causa DSc from the University of South Brittany, France, and University of Lorraine, France. Recently, he was awarded a Foreign Fellow of the European Academy of Sciences (EurASc) and was cited in the top 2% of scientists in India by Stanford University and was awarded "Honoured University of Professorship" by Siberian Federal University, Russia. He has received many national and international awards and has published over 1,000 peer-reviewed research papers, reviews, and book chapters. He has co-edited 100 books and is the inventor of more than five patents (granted-4, filed-11).

Contents

Contributors

Inci Albayrak
Department of Mathematical Engineering, Yıldız Technical University, Istanbul, Turkey

Saritha Appukuttan
Department of Chemistry, School of Arts and Sciences, Amrita Vishwa Vidyapeetham, Amritapuri, Kollam, Kerala, India

M. Banerjee
Nano Science and Technology Lab., School of Physics, Devi Ahilya University, Indore, Madhya Pradesh, India

Natrajan Chandrasekaran
Centre for Nanobiotechnology, Vellore Institute of Technology (VIT), Vellore, Tamil Nadu, India

Aakanksha Choudhary
Nano Science and Technology Lab., School of Physics, Devi Ahilya University, Indore, Madhya Pradesh, India

Haritha R. Das
School of Energy Materials, Mahatma Gandhi University, Kottayam

Ashok Garai
Department of Physics, The LNM Institute of Information Technology, Jaipur, Rajasthan, India

Mainak Ghosal
Coal Ash Institute of India, Dum Dum, Kolkata, India
School of Advanced Materials, Green Energy & Sensor Systems, Indian Institute of Engineering Science & Technology, Shibpur, Howrah, India

Józef T. Haponiuk
Department of Polymer Technology, Chemical Faculty, Gdansk University of Technology, Poland

Mert Akin Insel
Department of Chemical Engineering, Yıldız Technical University, Istanbul, Turkey

Tomy Muringayil Joseph
Department of Polymer Technology, Chemical Faculty, Gdansk University of Technology, Poland

Aruna Joshi
Nano Science and Technology Lab., School of Physics, Devi Ahilya University, Indore, Madhya Pradesh, India

G. Jayanthi Kalaivani
Department of Chemistry, Lady Doak College, Madurai, Tamil Nadu, India

Selcan Karakuş
Department of Chemistry, Faculty of Engineering, Istanbul University-Cerrahpasa, Istanbul, Turkey

Prachi Khamkar
Pharmaceutical Manufacturing Operations, CiREE Edutech, Pune, Maharashtra, India

Sarathi Kundu
Soft Nano Laboratory, Physical Sciences Division, Institute of Advanced Studyin Science and Technology, Paschim Boragaon, Garchuk, Guwahati, Assam, India

B. M. Madhu
V.T.U. Research Centre, Department of Electrical and Electronics Engineering, Siddaganga Institute of Technology, Tumakuru, Karnataka, India

Debarshi Kar Mahapatra
Department of Pharmaceutical Chemistry, Dadasaheb Balpande College of Pharmacy, Nagpur, Maharashtra, India

Meegle S. Mathew
School of Energy Materials, Mahatma Gandhi University, Kottayam

Siby Mathew
Department of Physics, Sacred Heart College, Thevara, Kochi, Kerala, India

S. Mathiya
Department of Chemistry, Lady Doak College, Madurai, Tamil Nadu, India

Amitava Mukherjee
Centre for Nanobiotechnology, Vellore Institute of Technology (VIT), Vellore, Tamil Nadu, India

G. S. Mukherjee
Defence Research & Development Organization, DRDO Metcalfe House DESIDOC Complex, Delhi, India

Swati Nagar
Nano Science and Technology Lab., School of Physics, Devi Ahilya University, Indore, Madhya Pradesh, India

Mereena Luke Pallikkunnel
Department of Polymer Technology, Chemical Faculty, Gdansk University of Technology, Poland

Rashmi
V.T.U. Research Centre, Department of Electrical and Electronics Engineering, Siddaganga Institute of Technology, Tumakuru, Karnataka, India

Preeti Sachdev
Nano Science and Technology Lab., School of Physics, Devi Ahilya University, Indore, Madhya Pradesh, India

R. R. N. Sailaja
The Energy and Resources Institute, Southern Regional Centre, Bangalore, Karnataka, India

Richa Shree
Centre for Nanobiotechnology, Vellore Institute of Technology (VIT), Vellore, Tamil Nadu, India

Virpal Singh
Department of Chemistry, M J P Rohilkhand University, Bareilly, Uttar Pradesh, India

Ranimol Stephen
Department of Chemistry, St. Joseph's College (Autonomous), Devagiri, Calicut, Kerala, India

S. K. Suja
Department of Chemistry, Lady Doak College, Madurai, Tamil Nadu, India

Rajan J. Sundara
V.T.U. Research Centre, Department of Electrical and Electronics Engineering, Siddaganga Institute of Technology, Tumakuru, Karnataka, India

Hrishikesh Talukdar
Department of Physics, Anandaram Dhekial Phookan College, Nagaon, Assam, India

Nevin Taşaltin
Department of Renewable Energy Tech., Maltepe University, Istanbul, Turkey
Department of Electrical & Electronics Eng., Maltepe University, Istanbul, Turkey

Jija Thomas
Department of Chemistry, St Mary's College, Sulthan Bathery, Wayanad, Kerala, India

John Thomas
Centre for Nanobiotechnology, Vellore Institute of Technology (VIT), Vellore, Tamil Nadu, India

Sabu Thomas
School of Energy Materials, Mahatma Gandhi University, Kottayam
International and Inter University Centre for Nanoscience and Nanotechnology, Mahatma Gandhi University, Kottayam, Kerala, India

Kiran John U.
Department of Physics, Sacred Heart College, Thevara, Kochi, Kerala, India

Jyothy G. Vijayan
Department of Chemistry, M. S. Ramaiah University of Applied Sciences, Bengaluru, Karnataka, India

Abbreviations

AAOx	aryl-alcohol oxidase
Abs	absorbance
AChE	acetylcholinesterase
AFM	atomic force microscopy
Ag-NPs	sliver nanoparticles
Ag-PS	silver polystyrene particles
AMIMC1	1-allyl-3-methylimidazolium chloride
APS	ammonium persulfate
ATH	aluminium trihydrate
Bcc	body-centered cubic
BE	binding energy
BET	Brunauer–Emmett–Teller
BMIMCl	1-butyl-3-methylimidazolium chloride
BNC	bacterial nanocellulose
BPE	bio-polyelectrolyte
BSA	bovine serum albumin
CAT	catalase
CD	circular dichroism
CDH	cellobiose dehydrogenase
CF	cellulose microfibrils
CMC	carboxy methylcellulose
CN	cellulose nano crystals
CNC	cellulose nanocrystals
CNF	cellulose nanofibrils
CNS	cashew nut shell starch
CNT	carbon nanotubes
CNT	carbon nanotube
CZT	cadmium zinc telluride
DCFH-DA	dichloro-dihydro-fluorescein diacetate
DLS	dynamic light scattering
d-NaPSS	deuterium-labeled NaPSS
DSV	dilute solution viscometer
EC	enzyme commission
EDX	energy dispersive X-ray

EMI	electromagnetic interference
FAD	flavin adenine dinucleotide
FE-SEM	field emission scanning electron microscopy
FT-IR	Fourier transform infrared spectroscopy
FWHM	full width half maximum
GDH	glucose dehydrogenase
GFRP	glass fiber reinforced polymer
GIXRD	grazing incidence X-ray diffraction
GMC	glucose-methanol-choline
GNC	glass-reinforced nanocomposite
GNPs	graphene nanoplatelets
GO	graphene oxide
GOx	glucose oxidase
HAS	human serum albumin
Hbg	hemoglobin
Hcp	hexagonal close packed
IS	impedance spectroscopy
IUB	International Union of Biochemistry
IUBMB	International Union of Biochemistry and Molecular Biology
LBL	layer-by-layer
LC50	lethal concentration 50
LG	locust gum
MALDI TOFF	matrix-assisted laser desorption/ionization time of flight
MB	methylene blue
MCC	microcrystalline cellulose
MFC	microfibrillated cellulose
MO	methyl orange
MOKE	Magneto Optical Kerr effect
MOx	methanol oxidase
MPA	mercaptopropionic acid
MPs/MP	microplastics
MWCNTs	multi-walled carbon nanotubes
MWS	Maxwell–Wagner–Sillars
NAD	nicotinamide adenine dinucleotide
NADP	nicotinamide adenine dinucleotide phosphate
NaPSS	poly(Na-4 styrene sulfonate)
NCC	nanocrystalline cellulose
NFC	nanofibrillated cellulose

NMR	nuclear magnetic resonance spectroscopy
NP	nanoparticles
OD	optical density
Oe	oersted
OTR	oxygen transfer rate
OV	ovalbumin
PAA	poly(acrylic acid)
PAH	poly(allylamine hydrochloride)
PAH	polycyclic aromatic hydrocarbons
PCBs	polychlorinated biphenyls
PCL	polycaprolactone
PDADMAC	poly(diallyldimethylammonium chloride)
PDH	pyranose dehydrogenase
PE	polyelectrolyte
PEG	polyethylene glycol
PEM	polyelectrolyte multilayer
PHB	poly(3-hydroxybutyrate)
PLA	polylactic acid
PMMA	poly(methacrylic acid)
PNC	polymer nanocomposite
Pox	pyranose 2-oxidase
PPC	protein-polyelectrolyte complex
PTFE	polytetrafluoroethylene
PU	polyurethaner
PVA	polyvinyl alcohol
PVDF	polyvinylidene fluoride
PVP	polyvinylpyrrolidone
QCM	quartz crystal microbalance
QD	quantum dot
RhB	rhodamine B
ROS	reactive oxygen species
SANS	small-angle neutron scattering
SAX	small-angle X-ray scattering
SBR	styrene-butadiene rubber
SEM	scanning electron microscopy
SLD	scattering length densities
SLS	static light scattering
SPEEK	sulfonated poly(ether ether ketone)
STMP	sodium trimetaphosphate

TEM	transmission electron microscopy
TEMPO	2,2,6,6- tetramethylpiperidine-1-oxyl
TGA	thermogravimetric analysis
UV-Vis	ultraviolet-visible light
VAO	vanillylalcohol oxidase
WAX	wide-angle X-ray scattering
WNC	walnut shell cellulose
XPS	X-ray photoelectron spectroscopy
XRD	X-ray diffraction
XRR	X-ray reflectivity

Symbols

AMP	amplitude of sonication
C	concentration
ci	concentration of the ith charge species
e	electronic charge
h	planks constant
Hc	coercivity
Hc‖	coercivity parallel to the film surface
Hs	saturation magnetization
kB	Boltzmann Constant
lB	Bjerrum length
li	separation between two neighboring charged ions
M	molarity
Mr/Ms	squareness of the loops
s	power-law exponent
N	normalized
T	temperature
T	absolute temperature
ts	sonication time
TG	Glass transition temperature
z	complex impedance
zi	valency of the ith charge species
z′	real part of impedance
z″	imaginary part of impedance
η	viscosity
ε	permittivity of solvent medium
ε0	permittivity of vacuum
ξD	Debye screening length
α	absorption coefficient
υ	frequency of incident photon
ε′	real part of dielectric constant
ε″	imaginary part of dielectric constant
ε*	complex dielectric constant
σAC	AC conductivity
ε0	permittivity of free space

$\tan\delta$	loss tangent
σT	total conductivity
$\sigma 0$	DC conductivity
θ	phase angle.
τ	relaxation time
ω	angular frequency

Preface

In the year 2020, macromolecular science reached the milestone of its 100th anniversary. According to Prof. Hermann Staudinger, during this period of 100 years, polymers have become a significant part of human lives. The importance of macromolecular science cannot, thus, be undermined. This book discusses the foundation of macromolecular science, its current challenges, its application in various fields, and future solutions. It looks at the present research scenario in the field of polymers and polymer composites. It also discusses the new applications of different polymers and its composites that have emerged recently and examining the application of polymer composites in EMI shielding, sensing, wound healing, etc.

The main goal of this book is to help researchers understand the significance of macromolecular science and research advances for sustainable development, for the benefit of humanity, to emphasize that macromolecular science is widely used in important fields of science such as chemistry, physics, engineering, and biology, and to stress on the never-ending nature of scientific research and discoveries. We thus believe that journeying through the various chapters will grant readers a thorough understanding of the status of research and development in the area of polymers and polymer composites.

CHAPTER 1

Introduction – Foundation and Growth of Macromolecular Science: Advances in Research for Sustainable Development

HARITHA R DAS[1], MEEGLE S MATHEW[1], SARITHA APPUKUTTAN[2], and SABU THOMAS[1]

[1]*School of Energy Materials, Mahatma Gandhi University, Kottayam*

[2]*Assistant Professor, Postgraduate and Research Department of Chemistry, Mar Athanasius College, Kothamangalam*

ABSTRACT

The success of polymers in consumer products lies in their capability and flexibility in providing tailor-made products for specific functions. This uniqueness of polymers will also help in solving the question regarding the sustainability of polymers in future. Of course, the challenges concerning sustainability needs to be addressed, and it is the responsibility of the research community to carry out research for empowering advanced applications and annulling the challenges in this field by carefully analyzing the past lessons. The future of polymer science is thoroughly dependent on the scientific advances and the willingness of the society.

1.1 INTRODUCTION

Research in polymer science has traversed across various dimensions and has amalgamated various emerging sciences, and this interdisciplinary

Foundation and Growth of Macromolecular Science: Advances in Research for Sustainable Development. Meegle S. Mathew, PhD, Józef T. Haponiuk, PhD & Sabu Thomas, PhD, DSc (Eds.)

nature of polymer science has paved the way toward the creation of novel materials called advanced functional polymers with diverse applications. Polymers are materials made up of long chains of repeating monomer units. Depending upon the type of molecules bonded and how they are bonded, the materials have distinct properties. They can be both natural and synthetic. Polymer nanocomposites can be defined as a mixture of two or more materials, where the matrix is a polymer and the dispersed phase has at least one dimension in the nanometer range. With the advent of nanotechnology, nanofillers are increasingly utilized in polymers. Nanoparticles are defined as particles with at least one diameter in the nanometer range. Their optical, electrical, and magnetic properties are all directly proportional to their size. In the nanoscale, as the particle size decreases, the ratio of surface area/volume increases, so that the surface properties become crucial.[1] The addition of varying concentrations of these nanofillers into the polymer can lead to improvements in their mechanical, thermal, barrier, and flammability properties, without affecting their ability to process.

The mechanical and thermal properties are strongly related to the morphologies obtained; therefore, evaluating the nanofiller dispersion in the polymer matrix is crucial. The nanocomposites thus obtained will have maximum reinforcement due to the large surface area of contact between the matrix and nanoparticles. This is one of the main differences between nanocomposites and the conventional composites.[2]

Different types of nanoparticle fillers can be used to tune the characteristics of a polymer. Various properties of the polymer matrix can be modified depending on the particle size, shape, specific surface area, and chemical nature, including electrical and thermal conductivity, polymer phase behavior, and thermal stability; mechanical properties such as stiffness, Young's modulus, wear, fatigue, flame retardancy, and density; and physical properties such as magnetic, optic, or dielectric properties.[1] Because high degrees of stiffness and strength are achieved with significantly less high-density material, polymer nanocomposites are lighter than conventional composites.[3]

Among the most often used functional nanofillers are magnetic nanoparticles. Magnetic nanoparticles are placed in a nonmagnetic or magnetic matrix in magnetic nanocomposites. Because of the high-surface area-to-volume ratio of the nanosized particles, these nanoparticles distributed in composites have a significant tendency to form agglomerates for energy reduction. Protection techniques have been devised to chemically stabilize magnetic nanoparticles by grafting or coating with organic species like surfactants or polymers, or covering with an inorganic layer, such as silica or carbon, to avoid aggregation. Magnetic polymer nanocomposite is formed when these functionalized magnetic nanoparticles are incorporated into polymer matrices.[4] Many factors influence the magnetic characteristics of the resultant nanocomposites, including nanoparticle dispersion/aggregation, interparticle interaction strength, and the effect of surface on nanoparticle magnetism.[5]or by generating the nanoparticles during the very synthesis of the embedding matrix. Two typical examples of these production methods are polymer nanocomposites and ceramic nanocomposites. The resulting magnetic properties turn out to be markedly different in these two classes of nanomaterials. The control of nanoparticle size, distribution, and aggregation degree is easier in polymer nanocomposites, where the interparticle interactions can either be minimized or exploited to create magnetic mesostructures characterized by anisotropic magnetic properties; the ensuing applications of polymer nanocomposites as sensors and in devices for Information and Communication Technologies (ICT magnetic nanocomposites are employed in a variety of medical applications, including cancer therapy, MRI diagnosis, and drug targeting.[6] Magnetic nanocomposites can also be used to remove heavy-metal contaminants, poisonous dyes, oils, and effluents from waste water.[7]

Cellulose is one of the most prevalent renewable polymers found in nature. When these cellulose chains are grouped together, highly organized regions are formed that can then be isolated as nanoparticles, referred to as cellulose nanomaterials or nanocelluloses. Nanocellulose possesses many of the same desired characteristics as cellulose, such as low density, nontoxicity, and high biodegradability. Because of its unique shape, size, surface chemistry, and high degree of crystallinity, they also have unique qualities such as high mechanical strength, reinforcing capabilities, and tunable self-assembly in aqueous conditions.[8] Nanocellulose can be modified through chemical, biological, and mechanical processes, resulting in improved qualities such as flame retardancy, transparency, and high flexibility. Cellulose nanocrystals (CNCs), nanofibrillated cellulose (NFC), and rigid bacterial nanocellulose (BNC) are some of the cellulose-based materials whose chemical and physical properties vary depending on their source and extraction process. The ability to process and function of the material to be produced are influenced by the surface chemistry. Nanocellulosic materials are used in a variety of interdisciplinary fields. They can be woven together to create highly porous and mechanically robust materials including nanopapers, nanocellulose films, aerogels, and hydrogels, which have applications in the paper industry, biomedicine, environmental remediation, optoelectronics, and engineering. Nanocellulosic materials are utilized to reinforce bionanocomposites because they generate strong networks with high tensile stiffness and strength. Great deal of interest has been directed to the development and implementation of polymer composite materials reinforced by stiff nanocellulosic particles derived from renewable sources.[9] The use of nanocellulose as a filler in polymer matrices has increased significantly in recent years. The possibility for utilizing the high rigidity of cellulose crystals is a primary reason for this. The mechanical performance of the resultant nanocomposites has been shown to be improved by the nanocellulose.[10]

Biopolymers are polymers which are derived from biological sources such as microorganisms, plants, or trees. Materials produced by synthetic chemistry from biological sources such as vegetable oils, sugars, fats, resins, proteins, amino acids, and so on can also be described as biopolymer. Biopolymers have complex molecular assemblies that adopt precise and defined 3D shapes and structures. Alternatives to fossil-fuel-based polymers have been developed using biodegradable biopolymers made from renewable resources. They are usually made from starch, sugar, natural fibers, or other organic biodegradable components in a variety of compositions. The biopolymers are degraded by exposure to bacteria in soil, compost, or marine sediment. Furthermore, compared to traditional incineration, subjecting biodegradable biopolymers to waste disposal by leveraging their characteristic of being degradable by bacteria in the ground greatly reduces CO_2 emissions. As a result, the usage of biodegradable biopolymers is being highlighted as a way to combat global warming.[11] Currently, there is an increasing trend to replace synthetic materials with biopolymers like cellulose, lignin, chitin, etc., in accordance with the "go green" motto.

1.2 APPLICATIONS OF POLYMERS

The polymer community has been holding brainstorming discussions on the decisive role to be played by polymers in several crucial areas of application. 3D printing has advanced rapidly in recent years with many areas of research now being translated into engineered products, especially in medical fields. Polymer printing is advantageous in many areas that benefit from the wide range of polymer material characteristics and processing approaches. 3D printing is a popular fabrication method because it allows creation of designs with complicated geometries and architectures that are impossible to achieve with traditional manufacturing methods. Also, polymers do not degrade in the body and cause mechanical issues such as stress shielding. Polymer printing is possible using extrusion, resin, and powder 3D printing processes that provide versatility for material selection and supporting designs with diverse architectures, responses, and layouts.[12]

Polymer-based technologies are the most promising for boosting the effectiveness of sensors and biosensors. Molecular imprinted polymers (MIP), conducting polymers and their composites, hydrogels, and other polymeric materials are employed in sensor devices. Polymer-based materials are frequently used in these types of sensors because they improve target molecule recognition, act as supports for functionalities immobilization (for instance, dyes, fluorophores, metal nanoparticles), and allow the detection of target analytes by changing their physical or chemical properties. Polymer-based sensors are of great advantage due to the possibility of modifying their chemical properties, tunability, biocompatibility, flexibility, and resistance to degradation.[13]

1.3 CONCLUSION AND FUTURE PERSPECTIVES

Smart polymeric materials that display rapid and on-demand biodegradation, programmability, reshapeability, adaptivity, and self-healing should be the focus of new developments. Stimuli-responsive as well as shape memory polymers are also gaining momentum in various applications. While most stimuli–responsive polymers investigated so far were intended to respond to one stimulus, it is also possible to create multiresponsive polymers that respond to different stimuli with more than one response. Apart from diagnostics, pharma, and biomaterials, other areas like optoelectronics, robotics, etc., would also be fitted into the application arena of polymers. Recent advances in 3D printing have caused a resurrection in rheological and molar mass analysis and will, undoubtedly, lead to the synthesis of novel polymers. Polymers and their composites are unavoidable candidates in space applications and find utility in almost all areas beginning from space craft, space station—thermal control coatings, adhesives, tapes, toughening/damping materials, seals, thermal insulations, etc. Thus, a future world without polymers is unconceivable not only because they have entered into the various walks of human life but also due to the vast job opportunities as well as economic impacts on the society.

1.4 OVERVIEW OF THE CHAPTERS

This introductory chapter provides a glimpse on polymers along with a summary of the various chapters that have been included in this book together. This present research scenario in the field of polymer, and polymer nanocomposite is discussed in this book. The initial sections of the book deal mostly with the various polymers, polymer nanocomposite, followed by application of polymer composite in EMI shielding, sensing, wound healing, etc.

The book also discusses about DNA Nanomechanics, its theory and applications. This chapter will give an insight into DNA double helix and its mechanical properties at the microscopic level. Further different theoretical and computational methods along with experimental techniques are discussed to determine these properties. The structural transitions of DNA under force, ionic effects, and protein binding effects on elastic properties of DNA will also be reviewed.

There is a chapter that also discusses the structure and the properties of protein–polyelectrolyte complexes. Another chapter discusses about the extraction of nanocellulose from nutshell and its applications in polymers as fillers.

Thus, we believe that the journey through the various chapters will grant the readers with a thorough understanding of the status of research and development in the area of polymers and polymer composite.

KEYWORDS

- **polymer**
- **polymer nanocomposite**
- **nanofillers**
- **biopoymers**
- **functional nanomaterials**

REFERENCES

1. Hanemann, T.; Szabó, D. V. Polymer-Nanoparticle Composites: From Synthesis to Modern Applications. *Materials* 2010, 3 (6), 3468–3517.

2. Dantas de Oliveira, A.; Augusto Gonçalves Beatrice, C. Polymer Nanocomposites with Different Types of Nanofiller. *Nanocomposites Recent Evol.* 2019. DOI: 10.5772/intechopen.81329.
3. Giannelis, E. P. Polymer-Layered Silicate Nanocomposites : Synthesis , Properties and Applications. *Appl. Organometal. Chem.* 1998, *12*, 675–680.
4. Kalia, S.; et al. Magnetic Polymer Nanocomposites for Environmental and Biomedical Applications. *Colloid Polym. Sci.* 2014, *292*, 2025–2052.
5. Barrera, G.; et al. Magnetic Properties of Nanocomposites. *Appl. Sci.* 2019, *9* (2), 212.
6. Świętek, M.; Tokarz, W.; Tarasiuk, J.; Wroński, S.; BŁazewicz, M. Magnetic Polymer Nanocomposite for Medical Application. *Acta Phys. Pol. A* 2014, *125*, 891–894.
7. Gu, J.; et al. Applied Surface Science Facile Removal of Oils from Water Surfaces Through Highly Hydrophobic and Magnetic Polymer Nanocomposites. *Appl. Surf. Sci.* 2014, *301*, 492–499.
8. Access, O. We are IntechOpen , the World ' s Leading Publisher of Open Access Books Built by Scientists , for Scientists TOP 1 % Preparation , Properties and Use Plant Materials.
9. Thomas, B.; et al. Nanocellulose , a Versatile Green Platform : From Biosources to Materials and Their Applications. *Chem. Rev.* 2018, *118* (24), 11575–11625. DOI: 10.1021/acs.chemrev.7b00627.
10. Bismarck, A. High Performance Cellulose Nanocomposites: Comparing the Reinforcing Ability of Bacterial Cellulose and Nano fi brillated Cellulose. *ACS Appl. Mater. Interfaces* 2012, *4* (8), 4078–4086.
11. Mohan, S.; Oluwafemi, O. S.; Kalarikkal, N.; Thomas, S.; Songca, S. P. Biopolymers – Application in Nanoscience and Nanotechnology. In Recent Advances in Biopolymers, 2016.
12. Arefin, A. M. E.; Khatri, N. R.; Kulkarni, N.; Egan, P. F. Polymer 3D Printing Review : Materials , Process , and Design Strategies for Medical Applications. *Polymers* 2021, *13* (9), 1499.
13. Alberti, G.; Zanoni, C.; Losi, V.; Magnaghi, L. R. Current Trends in Polymer Based Sensors. *Chemosensors* 2021, *9* (5), 108.

CHAPTER 2

Synthesis and Characterization of Chitosan–Starch and Glutaric Acid Films

VIRPAL SINGH

Department of Chemistry, M J P Rohilkhand University, Bareilly, Uttar Pradesh, India

ABSTRACT

The objective of this study is to investigate the synthesis of films of chitosan–starch–glutaric acid and its characterization. The films were synthesized using chitosan, starch, and crosslinked with and glutaric acid. The variation of concentration of chitosan-starch was 0.1/0.9 from 0.9/0.1. These films were characterized by percentage of swelling, FTIR. The swelling study was conducted in pH 2.2 and 7.4. The percentage of swelling was increased with increasing the concentration of chitosan and also increased with release time increasing. The percentage of swelling is more in basic medium at pH 7.4. The cryastalinity of films were increased by adding starch and also increased cryastalinity of films with crosslinking glutaric acid. FTIR study also supported that the films were crosslinked with different crosslinking agents.

2.1 INTRODCTION

Chitosan, (1, 4)-[2amino-2-deoxy-β-D-glucan] is a natural derivative of chitin, is obtained by its partial deacetylation. Chitosan has amine side group, which is responsible for its polycationic character, and formation of well-known intermolecular complexes with carboxylic acid and poly carboxylic acid. Chitosan is inert, hydrophilic, and biocompatible and biodegradable.[1]

Foundation and Growth of Macromolecular Science: Advances in Research for Sustainable Development. Meegle S. Mathew, PhD, Józef T. Haponiuk, PhD & Sabu Thomas, PhD, DSc (Eds.)

A number of researches have reported the preparation of membranes or films,[2] beads,[3,4] nanoparticles[5] for use in various fields such as metal–ion separation, gas separation, reverse osmosis, ultra filtration, drug release,[6,7] wound healing,[8] pervaporation, for affinity purification and packaging.[9,10] It possesses antimicrobial property and absorbs toxic metals like mercury, cadmium, lead, etc. In addition, it has good adhesion, coagulation ability, and immune-stimulating activity. The use of chitosan in pharmaceutical industry is still very limited because of its high cost, poor mechanical strength, fast dissolution in the stomach for oral administration, and limited capacity to controlled drug release.[11–13] Chitosan is insoluble in water, alkali, and organic solvents, but is soluble in acid because of the positive charge on C_2 of the glucosamine monomer at pH 6 or below. Some dilute inorganic acids, such as nitric acids, hydrochloric acids, and perchloric acids can also be used to prepare a chitosan solution.[14,15] The solution properties of chitosan depend not only on its average quantity of acetylation (DA) but also on the distribution of acetyl groups. The quantity of acid needed depends on the amount of chitosan to be dissolved. Chitosan is only soluble in acetic environment, and therefore unable of enhancing absorption in the small intestine.[16] There is a role of primary amino group, and the role of electrostatic attraction, hydrogen bonding, and hydrophobic effects going on aggregation of gastric mucin in the existence of chitosan. The theory behind the solubility of chitosan is that the amino groups on the chitosan molecule contain nitrogen atom, each with a single pair of electrons. These amino groups are the weak basic groups capable of taking up hydrogen ions, and consequently the chitosan molecule becomes a polycationic electrolyte. This is responsible for the dissolution of the chitosan molecule in an acidic environment.[17] Chitosan is a pseudoplastic material and an excellent viscosity-enhancing agent in acidic environments. The viscosity of chitosan solution increases with an increase in chitosan concentration and decreases with increase in temperature.[18] The film-forming qualities of chitosan found to be dependent on the homogeneity of the bulk material, the degree of acetylation, and acid degradation during dissolution. Hence, other biodegradable materials such as pectin, gaur gum, sodium alginate, and starch, etc., are used to reduce the cost. Starch can be used for making blends with other polymeric materials. Starch is water swellable excipient in nature. The main differences between starch and chitosan are the glucoside linkage: α (1, 4) for starch and β (1,4) for chitosan, and the hydroxyl group of the second carbon is replaced by the amine group. Commercialized starches are from cereals such as corn, wheat, rice, and from tubers or roots, potato, sweet potato, and tapioca starch.

Appearance and properties of the starch are varied on species. Starch, a polymer of α-D-glucose, composes of two forms of glucose polymers, linear (amylose), and branched chain (amylopectin). Starch occurs in many plant species in the form of spherical granules. Amylose is a linear homopolymer of α (1, 4)-linked glucose with a degree of polymerization of ~1000. Amylose makes up ~35% of starch (range of 11–36% depending on plant and organ), and amylopectin is highly branched form of "amylase," which has α(1, 6) glycosidic linkage.[19] It can form complex with fat or iodine because the core of the amylose helices is hydrophobic molecule. Amylopectin is large-sized polymer molecule of starch, which is the main component of the starch granules. It is highly branched and 4–5% of its glucose monomers contain a 1–6 linkage.[20] Keeping in view the above aspects, the present work aims to synthesis of chitosan–starch films using glutaric acid as crosslinking agent. These films are characterized by swelling and FTIR spectroscopy.

2.2 EXPERIMENTAL PART

2.2.1 *MATERIALS AND METHODS*

Chitosan of high molecular weight and having viscosity 800 cp is supplied by Sigma-Aldrich (Germany). Corn starch, procured from Himedia (India) and acetic acid (99.5%), is purchased from Merck (Germany). Glutaric acid is purchased from Loba Chemie (M.W = 132.11, purity 99%, and melting point 95–98 °C). Double distilled water is used for preparation of solutions.

2.2.2 *SAMPLE PREPARATION*

2.2.2.1 *PREPARATION OF GLUTARIC ACID CROSSLINKED CHITOSAN-STARCH FILMS*

For the preparation of films, the casting/solvent evaporation technique is employed. The solutions of chitosan/starch are prepared in a different weight ratios (0.1/0.1–0.9/0.1), separately. A known quantity of chitosan is weighed using a Citizen balance with a precision of 10^{-1} mg and dissolved in 2% acetic acid solution at room temperature (25 ± 2°C) with continuous stirring for 3 h, whereas, starch solution is prepared by dissolving it in distilled water at 95°C with constant stirring for 20 min. The resultant solution is cooled to room temperature (25 ± 2°C). Afterward, both the solutions are mixed together and kept for 24 h at room temperature in order to get a homogeneous solution. Chitosan-corn starch solution with constant stirring for 1 h to obtain homogeneous solution. 1 mL solution of 1% glutaric acid solution is taken, which acts as a crosslinking agent. The resultant solution is cast onto a flat, leveled, nonstick tray to set. Once set, the crosslinked film is dried at room temperature for 72 h before peeling the film off the tray. The film samples are stored in plastic bags and kept in desiccator at 60% relative humidity for further use. Crosslinked films are designated by four digits (CS19–CS91). The thickness of the films is measured by screw gauge. The thickness of the film is found to be 0.105 ± 0.005 mm.

The formulations of different synthesized crosslined films are presented in Table 2.1.

TABLE 2.1 Composition of Synthesis of Chitosan–Starch Crosslined Films.

Sample no.	Chitosan (g) in 20 mL (2% acetic acid solution)	Starch (g) in 20 mL distilled water
S	0.0	1.0
CS19	0.1	0.9
CS28	0.2	0.8
CS37	0.3	0.7
CS46	0.4	0.6
CS55	0.5	0.5
CS64	0.6	0.4
CS73	0.7	0.3
CS82	0.8	0.2
CS91	0.9	0.1
C	1.0	0.0

2.3 CHARACTERIZATION

2.3.1 FTIR SPECTROSCOPY

FTIR spectra of blend and crosslinked films are recorded by Perkin Elmer FTIR spectrophotometer using KBr pellets. FTIR spectra of the samples are taken in the range of 500–4000 cm^{-1} with a resolution of 2 cm^{-1}.

2.3.2 *SWELLING STUDIES*

Swelling studies are performed in solutions of pH 2.2 and pH 7.4 for understanding the molecular transport of liquid into chitosan–starch–glutaric acid films. A definite amount of dry sample is immersed in solutions of pH 2.2 and 7.4 at 37°C. At different time intervals, film is taken out and blotted off in between tissue paper (without pressing hard) to remove the surface adhered solution. The final weight of the samples is noted, and percentage of swelling is calculated as Ref. [21]

$$S = \frac{W_S - W_d}{W_d} \times 100 \qquad (2.1)$$

where W_s is the weight of swelled sample and W_d is the weight of the dry sample.

2.4 RESULT AND DISCUSION

2.4.1 *FTIR SPECTROSCOPY*

FTIR spectra of chitosan–starch blended (CS5) film and chitosan–starch-glutaric acid (CS55) film are shown in Figure 2.1. The broadband of chitosan film at 3288 cm^{-1} corresponds to –OH stretching, while the band at 2880 cm^{-1} represents -CH_2 group. The peak at 1646 cm^{-1} is due to C=O stretching of amide group. A small peak is observed at 1411 cm^{-1} corresponds to CH_3 symmetrical deformation mode. The peak at 1154 cm^{-1} indicates saccharide structure, and the band at 1067 cm^{-1} is due to C-O stretching vibration. The bending vibration C-H at 898 cm^{-1} represents β-linked chitosan molecule. In FTIR spectra of pure corn starch film, a peak at 3217 cm^{-1} indicates hydroxyl stretching vibration, and the appearance of other peaks at 929 cm^{-1} and 860 cm^{-1} indicate the characteristics of α–linkage in starch. Other peak at 1644 cm^{-1} is due to C=O stretching and a peakat 1114 cm^{-1} represents the presence of an ether group in the starch.[22] In the spectrum of chitosan–starch blended film, the peak corresponding to amino group of chitosan is shifted from 1656 cm^{-1} to 1643 cm^{-1}, which indicates the presence of interaction between the hydroxyl group of corn starch and amino group of chitosan.[23,24] In FTIR spectra (CS55) of chitosan–starch–glutaric acid film, no characteristic –COOH peak in 1698 cm^{-1} region is observed, which indicates that the added amount of glutaric acid to the blended solution reacted completely with chitosan and starch.[25,26]

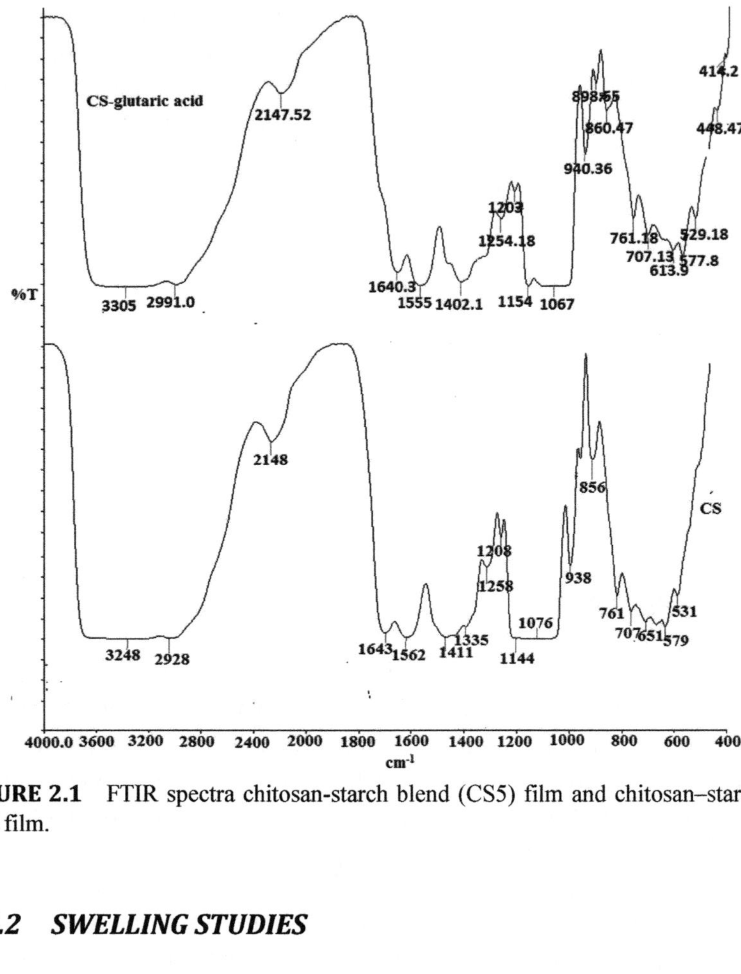

FIGURE 2.1 FTIR spectra chitosan-starch blend (CS5) film and chitosan–starch–glutaric acid film.

2.4.2 SWELLING STUDIES

The swelling capacity of the chitosan–starch–glutaric acid films is determined by submerging the films of known weight into solutions of pH 2.2 and pH 7.4 at room temperature. Figures 2.2–2.3 depict the percentage of film swelling with time in different solutions. The results indicate that the percentage of swelling increases with time. The rate of swelling of biodegradable crosslinked films increases linearly for first 4 h followed by almost a constant swelling, for all samples, for rest of the studied time period. Further, it is observed that the percentage of swelling is also influenced by the contents of chitosan and starch present in the film. The maximum swelling occurs when concentration of chitosan increases from 10 to –90%. The percentage of swelling is more in pH7.4 than of pH 2.2.

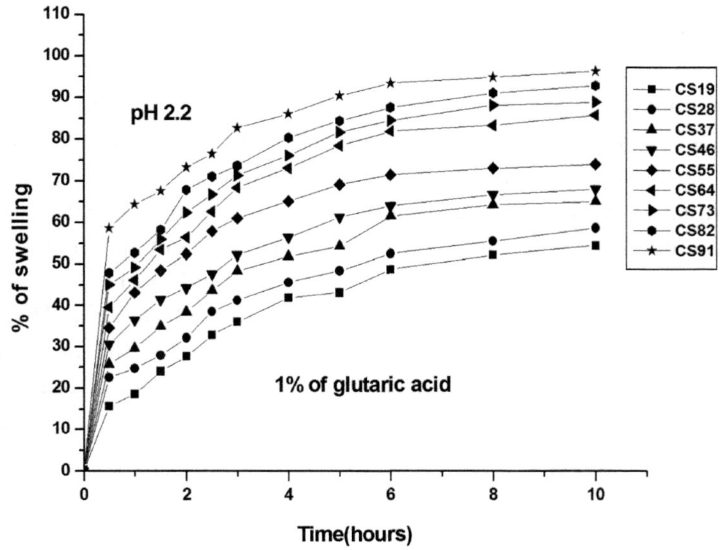

FIGURE 2.2 Swelling behavior of the crosslinked films as a function of time in pH 2.2.

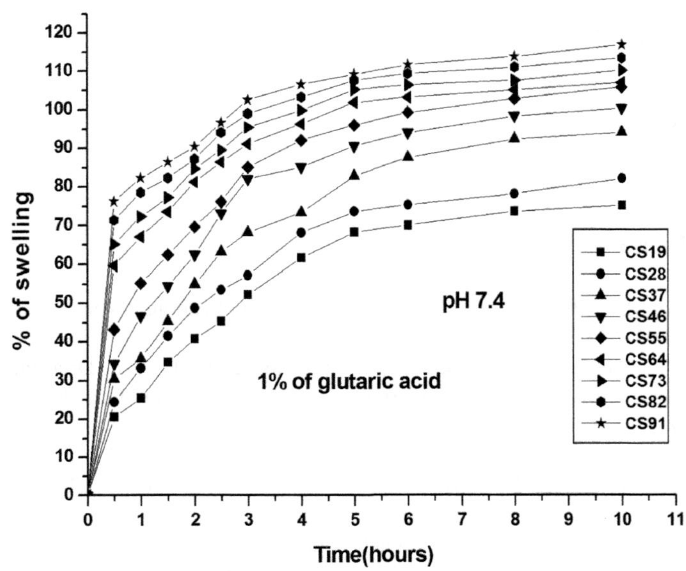

FIGURE 2.3 Swelling behavior of the crosslinked films as a function of time in pH 7.4.

2.5 CONCLUSIONS

pH sensitive chitosan–starch–glutaric acid films were prepared from variation of chitosan, starch, with crosslinked by glutaric characterized by FTIR. In FTIR spectra (CS55) of chitosan–starch–glutaric acid film, no characteristic –COOH peak in 1698 cm^1 region is observed, which indicates that the added amount of glutaric acid to the blended solution reacted completely with chitosan and starch. In this study, the influence of external stimuli such as pH of swelling media on equilibrium, swelling properties of crosslinked films were evaluated. It was observed that the swelling of chitosan–starch–glutaric acid films increased with increasing quantity of chitosan and also increasing with time. Percentage of swelling was more observed in pH 7.4. Percentage of cryastalinity increases with adding quantity of glutaric acid. These films were used in drug release studies applications due to the percentage of swelling of chitosan–starch–glutaric acid films are very good, and also drug release is measured as fuction of time at external stimuli pH 1.2 and pH 7.4.

KEYWORDS

- **chitosan**
- **starch**
- **glutaric acid**
- **swelling**
- **FTIR**

REFERENCES

1. Rinaudo, M. Chitin and Chitosan: Properties and Applications. *Prog. Polym. Sci.* **2006,** *31* (7), 603–632.
2. Puttipipatkhachorn, S.; Nunthanid, J.; Yamamoto, K.; Peck, G. E. Drug Physical State and Drug–Polymer Interaction on Drug Release from Chitosan Matrix Films. *J. Control. Release*. **2001,** *75*, 143–153.
3. Singh, V. Synthesis and Characterization Chitosan-Starch Crosslinked Beads. In *Advanced Polymeric Materials from Macro to Nano-Length Scales*; Chapter 1; Thomas, S., et al., Eds.; CRC Press: USA, 2015; pp 205–218.
4. Shu, X. Z.; Zhu, K. J. A Novel Approach Tripolyphosphate/Chitosan Complex Beads for Controlled Release Drug Delivery. *Int. J. Pharm.* **2000,** *201*, 51–58.

5. Wu, Y.; Yang, W.; Wang, C.; Hu, J.; Fu, S. Chitosan Nanoparticles as a Novel Delivery System for Ammonium Glycyrrhizinate. *Int. J. Pharm.* **2005,** *295*, 235–245.
6. Yu, Z.; Ma, L.; Ye, S.; Li, G.; Zhang, M. Construction of an Environmentally Friendly Octenylsuccinic Anhydride Modified pH-Sensitive Chitosan Nanoparticle Drug Delivery System to Alleviate Inflammation and Oxidative Stress. *Carbohydr. Polym.* **2020,** *236*, 115972.
7. Roberts, G. A. F. Solubility and Solution Behavior of Chitin and Chitosan. In: *Chitin Chemistry*; Roberts, G. A. F., Ed.; MacMillan: Houndmills, 1992; 274–329.
8. Aiba, S. Studies on Chitosan: 3. Evidence for the Presence of Random and Block Copolymer Structures in Partially Nacetylated Chitosan. *Int. J. Biol. Macromol.* **1991,** *13*, 40–44.
9. Muzzarelli, R. A. A.; Guerrieri, M.; Goteri, G.; Muzzarelli, C.; Armeni, T.; Ghiselli, R.; Cornelissen, M. The Biocompatibility of Dibutyryl Chitin in the Context of Wound Dressings. *Biomaterials* **2005,** *26*, 5844–5854.
10. Pelissari, F. M.; Yamashita, F.; Garcia, M. A.; Martino, M. N.; Zaritzky, N. E.; Grossmann, M. V. E. Constrained Mixture Design Applied to the Development of Cassava Starch–Chitosan Blown Films. *J. Food Eng.* **2012,** *108*, 262–267.
11. Silva-Weiss, A.; Bifani, V.; Ihl, M.; Sobral, P. J. A.; Gómez-Guillén, M. C. *Food Hydrocoll.* **2013,** *31*, 458–466.
12. Singh, V. Physicochemical Measurements of Chitosan-Starch –Glutaric Acid in Acetic Acid-Water Mixtures. *Green Chem. Technol. Lett.* **2016,** *2* (4), 180–184.
13. Campos, A. M. D.; Sánchez, A.; Alonso, M. J. Mucoadhesive Drug Delivery Systems. *Int. J. Pharm.* **2001,** *224* (1–2), 159–168.
14. Atyabi, F.; Manoochehri, S.; Moghadam, S. H.; Dinarvand, R. Cross-Linked Starch Microspheres: Effect of Cross-Linking Condition on the Microsphere Characteristics. *Arch. Pharm. Res.* **2006,** *29* (12), 1179–1178.
15. Chattopadhyay, D. P.; Inamdar, M. S. Aqueous Behaviour of Chitosan. *Int. J. Polym. Sci.* **2010,** *2010*, 1–8.
16. Wang, Q.; Dong, Z.; Du, Y.; Kennedy, J. F. Controlled Release of Ciprofloxacin Hydrochloride from Chitosan/Polyethylene Glycol Blend Films. *Carbohydr. Polym.* **2007,** *69* (2), 336–343.
17. Aldana, A. A.; González, A.; Strumia, M. C.; Martinelli, M. Preparation and Characterization of Chitosan/Genipin/poly (N-vinyl-2-pyrrolidone) Films for Controlled Release Drugs. *Mater. Chem. Phys.* **2012,** *134*, 317– 324.
18. Singh, V.; Kumari, K. Some Physicochemical Measurements of Chitosan/Starch Polymers in Acetic Acid-Water Mixtures. *J. Macromol. Symp.* **2012,** *320*, 81–86.
19. Yao, K. D.; Jing, L.; Cheng, G. X.; Lu, X. D.; Tu, H. L.; Silva, J. A.; Lopes, D. Swelling Behaviour of Pectin/Chitosan Complex Films. *J. Appl. Polymer Sci.* **1996,** *60* (2), 279–283.
20. Mathew, S.; Abraham, T. E. Characterisation of Ferulic Acid Incorporated Starch–Chitosan Blend Films. *Food Hydrocoll.* **2008,** *22*, 826–835.
21. Pelissari, F. M.; Yamashita, F.; Garcia, M. A.; Martino, M. N.; Zaritzky, N. E.; Grossmann, M. V. E. Constrained Mixture Design Applied to the Development of Cassava Starch–Chitosan Blown Films. *J. Food Eng.* **2012,** *108*, 262–267.
22. Silva-Weiss, A.; Bifani, V.; Ihl, M.; Sobral, P. J. A.; Gómez-Guillén, M. C. *Food Hydrocoll.* **2013,** *31*, 458–466.

23. Thakur, A.; Singh, V.; Wanchoo, R. K.; Singh, P. Synthesis and Characterization of pH Sensitive Poly (Acrylamide/Acrylic Acid) Hydrogels. *J. Polym. Mater.* **2006,** *23* (2), 185–196.
24. Singh, V.; Kumari, K. Synthesis and Characterization of Chitosan–Starch Crosslinked Film for Controlled Drug Release. *Int. J. Mater. Biomater. Appl.* **2014,** *4* (1),7–13.
25. Mitra, T.; Sailakshmi, G.; Gnanamani, A. Could Glutaric Acid (GA) Replace Glutaraldehyde in the Preparation of Biocompatible Biopolymers with High Mechanical and Thermal Properties? *J. Chem. Sci.* **2014,** *126* (1), 127–140.
26. Gong, R.; Li, C.; Zhu, S.; Zhang, Y.; Du, Y.; Jiang, J. A Novel pH-Sensitive Hydrogel Based on Dual Crosslinked Alginate/N-α-glutaric Acid Chitosan for Oral Delivery of Protein. *Carbohydr. Polym.* **2011,** *85*, 869–874.

CHAPTER 3

Bio-Based Polymer Science: From Past to Future in a Biodegradability and Eco-Friendly Context

TOMY MURINGAYIL JOSEPH[1], MEREENA LUKE PALLIKKUNNEL[1], DEBARSHI KAR MAHAPATRA[2], JÓZEF T. HAPONIUK[1], and SABU THOMAS[3]

[1]*Department of Polymer Technology, Chemical Faculty, Gdansk University of Technology, Gdansk, Poland*

[2]*Department of Pharmaceutical Chemistry, Dadasaheb Balpande College of Pharmacy, Nagpur, Maharashtra, India*

[3]*International and Inter University Centre for Nanoscience and Nanotechnology, Mahatma Gandhi University, Kottayam, Kerala, India*

ABSTRACT

Renewable resources are used to make bio-based polymers, also known as bio-based resins (algae, bacteria, microorganisms, plants, etc.). Because of their biodegradability and improved performance after reinforcing, bio-based polymers have received much attention as a possible replacement for conventional polymers. With the global depletion of crude oil reserves and rising environmental concerns, efforts are being made on a global scale to find viable alternatives to petroleum-based polymeric materials for a sustainable and green civilization. Innovative technologies for converting these natural resources into value-added chemicals, as well as revolutionary polymerization processes for producing high-performance, low-cost polymers with

Foundation and Growth of Macromolecular Science: Advances in Research for Sustainable Development. Meegle S. Mathew, PhD, Józef T. Haponiuk, PhD & Sabu Thomas, PhD, DSc (Eds.)

customizable architectures and functions, are critical components of long-term sustainability. Furthermore, developing state-of-the-art characterization methods for these novel materials is critical to exposing their unique structures and facilitating their implementation in new domains including sensors, structural components, and biological devices. Despite recent advances in renewable polymers such as PLA, most suggested bio-based materials are still far from commercialization and are unlikely to replace petroleum-based goods. Nanofillers and fibers have been added into bio-based polymer matrices to increase physical and thermomechanical characteristics. This special issue aims to provide a great platform for scientists and researchers working on bio-based polymers and composites to share their cutting-edge research on the conversion of natural resources into value-added chemicals, the modification and preparation of new bio-based polymers, as well as the characterization and application of bio-based polymers and composites.

3.1 INTRODUCTION

Renewable resources are used to make bio-based polymers, also known as bio-based resins (bacteria, algae, plants, microorganisms, etc.). They may be made directly or by the synthesis of monomers, which must be followed by polymerization. Polylactic acid, polyhydroxybuturate, poly-*L*-lactide, polyamide, polypropylene (PP), polyhydroxyalkalonates (PHAs), etc., derived from bio-based ethylene produced by converting ethanol, polyethylene terephthalate, and all other thermoplastic materials are accessible on a variety of markets. Other thermosetting bio-based resins include polyurethanes including bio-based diols and bio-based epoxy epichlorohydrine derived from glycerol. Bio-based polymers are mostly employed in packaging; nevertheless, their use in more demanding applications remains a significant issue. This can be explained by the following evidence:

- Bio-based matrices have poor durability in long-term applications such as transportation and construction;
- Significant modifications to existing processes are required;
- Bio-based polymers are more expensive than oil-based polymers; and
- Bio-based matrices have significantly lower performance than oil-based matrices.

In the fast-growing and busy interdisciplinary field of biomaterials, where engineers, scientists, and biologists integrate their expertise, bio-based polymer hybrid composites have a unique position. Advances in the science and engineering of biomaterials are widely recognized for their importance in tackling complicated medical and biological challenges requiring tissue replacement, repair, and regeneration. The definition, significance, classification, and use of polymer hybrid composites in the biomedical area are all covered in this chapter. It is divided into two sections. The first section contains information on the classification of biomedical polymer hybrid composites, as well as compositions and specifications of nanocomposites based on polymer hybridization with natural polymers, biopolymers, minerals, and metals. This chapter looks at biological applications of polymer hybrid composites, including drug delivery, tissue engineering, bioadhesion, and so on.[1]

3.2 BIOMASS

According to science, it is described as the total dry matter of an entity with a certain area or period, except water. Biological origins, on the other hand, are referred to as biomass. They are simply living beings that create other organisms. Nonfossilized species that were newly harvested from live creatures are referred to as nonfossilized animals. When consumers move to blended biomass, pyramid food producers are shrinking. Since essential functions including metabolism, discharge, death, and energy conversion necessitate constant contact between manufacturers, consumers, and separators, this is the case. Both internationally and locally, the concept may be expressed. Organic matter includes barley, shrubs, tundra, savanna, tropical monsoon plants, and rainforests. Photosynthesis is the conversion of solar energy into chemical energy by green plants. It acts as a safety net in the face of global oil scarcity. Biomass enhancement innovations are being explored and implemented in agricultural practices to increase and track energy efficiency. Hemp, wheat, rice, and sugar beet are all extensively cultivated crops in the United States. As a consequence, it aspires to be used as a source of energy while still producing nutrients. Biomass is a renewable energy source. Biomass is processed and used in industry to provide food and fuel. The carbon cycle provides plant-based energy.[2]

Biofuels are made by converting biomass into methane gas and carbon dioxide in an oxygen-free environment, with the help of microbiological flora. The bulk of biological matter is made up of farm and animal waste. Methanol and an alkali catalyst reaction are used to produce vegetable oils (soybean, sunflower, rapeseed, saffron, and waste frying oils) as well as methanol and waste frying oils.[3] Animal manures are fermented during the biogas reaction. As a consequence, it is becoming more common. Poultry manure is commonly used in agriculture because it produces saltwater, which reduces soil yield. Such agricultural wastes, field wastes, forest bio-wastes, and urban effluents are also used as raw materials. In addition to conventional combustion processes, it can be used to make thermal and electrical fuels, as well as solid, liquid, and gaseous fuels. The basic turkey has a fair probability of fulfilling the resource requirements.[4] It includes biofuel imports, energy cultivation development, rural socioeconomic frameworks, and natural energy supply and atmosphere protection. Biodiesel is a biofuel that encourages green fuels while still providing lubricant benefits to the engine. It has a large flash point and can be shipped and stored safely. This generates employment, particularly in rural areas, while also providing energy and cost savings at manufacturing facilities that are not linked to the power grid. The first time methane gas from septic tanks was used to fuel street lights was in England.[5]

There are no harmful environmental consequences. It does, though, have certain environmental implications. The waste will be recovered after it has been burned. This causes both acute and long-term visual emissions. In the absence of other renewable sources, it is a renewable resource that can provide a reliable supply of electricity. Since biomass is a local fuel with no greenhouse gas emissions, storage conditions are easy and convenient, improving biomass' flexibility. For combustion, it may be used for fossil fuels. As a consequence, fossil-fuel-related emissions are reduced, but agriculture and processing still necessitate long-term facilities and manpower. The stage in which fossil fuels pass through and perish justifies these shortcomings.[6]

3.3 RAW MATERIALS AND METHODS

There are two types of materials that can be used to generate biomass energy: classical and modern. Forests include traditional biomass, which consists of plant and animal wastes.[7] Agro-based industries, urban, domestic, vegetable (bark, root, straw, and stalk), woodland industrial effluents, and energy afforestation both utilize modern biomass energy. Protein plants (wheat, beans, and peas), fiber plants (muscantus, sorghum, hemp, and flax), starch plants (artichoke, beet, maize, wheat, and potatoes), and oilseed plants (soy, sunflower, and rapeseed) are all forms of vegetable waste. Carbon processing employs both physical and transformational methods. Sizing, crushing, processing, washing, extraction, and brewing are examples of physical processes. Thermochemical and biochemical systems are examples of transformation processes. The conversion processes and end products that have been proven are mentioned below.[8] Biomethanization is the mechanism through which hydrogen is generated via biophotolysis, bioethanol is produced through fermentation, the pyrolytic liquid is produced through pyrolysis, gas fuel is produced through gasification, biochar is produced through carbonization, and biodiesel is produced through esterification.[9]

3.4 ECO-FRIENDLY

Humans, and all living organisms, are affected by abiotics and antibiotics. These environment-creating variables are critical for maintaining and restoring ecological equilibrium. The use of modern research and technology in agricultural practices by the man at the top of the diet pyramid causes environmental issues. Disrupt the ecosystem equilibrium and emission loop caused by humans. People need to create eco-friendly goods and start consuming them as a consequence of rising usage. Eco-friendly goods are those that do not damage the atmosphere, do not result in the usage and waste of fossil fuels, and do not disturb the existing ecosystem order and cycles. Eco-friendly heart materials are reusable and recyclable, do not produce harmful contaminants, conserve water and electricity, and cause less damage to nature and animals during the manufacturing phase. The planet is a house away from home thanks to transportation, schooling, healthcare, and other favorite things. Pollution may be decreased, lowered, and eventually eliminated by using environmentally safe heart goods.[10]

Textiles, animal, flower, or wood materials that are either natural or biodegradable conserve resources or consume less energy are examples of eco-friendly products. Eco-friendly heart drugs are those that are 100% natural in nature and have no negative effects on health or the climate. Agricultural goods are used in ecological products, also known as sustainable products. However, due to its high price and low production, it does not have enough users. Washing machines, dishwashers, hybrid motors, and motorcycles are examples of items made with emerging technologies today.[11] Fujitsu was the first to market for an environmentally safe heart machine in 1993.[12] The first leadless motherboard was installed in 2002. Devices are approved under special tests that are eco-friendly heart products with a focus on current environmental concerns. Fujitsu, which also controls the Blue Angel, Nordic Swan, and Ecomark sustainable brands, strikes a compromise between low oil costs and innovative energy, reliable manufacturing, and component usage. Use of low-risk and volatile products, use of natural chemicals such as vinegar and carbonate, blockade of lime such as lime salt and vinegar, use of abrasive soap-based chemical agents instead of chemical oils used for cleaning; use ventilated areas, use cotton bags instead of plastic bags, use rechargeable batteries instead. Wooden toys should be favored over plastic toys, and wood goods should be preferred over furniture. As a consequence of anthropogenic causes, today's climate change is being felt, and governments have implemented a variety of solutions, such as waste separation and plastic bag disposal. These applications would be improved and added to in the future. The cry for biodiversity, the preservation of man–environment harmony, and the conservation of endangered natural resources both hinge on each citizen taking personal responsibility. Manufacturers can increase visibility by the supply of eco-friendly cardio goods and increasing market appetite for such products by matching pricing policies. The use of an eco logo will help raise visibility and promote the sale of environmentally sustainable goods. In this way, a comparative edge can be gained in a competitive environment.

3.5 SUSTAINABILITY

Sustainability is "ensuring the consistency of variety and competitiveness while maintaining the potential to last indefinitely."[13] "Our collective potential rapper study has entered the literature," wrote the World Environment Development Commission in 1987 under the auspices of the United Nations. According to the study, humans have the potential to support growth by meeting everyday needs without jeopardizing nature's ability to meet future generations' needs. Natural resources are essential to its sustainability, but there are no limitless reserves of market oil. Nonrenewable natural resources are causing ecosystem loss, and research is ongoing to avoid this. The usage of green energy sources necessitates an appreciation of sustainability in general. Sustainability and a healthy equilibrium between nature and man, on the other hand, will only be accomplished if citizens make the right decisions. This is because anthropogenic effects are posing a threat to the long-term viability of habitats. The advent of skewed urbanization is followed by the degradation of land, the introduction of toxic pollutants into the environment and water as a consequence of increasing levels of waste, improper soil agriculture, the opening up of natural ecosystems owing to dramatic increases in population, depletion of biodiversity, and loss of biodiversity.

With the right knowledge, the whole past of society and life is explored, and a rational and sustainable lifecycle is reached. This is an indication of a long-term human commitment. To preserve the ecosystem, reduce human interaction with nature, and preserve ecological equilibrium, an international obligation is needed. As a result, long-term strategies must be developed, and infrastructure, energy supplies, the atmosphere, and climate must all be accepted. The best direction to accomplish priorities and priorities determines the long-term viability of sustainability within the development cycle. We must develop a vision for the global framework, including energy, the economy, and the climate, while still meeting all personal and social commitments. It will only satisfy its needs for shelter, water, and restoration, which involves essential tasks, if ecological conditions appropriate for its structure are present. These elements can be found in natural habitats. The existing biodiversity and ecosystem roles are supported. Each disruption of the ecological equilibrium has a detrimental impact and disrupts a variety of critical energy activities.

Climate change and global warming are being exacerbated by rising urbanization, air pollution, and industrialization, all of which are disrupting the environment.[14] The loss of ecosystem diversity has the ability to wipe out all habitats. The transfer of systems that meet the requirements that can ensure the survival of all living things by creating an integrated system is referred to as ecological sustainability. The term "sustainability" refers to the need to safeguard the climate. Instead of overconsumption, being in a position to protect production and nature prevents natural resources from being depleted. The aim of artificial environments is to construct artificial habitats. The concept of sustainability, which is high on the scientific agenda, is being investigated in a variety of areas, including urbanization, tropical habitat protection, marine pollution prevention, and flora conservation. In addition to the elements that have been tested, the utilization of natural technologies like solar, wind, biomass, and hydropower should be maximized rather than polluting nature. Toxic substances can be stored using a recycling ecology to restore a habitat that has lasted hundreds of years. Endangered species can be protected with legislative and nongovernmental organization (NGO) assistance. Increased landscaping and cleaning facilities are possible.[15]

The amount of energy required benefiting from an emerging technology in uses like heating, ventilation, viewing TV, listening to music, and enlightenment is enormous. The amount of energy used by air conditioners, particularly in hot countries, necessitates EU-wide action. Private land consumes two-thirds of the electricity utilized in Europe as a function of construction use. Establish building design codes and measure building energy efficiency. Energy efficiency certifications are becoming more popular. Various projects in the Asia-Pacific area fund intensive campaigns in the United States and Canada to curb greenhouse gas pollution and electricity use. Individual studies alone would not be enough to restore the ecological equilibrium, so the production and usage of applications that ensure biodiversity in agricultural plants, structures, transportation, and personal automobiles is needed.[16]

3.6 BIODEGRADABILITY

A biodegradable commodity is the one that degrades biologically and has a biological impact on the environment. In this normal cycle, the natural gases CO_2 and N_2 decompose to release toxic salts and water. Over time, the usage of biodegradable products has grown in popularity. The pharmaceutical and agriculture sectors both utilize biodegradable polymers. In addition, petroleum-based biodegradable materials are used to mitigate the harm caused by goods that utilize oil as a raw material. We use plastic in nearly every aspect of our lives. It is a common raw material because it is cheap and simple to work with. Plastic made from fossil oil includes toxic chemicals. As a result, the latest conversation around cancer drugs that we learn on a daily basis is pervasive in this sense. This plastic, which has a negative impact when exposed to sunlight, decreases the rate of long-term breakdown in the raw material and is hazardous to human health. Rescuers are bioplastics with biodegradable properties.[17]

3.7 BIOPLASTICS

Thermoplastics such as polyethylene, polyethylene terephthalate, polystyrene, polypropylene, etc. currently account for 60% of Europe's total plastic market.[18] While these plastics are historically derived from petrochemicals, there is an increasing market for natural resource-based plastics (also known as "bioplastics") as alternatives to their petrochemically derived equivalents. Both bioplastics are made from renewable natural resources. Not all, however, are biodegradable.[19] Bioplastics are a form of plastic that is classified as either biodegradable or nonbiodegradable. The variety in biodegradation rates and routes characterizes biodegradable bioplastics. Polyhydroxyalkanoates, polylactic acid, starch, cellulose, etc. are examples of such plastics. Bio-based polymers that are biodegradable may be reprocessed or reduced to ashes in the same way as oil-based plastics can, but they are not commonly recycled since they are considered pollutants in the modern recycling method. Plastics may even be microbially destroyed depending on their form, providing for alternative end-of-life disposal alternatives such as industrial and domestic composting, as well as anaerobic digestion, fostering the establishment of a circular economy.[20] Polyol-polyurethane, bio-polyethylene, and Bio-polyethylene terephthalate (Bio-PET) are examples of nonbiodegradable bioplastics (Bio-PE).

The raw material is plastic, which can induce cancer when heated and by removing these results, biodegradable plastics are produced, allowing us to use healthier materials. The sector has gained importance for producing biodegradable goods, which we refer to as eco-friendly owing to the long-term breakdown rate in human health and nature, as well as the damage to the atmosphere. We also encounter terms like polymer and bioplastics as we hear the word biodegradable. It has the property of being a biodegradable bioplastic. It has recently been substituted for the idea of bioplastic as a "biodegradable plastic yarn." Bioplastics incorporate edible organic sources such as vegetable oil and vegetable starch; biodegradable properties lead to the manufacture of materials that degrade in nature in a limited period of time, reducing environmental damage. It is possible to recycle and refine biodegradable plastics. Waste collection and emissions in the forest are also avoided in this manner. This cuts down on the usage of fossil fuels, lessens the harm to the earth and climate, and speeds up the decomposition process. The manufacture of bags that decompose in nature has also become widespread. A natural quadrilateral's dissolution period is reduced to 45 days thanks to this biodegradable content. Cleaning our homes with detergents and cleaning supplies has also improved. It should be noted that utilizing this content is more costly and increases the price of the goods for sale by a factor of two. Because of the market's demand for this low-cost commodity, hazardous compounds are also being produced.

Supporters of her case have been trying to make the full text of her speech accessible on the internet. The organizations that awarded the certificates were founded after special training. China, India, the United States, Belgium, and Sweden were among the countries tested by JBA in 2005. Projects in Europe are born out of a sense of understanding. Forbioplast, Hydrus, and Biopack are a few of them. Biodegradable plastic has spawned fresh industry prospects, spawned a slew of production plants, and sunk millions of dollars through science around the globe. It was able to reach an annual market value of 110,000 tonnes thanks to fast and strong expansion. With health-conscious individuals, commitment and confidence in fossil fuels have waned. Consumer values fluctuated as a result.[21] As the market desire for renewable products grows, bioplastics, which may lessen our reliance on fossil fuels and greenhouse gases, will become more ubiquitous. Bioplastics output is projected to increase by as much as 20% by 2022, and customer awareness of bioplastics would need to keep pace.

3.8 BIOPLASTICS: BIODEGRADABLE VS BIO-BASED

To figure out these three concepts (compostability, oxo-degradability, and biodegradability), it is first necessary to comprehend bioplastics, which apply to a vast family of plastics that are bio-based at the start of their lives, biodegradable at the end of their lives, or both.

Bioplastics may be divided into three categories as a result of this:

1. Nonbiodegradable and bio-based (e.g., bio-based PET, bio-based PE).
2. Petroleum-based, biodegradable (e.g., PCL).
3. Fully or partly biodegradable and bio-based (e.g., PLA or starch blends).

3.9 BIODEGRADABLE VERSUS COMPOSTABLE VS OXO-DEGRADABLE PLASTICS

Compostable plastics have been examined and certified by a third party to meet international biodegradation criteria in an industrial composting facility environment, such as ASTM D6400 in the US or EN 13432 in Europe . Materials certified to ASTM D6400 or EN 13432 may decompose in 12 weeks and biodegrade at least 90% in 180 days in a municipal or commercial composting facility. Approximately 10% of the solid material would be left at the conclusion of the 6-month period as useable manure, water, wood, etc. These requirements often guarantee that all remaining manure is clear of pesticides, ensuring that the compost is safe to use when sold for planting or agricultural purposes. Unless otherwise stated, certified compostable goods must be disposed of in a licensed municipal composting unit, not at home. Many approved compostable products, to biodegrade rapidly or at all, need the higher temperatures of industrial environments. Since industrial composting is seldom collected curbside in the United States, approved compostable goods are better used in closed structures like stadiums, colleges, theme parks, etc., where organic waste and compostable is closely supervised and managed to ensure effective disposal in a commercial composting facility. Composting is being used by companies including San Francisco International Airport and Safeco Field in Seattle to reduce their carbon emissions and divert plastic waste from landfills.[24]

Oxo-degradables are often mistaken for biodegradable plastics, but they are a separate group. They are not bioplastics or biodegradable plastics, but rather traditional plastics combined with an ingredient to mimic biodegradation. Microplastics are formed as oxy-degradable plastics fracture into smaller and smaller fragments; however, unlike biodegradable and compostable plastics, they do not degrade at the molecular or polymer level. The consequential microplastics are released into the atmosphere and remain there forever until they degrade completely. If bioplastics are to gain market share in the coming years, the environmental benefits must be emphasized in food and material sales. The Green Guides of the Federal Trade Commission (FTC) are a decent place to start. The Green Guides set forth best practices for labeling and marketing green products in a way that meets client demands. Transparency would not only enable customers to make better buying choices, but it will also guarantee that bioplastics are disposed of properly. Bioplastics' sustainability benefit proposition of diverting agricultural waste from landfills, reducing carbon gases, and maintaining safe resource use, is strengthened by better end-of-life disposal. Green Dot specializes in alloying biodegradable polymers to satisfy the durability standards of customer-specific parts. When it comes to our bioplastic material choices, we strive to be open and honest. That's why, when we say a substance is "biodegradable," we are referring to the ASTM D6400 and EN 13432 compostability criteria. All will be on the same page this way. It provided enough time, almost every substance can biodegrade. Nevertheless, because the period of the biodegradation phase is strongly affected by other factors including such humidity and temperature, announcing a plastic to be "biodegradable" without offering extra context (i.e., in what timeframe and under what environmental conditions) can be perplexing to customers. Trustworthy companies are more likely to offer specific claims, such as certifying that their bioplastics are biodegradable.

Compostable plastics are a subclass of biodegradable polymers that meet the standard biodegradation standards and timetable. Although all biodegradable plastics decompose, not all compostable plastics decompose.

3.10 RECYCLING

Owing to technical advancements, the rate of industrialization and urbanization is speeding up. Human tension of nature is on the rise as the world's population grows. Because of their amount and quality, waste from manufacturing and marketing practices, as well as unconscious resource usage and use, pose an environmental hazard. This risk to the climate, life, cities, households, and even individuals necessitate waste collection, recycling, and usage. The first time Environmental Law No. 2872 was established to incorporate waste legislation was in 1983. It is known as "harmful substances discarded or tossed into the atmosphere as a consequence of some activity."[25–32] Good, liquid, gaseous, and processing wastes are categorized based on environmental, physical, output, and use characteristics. It may have come from an agricultural, commercial, or domestic environment. In 2008, our Waste Disposal Act included general waste management standards, as well as the regulation of medical waste, radioactive waste, vegetable waste oils, and packaging waste in 2005. The United Nations Environment Programme (UNEP) defines solid waste as "materials that the owner does not use, use, uses, disposes of, and disposes of." Raw products, oils, and water are both wastes that have lost their use and economic worth since they have been used. They can be removed in a manner that would not endanger the ecosystem. Domestic solid waste, radioactive waste, medicinal waste, factory waste, agricultural and garden waste, unique waste, building waste, and waste are the seven types. It has a clear or indirect link to living things. It is a means of nutrients and replication for certain species and brings infectious or viral diseases (malaria, rabies, measles, vomiting, cholera, plague, leprosy). Recycling is the conversion of existing waste into secondary raw materials using a variety of methods. It is a set of discarded products that can be reused and reused in practice once they have been employed in some form. The first justification for reuse is the war's resource constraints. To counter resource waste management, waste management, energy crisis, and environmental degradation, various waste recycling approaches have been created. High productivity can be attained in the processing of waste that has been removed from its source, as well as in the removal of waste that remains after treatment. Foreign currency saves a significant sum of money on the cost of manufactured recycled products and the resources used in manufacturing, in addition to eliminating certain issues. It is a significant investment in both the economy and the future. Prevention, reduction, reuse, reuse, energy recovery, and elimination designation are also included in the waste management range. The defined basic steps of the method are listed below, but they vary depending on the waste.

1. Separate disposal at the site: Recyclable waste is processed at the source (without contamination). By that the phases of the operation and avoiding waste, it saves time, water, and electricity.
2. Sorting: This entails sorting items gathered individually at the source into different recycling groups. It would be possible to provide direct connections to the required recycling scheme. This saves money in terms of transportation, labor, and time. Paper, metal, plastic, glass, and electronic waste are separated for a special collection at the source. Electronic waste, end-of-life cars, end-of-life tires, waste battery storage, waste oils, manufacturing waste, medical waste, landslide waste, radioactive waste, and domestic waste are among the other items used, depending on the source. Waste, porcelain waste, stone, and clay waste are also examples of waste (cloth waste).

Plastic recycling, like many other forms of pollution, is a serious issue in Russia. One of the more serious issues is that we lack a basic grasp of what plastic is made of and how it is classified. It ignores maintenance problems, technological limitations, and regulations. Meanwhile, Russia continues to make progress in the battle against plastic pollution. Scientists at Samara University, for example, have created a method for making bioplastics from organic waste, plants, and fruits. A study on a genetically engineered plant focused on Tefroseris (field cross) capable of decomposing plastic was carried out at the University of Kemerovo. A plant for producing paving slabs from recycled plastic is located in the city of Imwa in the Komi Republic. Plastic containers are thrown away in special trash dumps in the area. Every day, 30 m^2 of plastic paving slabs are created as a result. One of the big issues of the 21st century is polymer waste. Various nations war in various forms. But one thing is certain: waste recycling is quickly becoming one of the most exciting fields of industry for IT and gadgets, even on par with virtual reality.

Most of the most polluting products in the world are plastic. Polymers are cheap and durable, and they can be used almost everywhere. As a consequence, polymers make up almost half the human waste. They decompose for hundreds of years under normal environments. Harmful compounds such as styrene, phenol, and formaldehyde are emitted during the fermentation phase. Plastic recycling, on the other hand, is complex and unprofitable. As a result, just 10% of the world's plastic waste is processed. One of the global alternatives in the battle against plastics is the development of biopolymers. Many of them are still in usage in different aspects of life. Water-soluble polymers, which are assimilated without harming the human body, are used in medicine during surgery. There are not enough of them in other places. Bioplastics, on the other hand, are becoming more used in conventional packaging and household appliances as technology advances. This is because producers have historically found it unprofitable to invest in this market. Bioplastics are more costly to produce. However, as technology advances, barriers are increasingly eliminated. The biopolymer industry was worth just under US$65 million in 2013. It has almost tripled in size already. Bioplastics are expected to account for 5–7% of all polymers by 2020. It is currently about 1%. Polylactide is one of the most common biopolymers today. Lactic acid is used to make it. Sulsar, a Swiss business based in the Netherlands that produces around 5000 tons of biopolymers each year, has established a facility to make such plastics. Surprisingly, the firm would not need to adjust the technology entirely. It was sufficient to marginally change the plant for the manufacture of traditional polymers to produce bioplastics. Renault, a Russian financial firm, is one of the company's major shareholders.

Plastic is manufactured in Switzerland. Separating waste not only by nature but also by color is standard practice in the country to make the operation easier. The bottle lids are held in a different container in this situation. In the United States, the fight against polymer waste is handled differently. Food sold in plastic packaging, except manufactured with biopolymers, is banned in Minneapolis and St. Poe, for example. For polymer waste, states have a government-sponsored segregation scheme. People get a range of incentives for collecting bottles, ranging from cash to perks and bonuses. They developed technology at one of the United States' universities that might, in theory, help eradicate plastic in the future. The plastic is cooked for 3 h at 70°C in a barrel with a catalyst. After that, the plastic decomposes into carbon, and is used to power the batteries. They are said to perform significantly easier and over a considerably longer length of time.

Laws were enacted in Japan 20 years ago that specifically restricted the usage of hydrocarbon polymers. Legal companies can incur a very low tax whether they segregate or reuse such pollution. Individuals enjoy varying benefits, such as lower energy rates. The dilemma was approached differently in Germany. German clothes brands utilize recycled plastic in addition to processing and segregating waste. Puma brand in cycle was created as a one-of-a-kind ensemble. Traditional sportswear manufactured from natural fibers divided by polyester obtained from recycled plastic bottles is included in the German "cycle," as the name suggests. The set was made entirely of biodegradable items. In their shops, the organization also placed special garbage bins where rotting shoes can be thrown. The nonbiodegradable portion is used to make fresh clothes. The other is a polyester granulate that is said to be environmentally friendly by the producer. In Edmonton, Canada, they studied how to produce biofuels from plastic waste. It is mostly used in race cars. Methanol is made from waste that helps the vehicle to travel at high speeds. The city is heated with processed materials. Scientists in China experimented with the decomposition of plastics using petroleum ether and iridium. This catalyst is used to heat the plastic to 150°C. Fermentation may be used as a source of energy. The downside is that a single component of the catalyst will decompose into 30 pieces of plastic. Despite the fact that iridium is a precious metal, its industrial use is currently unprofitable. Scientists are also working on lowering the cost of technology.[33,34]

3.11 CONCLUSIONS

Natural decomposition occurs only in a few biopolymers such as sugar, cellulose, and lactic acid. In terms of longevity and power, they sometimes fall short of man-made polymers. As a result, the market's existing growth is minimal. Their usage is often limited due to their high expense. Despite the fact that substantial study is being conducted in this region, the current scenario is not encouraging. Microorganisms that may decompose plastic are now being researched. Around the world, efforts are also being made to effectively generate fuel from plastic waste by managed combustion. There are already three methods to reduce consumption: reuse as frequently as possible, compost, and recycle again. Currently, only around 15% of the overall waste generated is recycled. Germany, Austria, and South Korea are leading the way in this region. More than half of the plastic waste they produce is recycled. Despite boasts of a 60% recycling record, India does not even rank in the top ten countries in global reports. Plastic waste has been scattered due to a shortage of organized waste collection schemes and trash cans in public areas. New advances in this region are still hampered by the government's insufficient assistance to waste management facilities.

KEYWORDS

- **polymer**
- **biomass**
- **eco-friendly**
- **biodegradability**
- **sustainability**
- **bioplastics**
- **recycling**

REFERENCES

1. McKendry, P. Energy Production from Biomass (Part 1): Overview of Biomass. *Bioresour. Technol.* **2002,** *83* (1), 37–46, ISSN 0960-8524. https://doi.org/10.1016/S0960-8524(01)00118-3.
2. Alluvione, F.; Moretti, B.; Sacco, D.; Grignani, C. EUE (Energy use Efficiency) of Cropping Systems for a Sustainable Agriculture. *Energy* **2011,** 36 (7), 4468–4481, ISSN 0360-5442. https://doi.org/10.1016/j.energy.2011.03.075.
3. Kwietniewska, E.; Tys, J. Process Characteristics, Inhibition Factors and Methane Yields of Anaerobic Digestion Process, with Particular Focus on Microalgal Biomass Fermentation. *Renew. Sustain. Energy Rev.* **2014,** *34*, 491–500, ISSN 1364-0321. https://doi.org/10.1016/j.rser.2014.03.041.
4. Demirbaş, A. Biomass Resource Facilities and Biomass Conversion Processing for Fuels and Chemicals. *Energy Convers. Manag.* **2001,** *42* (11), 1357–1378, ISSN 0196-8904. https://doi.org/10.1016/S0196-8904(00)00137-0.
5. Hanff, E.; Dabat, M. H.; Blin, J. Are Biofuels an Efficient Technology for Generating Sustainable Development in Oil-Dependent African Nations? A Macroeconomic Assessment of the Opportunities and Impacts in Burkina Faso. *Renew. Sustain. Energy Rev.* **2011,** *15* (5), 2199–2209, ISSN 1364-0321. https://doi.org/10.1016/j.rser.2011.01.014.
6. Balat, M.; Balat, M. Political, Economic and Environmental Impacts of Biomass-Based Hydrogen. *Int. J. Hydrog. Energy* **2009,** *34* (9), 3589–3603, ISSN 0360-3199. https://doi.org/10.1016/j.ijhydene.2009.02.067.
7. Carating, R. B.; Galanta, R. G.; Bacatio, C. D. Soils and the Philippine Economy. In *The Soils of the Philippines. World Soils Book Series*; Springer: Dordrecht, 2014. https://doi.org/10.1007/978-94-017-8682-9_5
8. Doliente, S.; Narayan, A.; Tapia, F.; Samsatli, N. J.; Zhao, Y.; Samsatli, S. Bio-aviation Fuel: A Comprehensive Review and Analysis of the Supply Chain Components. *Front. Energy Res.* **2020,** *8*, 110. https://doi.org/10.3389/fenrg.2020.00110
9. Rodriguez Iglesias, J.; Castrillón Pelaez, L.; Marañon Maison, E.; Sastre Andres, H. Biomethanization of Municipal Solid Waste in a Pilot Plant. *Water Res.* **2000,** *34* (2), 447–454, ISSN 0043-1354. https://doi.org/10.1016/S0043-1354(99)00176-1.

10. Han, H.; Hsu, L. T. J.; Lee, J. S.; Sheu, C. Are Lodging Customers Ready to go Green? An Examination of Attitudes, Demographics, and Eco-Friendly Intentions. *Int. J. Hosp. Manag.* **2011,** *30* (2), 345–355, ISSN 0278-4319. https://doi.org/10.1016/j.ijhm.2010.07.008.
11. ThyavihalliGirijappa, Y. G.; Mavinkere Rangappa, S.; Parameswaranpillai, J.; Siengchin, S. Natural Fibers as Sustainable and Renewable Resource for Development of Eco-Friendly Composites: A Comprehensive Review. *Front. Mater.* **2019,** *6* (226), 2296–8016. https://doi.org/10.3389/fmats.2019.00226.
12. Anchordoguy, M. Japan's Software Industry: A Failure of Institutions? *Res. Policy* **2000,** *29* (3), 391–408, ISSN 0048-7333. https://doi.org/10.1016/S0048-7333(99)00039-6
13. Kuhlman, T.; Farrington, J. What is Sustainability? *Sustainability* **2010,** *2* (11), 3436–3448. https://doi.org/10.3390/su2113436
14. Patella, V.; Florio, G.; Magliacane, D.; et al. Urban Air Pollution and Climate Change: "The Decalogue: Allergy Safe Tree" for Allergic and Respiratory Diseases Care. *Clin. Mol. Allergy* **2018,** *16*, 20. https://doi.org/10.1186/s12948-018-0098-3
15. Curtin, R.; Prellezo, R. Understanding Marine Ecosystem Based Management: A Literature Review. *Marine Policy* **2010,** *34* (5), 821–830, ISSN 0308-597X. https://doi.org/10.1016/j.marpol.2010.01.003.
16. Khan, M. A.; Khan, M. Z.; Zaman, K.; Naz, L. Global Estimates of Energy Consumption and Greenhouse Gas Emissions. *Renew. Sustain. Energy Rev.* **2014,** *29*, 336–344, ISSN 1364-0321. https://doi.org/10.1016/j.rser.2013.08.091.
17. Tokiwa, Y.; Calabia, B. P.; Ugwu, C. U.; Aiba, S. Biodegradability of Plastics. *Int. J. Mol. Sci.* **2009,** *10* (9), 3722–3742. https://doi.org/10.3390/ijms10093722
18. PlasticsEurope. Plastics—The Facts [Online]. https://www.plasticseurope.org/en (accessed on Apr 15, 2020).
19. Narancic, T.; O'Connor, K. E. Microbial Biotechnology Addressing the Plastic Waste Disaster. *Microb. Biotechnol.* **2017,** *10*, 1232–1235.
20. Narancic, T.; Verstichel, S.; Reddy Chaganti, S.; Morales-Gamez, L.; Kenny, S. T.; De Wilde, B.; Babu Padamati, R.; O'Connor, K. E. Biodegradable Plastic Blends Create New Possibilities for End-of-Life Management of Plastics but They Are Not a Panacea for Plastic Pollution. *Environ. Sci. Technol.* **2018,** *52*, 10441–10452. [Google Scholar]
21. Ren, X. Biodegradable Plastics: A Solution or a Challenge? *J. Clean. Prod.* **2003,** *11* (1), 27–40, ISSN 0959-6526. https://doi.org/10.1016/S0959-6526(02)00020-3
22. Rujnić-Sokele, M., Pilipović, A. Challenges and Opportunities of Biodegradable Plastics: A Mini Review. *Waste Manag. Res.* **2017,** *35* (2), 132–140. DOI: https://doi.org/10.1177/0734242X16683272
23. Lambert, S.; Wagner, M. Environmental Performance of Bio-Based and Biodegradable Plastics: The Road Ahead. *Chem. Soc. Rev.* **2017,** *46*, 6855–6871. DOI: https://doi.org/10.1039/C7CS00149E
24. Napper, I. E.; Thompson, R. C. Environmental Deterioration of Biodegradable, Oxo-biodegradable, Compostable, and Conventional Plastic Carrier Bags in the Sea, Soil, and Open-Air Over a 3-Year Period. *Environ. Sci. Technol.* **2019,** *53* (9), 4775–4783. DOI: 10.1021/acs.est.8b0698
25. Al-Salem, S. M.; Lettieri, P.; Baeyens, J. Recycling and Recovery Routes of Plastic Solid Waste (PSW): A Review. *Waste Manag.* **2009,** *29* (10), 2625–2643, ISSN 0956-053X. https://doi.org/10.1016/j.wasman.2009.06.004.

26. Arena, U.; Mastellone, M. L.; Perugini, F. Life Cycle Assessment of a Plastic Packaging Recycling System. *Int. J. LCA* **2003,** *8*, 92–98. https://doi.org/10.1007/BF02978432
27. Santos, A. S. F.; Teixeira, B. A. N.; Agnelli, J. A. M.; Manrich, S. Characterization of Effluents Through a Typical Plastic Recycling Process: An Evaluation of Cleaning Performance and Environmental Pollution. *Resour. Conserv. Recycl.* **2005,** *45* (2), 159–171, ISSN 0921-3449. https://doi.org/10.1016/j.resconrec.2005.01.011.
28. García, M. T.; Gracia, I.; Duque, G.; de Lucas, A.; Rodríguez, J. F. Study of the Solubility and Stability of Polystyrene Wastes in a Dissolution Recycling Process. *Waste Manag.* **2009,** *29* (6), 1814–1818, ISSN 0956-053X. https://doi.org/10.1016/j.wasman.2009.01.001.
29. Shen, L.; Worrell, E. Chapter 13 - Plastic Recycling. In *Handbook of Recycling*; Worrell, E., Reuter, M. A., Eds.; Elsevier, 2014; pp 179–190, ISBN 9780123964595. https://doi.org/10.1016/B978-0-12-396459-5.00013-1.
30. Pacheco, E. B. A. V.; Ronchetti, L. M.; Masanet, E. An Overview of Plastic Recycling in Rio de Janeiro. In *Resources, Conservation and Recycling*, 2012; vol 60, pp 140–146, ISSN 0921-3449. https://doi.org/10.1016/j.resconrec.2011.12.010.
31. Siddique, R.; Khatib, J.; Kaur, I. Use of Recycled Plastic in Concrete: A Review. *Waste Manag.* **2008,** *28* (10), 1835–1852, ISSN 0956-053X. https://doi.org/10.1016/j.wasman.2007.09.011
32. Achilias, D. S.; Roupakias, C.; Megalokonomos, P.; Lappas, A. A.; Antonakou, E. V. Chemical Recycling of Plastic Wastes made from Polyethylene (LDPE and HDPE) and Polypropylene (PP). *J. Hazard. Mater.* 2007, *149* (3), 536–542, ISSN 0304-3894. https://doi.org/10.1016/j.jhazmat.2007.06.076.
33. Muringayil Joseph, T.; Murali Nair, S.; Kattimuttathulttara, S.; Haponiuk, J. T.; Thomas, S. Copolymerization of Styrene and Pentadecylphenylmethacrylate (PDPMA): Synthesis, Characterization, Thermomechanical and Adhesion Properties. *Polymers* **2020,** *12*, 97. https://doi.org/10.3390/polym12010097
34. Joseph, T. M.; Mahapatra, D. K.; Luke, P. M. Perspectives of Cashew Nut Shell Liquid (CNSL) in a Pharmacotherapeutic Context. In *Advanced Studies in Experimental and Clinical Medicine*, 2020; pp 123–133. https://doi.org/10.1201/9781003057451-7
35. Suresh, K. I.; Nutenki, R.; Joseph, T. M.; Murali, S. Structural, Molecular and Thermal Properties of Cardanol Based Monomers and Polymers Synthesized Via Atom Transfer Radical Polymerization (ATRP). *J. Macromol. Sci. Part A* **2022,** *59* (6), 403–410. DOI: 10.1080/10601325.2022.2053288

CHAPTER 4

Zein-Based Composites: Synthesis, Characterization Properties, and Applications

JYOTHY G VIJAYAN

Department of Chemistry, M. S. Ramaiah University of Applied Sciences, Bengaluru, Karnataka, India

ABSTRACT

Zein is an edible, biodegradable prolamine corn protein. It is a promising polymer, which has potential applications in different fields. Zein has low water solubility and higher barrier properties. Zein has many applications in industries including coating, textiles, plastics, adhesives, etc., due to the insolubility of zeins in water. Zein has huge application in biomedical field also. It helps to support growing cells and delivery drugs. This review covers synthesis, characterization, and modification of zein and zein-based composites for various applications. There are many methods used for zein characterization that includes nuclear magnetic resonance spectroscopy NMR, DLS dynamic light scattering, MALDI-TOFF, STD, etc. Sponge-like structure scaffolds on zein with enough thickness and porosity helps it to be used in tissue engineering applications. The aim of the review is to make a comprehensive analysis of the synthesis, characterization, and application of zein and its composites. The current trends in research work and future directions are also described.

Foundation and Growth of Macromolecular Science: Advances in Research for Sustainable Development. Meegle S. Mathew, PhD, Józef T. Haponiuk, PhD & Sabu Thomas, PhD, DSc (Eds.)

4.1 INTRODUCTION

Zein is an amorphous polymer, which is thermally stable and viscoelastic in nature. It is a water-insoluble protein derived from maize. It has wide applications in packaging and food industries. Zein is used in designing biopolymer-based nano liposome, post hydrolysis with enzymes such as papain and alcalase. It exhibits excellent thermoplastic, barrier, and hydrophobic properties, which are used for the fabrication of biodegradable composites. Researchers found that zeins displayed characteristics like carrier of antioxidants that are used widely among others. Wilson et al. (1988) discovered that α-helix was the main and only Zein present in the synthesized industry. It has 50% more amino acid residues and near to 10 tandem repeats of helical segment of nonpolar residues.[1] Wang et al. (2008) suggested the 3° tertiary structure of hydrogen α-helices and the observed tandem repeats which is analyzed by surface adhesion with the sample Zein.[2] Mass spectroscopy MALDI-TOF has become the most important spectroscopy used nowadays to identify the molecular weight of Zeins. Through MALDI, nine different α-zein proteins were analyzed. It includes Z19 and Z22. Compared to α-zein, Υ-zein degrade faster when placed outside the test protein body.[8] β-zeins were the slowest among the three. Alpha and gamma zeins degrade very slowly. Larkins et al. (1989) used immune staining and microscopic methods to identify the origin of zein proteins within the protein bodies.[9] Zein is used as a pharmaceutical excipient in tissue engineering. Being natural, it is the most safely used protein compared to other synthetic and individual products. Zein also exhibits a controlled drug-releasing characteristic. Due to that it is used in micro-encapsulation, films, and hydrogels and in drugs related to anticancer activities. Commercial zein, which is very hydrophobic in nature, helps to develop oral-controlled drug delivery systems.

This review explains about zein and zein-based composites, extraction, properties, characterization, and applications. Biodegradation and biocompatibility are the two main factors responsible for the high performance of zein used in different areas. It includes food packaging, biomedical, and pharmaceutical fields. Zein contains hydrophobic and hydrophilic groups so it is important to mix zein with other polymeric moieties to get new derivatives with good applications. Electrospinnability of zein helps to produce zein and zein-based fibers, which have wide applications in different areas.

4.1.1 PROCESSING OF CORN WITH HIGH PROTEINS

4.1.1.1 DRY MILLED CORN

Here, the corn endosperm is differentiated from the pericarp and germ through the milling process. It is water-tempered corn grit and is not exposed to high heat. Dry milled corn (DMC) contains low quantity of proteins, that is, 6–8%. Commercially DMC is processed by three techniques: (1) full-fat milling process, tempering-degerming milling process, and bolted milling process. (2) In tempering milling, water is added more to increase the moisture. This method is used to separate corn into bigger grits, fines, smaller grits, germs, and pericarp. In DMC, corn is dried in low temperature, and then physical grinding continues. This process should not harm the zein physiology and protein. Then the extraction is done with the use of 70% ethanol and the absence of reducing agents. Alpha and beta zein in small quantities were also extracted.[12]

4.1.1.2 WET CORN MILLING

This process helps to produce a side product that contains maximum percentage of zein. Through wet milling, gluten meal is obtained which contains the high percentage of zein and proteins.[11] Steeping techniques of corn enhance the property of protein and help to separate them into the fiber and germ. Here, addition of reducing agents such as sulfur dioxides is used to break the disulfide linkages. Presence of red color of zein by CGM is due to the effect of degree of drying. Duration of drying also causes the reduction in the yield of zein.

4.1.1.3 DRY GRIND ETHANOL

The dry grind ethanol method is used to produce ethanol from corn. Its coproduct is called DDGS. This is typically used as fuel ethanol and beverage alcohol. The traditional dry grind ethanol process is complicated and can cause damage to zein. Water is added with milled corn, lime, and ammonia to form slurry. Next phase is liquefaction. Finally, it goes to high-temperature heating. Then it is cooled to 60°C and glucoamylase is added to synthesize glucose.[8]

4.2 PROPERTIES AND COMPOSITION OF ZEINS

- Zein has 5–12% protein based on its dry weight
- Major 75% of protein is in the tissues of endosperm and 25% in the germ and barn
- Zein contains glutamic acid, leucine, proline, and alanine that are hydrophilicin nature
- Zein in its pure form is hydrophobic, colorless, odorless solid, nonpolar soluble, biodegradable, and edible
- Depending on the solubility, corn products are divided into four groups shown in Table 4.1.

TABLE 4.1 One Solubility of Corn Proteins.

Proteins	Solubility
Albumins	Water soluble
Globulins	Aqueous-saline soluble
Prolamines	Water-alcohol soluble
Glutelins	Water-alcohol insoluble

Due to its hydrophilic and hydrophobic properties, zein is amphibilic in nature. Extracted from corn gluten, zein is not widely used in large-scale quantities. The conventional process of extraction of zein requires high cost. After extraction, zein is yellowish in color. Zein is categorized into three types based on their molar mass. These are alpha, beta, and gama zeins. α-zeins is the major product in the commercial preparation. Researchers found that in maize α-zeins fraction is decreased and ϒ-zein fraction increases as the plant grew at atmospheric carbon dioxide concentration. Composition and structure of zeins depends on the external conditions like solvent, concentration, temperature, reagent, raw materials, and drying.

Zeins are grafted by different processes such as amination, carboxylation, and oxidation. Presence of different functional groups like phosphates, amines, amides, sulfates, hydroxyls, carboxylates etc. on the surface of the fibers enhances the solubility and chemical reactivity of the zeins. Surface functionalization helps to increase the properties of zeins. Zein is an amorphous polymer that exhibits glass transition temperature and viscoelasticity.

4.3 EXTRACTION OF ZEINS

The method of extraction of zein is summarized in the following[11]

***Nonaqueous solvent process*:** Here, in this process, a combination of organic solvent with water or a mixture of 2 organic solvents was used in the oxidation. In this process, 70 solvents comprised the extraction. Temperature can be increased to get more than 60% yield of zein. Solvents like binary or tertiary containing acetone, dioxane, etc. can also be used for zein extraction.

***Aqueous solvent method*:** Here, acidified or alkalized water is used to transfer insoluble amine to soluble salts. Arranging pH more than 7 helps in the degradation of zein using high amount of solvent.

***Enzymatic hydrolysis*:** This process of zein with alcalase was used to generate water-soluble zein compared with the use of other enzymes.

Gelation: Here, solvent selection of zein extraction was done by variation of concentration, temperature, pH, agitation, etc.

4.3.1 ZEIN SOLVENTS

Zein is well known for its solubility in different binary solvents like aliphatic and water or aqueous isopropanol, etc. But zein is not soluble in water. It was found that zeins perfectly dissolve in glycerol at high temperature. Zein/glycerol solution can be heated up to 200°C without any structural change. Zein is also soluble in glacial acetic acid and phenol. Propylene glycol is found to be a better solvent for zein.[12] Ratio of the solvent mixture and temperature are the key factors, which has an effect on the cloud point of zeins. Binary and ternary solvents of zein are shown in Tables 4.2 and 4.3.

TABLE 4.2 Binary Solvents for Zeins.

Secondary solvents	Aliphatic alcohol
Acetone	Acetone
Acetonyl acetone	Benzene
n-Butanol	Ethyl lactate
Ethanol	Acetaldehyde
t-Butanol	Butyl lactate
Methanol	Toluene
Isopropanol	Ethylene glycol
Isobutanol	Dichloro methane
	Nitromethane
	Nitro ethane
	Furfural
	Propylene glycol

TABLE 4.3 Ternary Solvents for Zein.

Water	Lower aliphatic alcohol	One of the following
Water	Acetone	1,3 butanediol
	Acetaldehyde	Diethylene glycol
	Benzene	Ethylene glycol
	Acetonyl acetone	Propylene glycol
	Nitromethane	Dipropylene glycol
	Nitroethane	2,3 butanediol
	Dioxane	1,4-butanediol
	Methyl acetate	
	Formaldehyde	

4.3.2 GELATION

Zein solutions and dispersions get gelled with the variation of heat and time. This property increases with increasing concentration of zein. Zein in alcohol get gelled slower when the quantity of water is less. If the percentage of water is above 5, then gelling stops in the zein solutions. Addition of resins reduces the rate of zein gelation. Addition of aldehydes above 20% also retards the gelation process. Zein solution can be stabilized against gelation by adding the solvents by different processes and steps. Isolation and other process conditions also retard gelation in zeins.[18]

4.3.3 PLASTICIZERS

Along with the solvents, plasticizers also help to improve the properties of zeins. Plasticizers have to be polar or nonpolar in nature, for example, lactic acid, oleic acid, acetanilide, dibutyl tartnate, etc. Water is considered as the most effective plasticizers for the preparation of zein. Primary plasticizers include PEG, PPG, TEG, oleic acid, stearic acid, acetamide, urea, acetanilide, diethanol amine, triethanol amine, glycerol monooleate, glycerol monophthalate, monomethyl azelate, etc.[22] Plasticizers are listed in Table 4.4.

TABLE 4.4 Zein in Plasticizers.

Primary plasticizers	Secondary plasticizers
Glycols	Dibutyl phthalate
Fatty acids	Glycerol
Amides	Dibutylsebacate
Amines	Sorbitol
Sulfonamides	
Glycol ethers	
Esters	

4.4 CHARACTERIZATION OF ZEINS

Zein is a fiber product from the corn industry and is a biopolymer. Zein is a unique and complex material in its purest form. Studies and research has been done on Zein since the last fifty years.. There are many methods of structural characterization of zein before and after its extraction and purification processes. SAXS is used to find the folded helical segment of zein. There are many methods used for zein characterization that includes nuclear magnetic resonance spectroscopy NMR, DLS dynamic light scattering, MALDI-TOFF, STD, etc. Sponge-like structure scaffolds on zein with enough thickness and porosity helps it to be used in tissue engineering.[11,19,24,60,72]

Zein is a biocompatible, renewable, and biodegradable resource that contains major part of corn proteins. Zein and zein-based materials, that is, biodegradable polymers are used in different applications that exhibit their biocompatibility. Zein is a plant-sourced, protein-based biodegradable protein. It has originated from the prolamine protein group and is insoluble in a nonaqueous environment. The main factor to find the solubility of zein is to identify its amino acid composition.

Fourier transform spectroscopy (FT-IR) is used to find chemical changes occurring in the molecule during the reaction process. Another technique generally used for analysis is circular dichroism CD. This method is used to find the secondary structure and is usually used with FT-IR.[7,10,62,73] The tertiary structure also can be analyzed with the two. For secondary structure analysis, CD is considered more reliable than FT-IR.

FT-IR is considered the main option for researchers to have a better understanding of various processes like extrusion, electrospinning, plasma treatment, and other processes of preparation of zein.[45,48,51,55]

Dissolving zein in different solvents is very essential for the majority of its applications. FT-IR is used to analyze the effect of different solvents on neat zein and zein-based products. The influence of different alcoholic solutions as a synthesizing medium and the influence of alcohols on mechanical properties and surface properties were identified by FT-IR.[17,33] To reduce the percentage of a random coil structure, zein has to dissolve in ethanol or isopropanol. It proves that zein films synthesized with isopropanol solutions are highly hydrophilic than ethanol-made zeins. Addition of plasticizers as fillers, for example, glycerol, decreases the beta turn fraction and produce a more volume coil structure formation. Increasing the concentration of plasticizers in zein composites or film preparation leads to elongating and rupturing the structure from 20% to 40%.[41] The association of zein with chitosan, forming microparticles, was investigated and found the application of such biomaterial as a carrier for controlled drug release (Fig. 4.1).

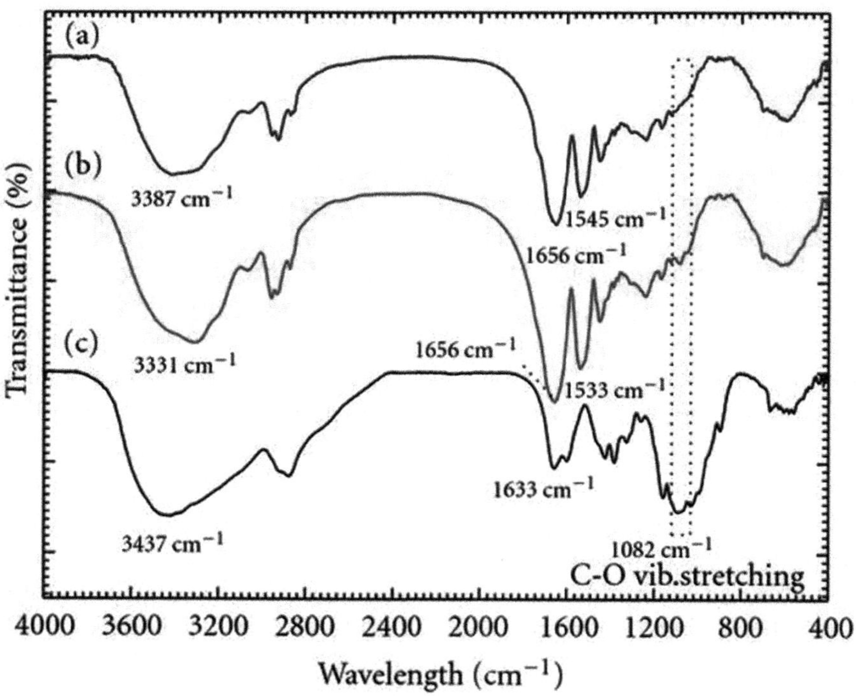

FIGURE 4.1 FT-IR spectra of (a) zien (ZN) microspheres, (b) ZN/CHI microspheres, and (c) pure chitosan (CHI). 2011, MDPI.
Source: Reprinted from Ref. [56]. https://creativecommons.org/licenses/by/3.0/

Mainly 2° (secondary) amides of zeins were influenced by elasticizers that were analyzed by FT-IR. It is easy to identify the bond that is formed in zein-based complexes and microspheres formed from zein protein blend through FT-IR. Pristine zein film exhibits higher brittleness when plasticizers were added.[3,40,53,42] It also helps the researcher to decide the type of plasticizer that needs to be added to the zein. It is considered as the most challenging part.

Sysnthesis of zein fibers by the electrospun method is used to analyze by FT-IR. Compared to the solvent casting method, electrospun was more good in the evaporation of the solvent to increase the mechanical properties of the zeins; the crosslinking technique is usually selected. FT-IR analysis is used to study and analyze the zein-based products. In zein dough, complex formation is attained and also zein is used as a substitute for gluten in many cases. FT-IR is also used for the analysis of molecules on the basis of varying tensile properties of dough with the addition of zein. Researchers also studied the rheology of zein dough and found that a sudden increase in the beta sheet content will increase the viscoelasticity of zeins.[15,47,55,63]

While preparing the films, transparency was considered the most challenging issue. To improve transparency zein fibers were processed with heat and moisture. Column-purified decolored zein was studied by using FT-IR to find the changes in the zein, column filteration caused to hike the α-helical content of zein.[11] The circular dichroism technique also proved the findings of FT-IR, which indicate that the column purification increases α-helical content by 5%. Compared to other spectroscopic techniques FT-IR was efficiently used to find the cleavage of amino acids and the disappearance of primary structure of zein after the extrusion.[45]

FT-IR spectra of zein has the following bands:

TABLE 4.5 FT-IR Spectra Details of Zein.

Wavelength cm^{-1}	Groups
2800–3500	N-H and O-H stretching
1650	C=O carbonyl stretching
1540	Amide II angular deformation of N-H band
1230	Axial deformation vibrations of the C-N bonds

Raman spectroscopy helps to find the structural characterization of different organic materials. Raman spectroscopy is similar to IR. Two are used for the rotational, vibrational, and other motions present at the molecular level of zein samples. Different types of solvents were used to study the secondary and tertiary structures of dissolved zein and Raman spectra was effectively used to analyze the structure. From the most-cited literature, it was found that the plane-polarized Raman result availed from the perpendicular and parallel directions of zein fibers did not reveal any difference. It was found that the Raman spectra of edible zein–chitosan film prepared were not shown any new peak compared to pure samples.[13] A hydrogen bond was developed from the amide groups of zein glutamine and hydroxyl groups of chitosan. It was observed that increasing the content of zein in the film increases the elasticity of the film due to the increase in the discontinuity of the films. It caused the composites to become more extensible. Raman spectroscopy is used to analyze the encapsulation of beta-carotene. Some studies also discovered that when Raman spectroscopy is coupled with Raman microscopy, then it will turn into a powerful tool.[21,65]

Zein fibers can decrease fluorescence characteristics and it can be easily characterized using FT-Raman spectra. By using FT-Raman spectroscopy it was proved that the secondary structure of zein protein is not changed by applying compression to its maximum. FT-Raman spectroscopy is an effective technique to identify the number of specific bonds between materials and also the validation of the encapsulation procedure.[17] Structural data derived from the FT-Raman and FTIR analyses were used to study the zein protein structure. This gives an accurate method to analyze a protein secondary structure based on the amide I band in vibrational spectra (Fig. 4.2).

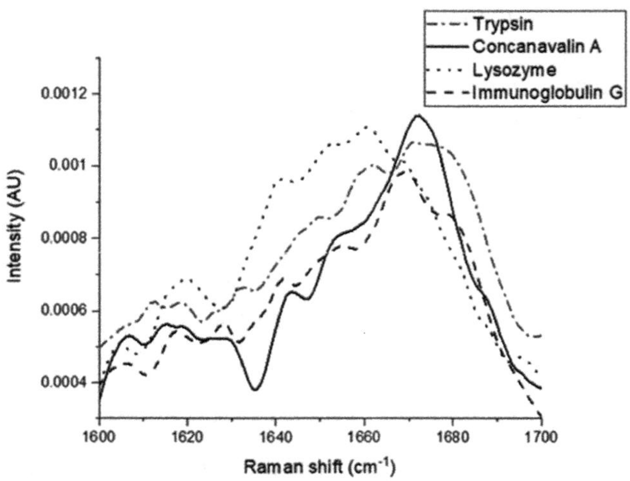

FIGURE 4.2 FT-Raman spectroscopy of standard proteins in aqueous solution (20% w/v) 2020, MDPI.

Source: Reprinted from Ref. [57]. https://creativecommons.org/licenses/by/3.0/.

In circular dichroism studies, the main source to generate absorption is the circularly polarized light. After analyzing the zein sample and its chirality, some well-polarized light gets absorbed by the zein sample and some is not absorbed at all. CD spectroscopy is equally used with some spectroscopy like FT-IR. FTIR-Raman combo to evaluate the secondary structure of zein is highly effective. Purity of neat zein is studied effectively by using CD spectra. Major CD experiments were usually conducted by using the zein–ethanol mix. CD also explain the crosslinking in zein. The self-assembly mechanism of the zein sample was well explained by CD with ELSA.[17,31]

To find the crystallinity of zein XRD is highly recommended. XRD is the most suitable method to identify the organizational structure of zein. SAXS and WAXS are used to find the structure of zein and modified zeins. From SAX, it was proved that there was a major difference between the different morphologies of neat zein and modified zein prepared by the casting method, resin formation, and fiber formation methods.[43] XRD helps to analyze the effects of different parameters on the crystalline index. XRD is used to analyze how the different ways of synthesis effect the structure of zeins. XRD also used to analyze the secondary structure of zeins. From the literature, it was found that the optimum conditions for the fabrication of zein-based composite nanoparticles from ethanol/H_2O cosolvents using electrospinning and the properties of the composite showed good antibacterial activity. The zein/Ag nanoparticles were characterized using X-ray diffraction (XRD) analysis. The antibacterial activity of the zein/Ag composite nanoparticles was also investigated. The XRD patterns and the TEM images indicate the coexistence of a zein matrix and well-distributed Ag nanoparticles.

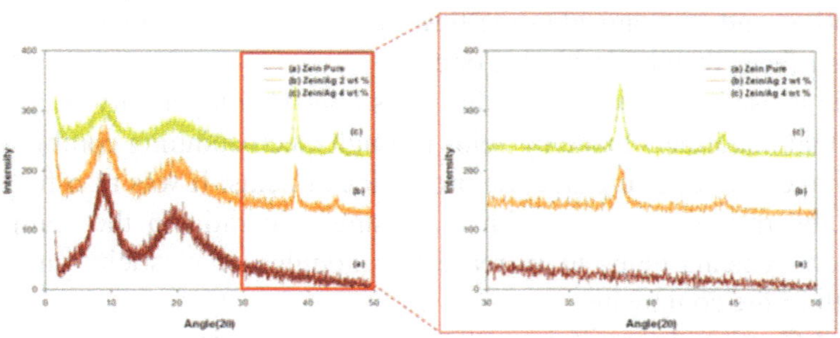

FIGURE 4.3 Three X-ray diffraction patterns of zein/Ag nanoparticles electrospun from ethanol aqueous solutions with an ethanol/water ratio of 8/2 (v/v) at Ag concentrations of (a) 0, (b) 2, and (c) 4 wt% (polymer concentration = 10 wt%, TCD = 15 cm, and applied voltage = 15 kV). 2016, MDPI.

Source: Reprinted from Ref. [74]. https://creativecommons.org/licenses/by/3.0/.

Atomic force microscopy (AFM) is an imaging microscopic technique, which helps to identify the surface topology in micro- and nanoscale dimensions. AFM is a characterization technique that helps to understand and identify the morphology of surface of pure zein and zein-based composites. It also helped to understand the interaction between zein and other molecules. The affinity of zein to the hydrophilic and hydrophobic surfaces was targeted by AFM and DPN (Dippen nanolithography AFM) is used to study the effect of solvents on the structure of films synthesized by the spin coating method. Generally, the AFM image showed smoother surfaces when films were coated with CH_3COOH solutions than Et-OH (95%) solutions. The difference in roughness shown by zein molecules was the result of the net electrical changes of the molecules. AFM helps to identify the surface topography and surface rheology of zeins and its composites[38,44,56]

4.5 APPLICATIONS OF ZEINS

***Film Formation*:** Zeins possess properties like good degree of polymerization and its chemical structure, molecular weight, and other properties, which help to make it in the use of film preparation. By wet and dry extraction methods, film can be prepared. The zein films prepared are hard, brittle, and tough.[33,51] The effective tuning of zein films addition of plasticizers is essential.

Plasticizers, which have nonpolar and polar groups, are most effective in the preparation of zein films. As it is alcohol soluble, it can be effectively used for making films. Films display properties like biodegradability, water absorption, and mechanical properties. Zein has more affinity toward hydrophilic surfaces than hydrophobic ones.

Zein is used as a textile fiber due to its fiber-forming ability. Researchers found that nanofibers made by zeins through elecrospinning combined with cross linkers help to improve the properties of zeins. It was also found that water absorption decreases with increasing zein ratio in the composites. Addition of zeins helps to increase the modulus of elasticity and tensile strength of the composites.

***As A Coating Agent*:** Zein can be used as an efficient coating agent. But when zein is combined with other materials like shellac, it can be used for coating together or individually.[15] Adding compounds like shellac enhanced the properties of zein by making it more water resistant, tough, and scuff resistant. In ancient times, tracel quantities of zeins were used as a substitute for shellac in varnishes. Zein is a component with resins in varnishes in the ratio zein 45: resin 55. Zein gelation tendency can be stopped by decreasing the pH. For that, hydrochloric acid can be added. In resin, packaging zein has found to be an efficient coating agent. Coating can be made by a mix of zein, resin, and plasticizers. These coatings can be used in fiber board containers because of its high grease resistance, printability, and dispersing effect.

***In Paper Industry*:** Zein is used as a pigment coating agent in paper industry also. Here, zein actsas an adhesive to bond, the pigment to the base paper.[66] Zein is of less cost and fast processing compared to casein. But zein-coated paper is proved less efficient in the waxy pick test than casein-coated paper. Zein-coated paper is high in demand in the food industry. Zein-coated paper can be used in quick service packaging. Zein is a natural and best choice for tablet coating. It is rapid and resisted to heat, humidity, and abrasion. These zein coatings concealed the taste and odors of the original tablets without interfering with solubility.

***Zein Films*:** Freestanding films are used in the mid-1930. The factors in the zein production are solvent type, content of plasticizers, substrate type for casting and drying condition. Zein is added with solvent, plasticizers, cross linkers and cast into nonstick substrate, then allow the solvent to evaporate to get zein-based film and can be peeled off from the nonstick mold. Zein sheets also can be prepared by adding zein with the plasticizer and then kneading the mixture. Heat is applied to the mixture, depending upon the plasticizer content. Finally, the softened zein mixture is compressed and rolled into sheets. Zein films are made depending upon the type and quantity of the plasticizer they include.[51] It will also effect the tensile strength and elongation of the film. Plasticizers like glycerols can cause a varied effect for zein film preparation. Zein is hygroscopic in nature. It can absorb water and also lose water from the surrounding environment depending upon the humidity of the environment. Researchers found out that zein films stored at different humidity levels have different tensile properties. Tensile property will decrease in accordance with the increase in humidity.

***Fiber*:** Zein fibers were synthesized by extruding a solution containing zein, urea, and formaldehyde in aqueous methanol. Zein fibers were comparable to wool fiber based on its properties and strength. The formaldehyde solution process could generally reduce the shrinkage of the fiber. Acetylation of the zein fiber helps to increase the water resistance and makes the fiber soft. Curing of zein fibers makes to sustain in alkaline and acidic conditions. Curing mixture includes an inorganic solvent, aldehyde, and strong acid.[52]

***Molded Articles*:** Freshly precipitated zein solutions with plasticizer as water are ductile, soft, and elastic molded in different shapes. Zein plastic replaced casein-type plastic due to its slower curing. Compared to casein plastics, zein plastics are more stable and water resistant. Zein molds were manufactured by using 20% water, 5% aldehyde, and zein along with other materials. Zeins are used now as molded articles.[22] Zein is used as an adhesive to bond wood. Here zein with formaldehyde were used. Zein-based adhesives are strong and better choice for adhering veneers. It is used in cork melting. It is used in gaskets, shoe fillers, etc. because it can resist mold growth.

***Zein-Based Inks*:** Zein is used in printing inks as well. Three types of inks are heat set, vapor set, and flexographic. In heat set and vapor set ink, zein is dissolved in organic solvents. In heat set zein inks, zein and pigment were dissolved in low boiling glycols. Vapor set zein inks are analogics to heat set ink. Here, the selection of a solvent is essential, because the solvent must dissolve in zein and modifying resin. To harden the ink, modifying resins are used. It is used in letter press equipment. Zein ink is also used in flexographic printing. It also has the ability to print on a wide variety of substrates.

***As a Drug Carrier*:** Zein-based nano composite shows excellent application in the drug industry due to the excellent compatibility, degradability, and its surface area.[5,52] It can create an active site with hydrophobic and hydrophilic drugs. It contains high percentage of hydrophobic amino acid residues; it can be used for the release of encapsulated compounds. Zein-based materials are used mainly for carrying hydrophobic drugs, which cannot be absorbed by the human body. Zein acted as an intermediary agent between hydrophobic drugs and epithelial cells of column. The main advantage is that Zein develops hydrophobic drug permeation and absorption, acting as a targeted delivery and controlled release agent. Protein-based polymers are highly used for biomedical applications due to its biocompatibility, biodegradability.[13,25,27,35,37,51,54] Among plant-based protein polymers, zein is attracting researcher's attention due to the production of zein-based film, sheets, nanoparticles, microphores, and nanofibers. Zein-based biomaterials' mechanical and physical properties can be enhanced by crosslinking and plasticizing by the combination with biopolymers or inorganic materials. Zein materials are effective compounds, which possess antibacterial, antioxidant, and anti-inflammatory properties.[57,64,69,70]

Zein–silver composites exhibit effective antibacterial properties and are analyzed in different pH revealed antibacterial and antiseptic properties. It is because the silver prevents the growth of bacteria. Zeins attained this property by the formation of silver–nitrogen covalent bond. Zein–Ag compound hinders bacterial growth. It made zein more biocompatible and leads to wound care. Carbon–zein composites are also developed by Dhandauth et al., through electrospinning. The biocomposite exhibited antithrombiotic properties. Polyurethane zein composite also exhibited a good wound-dressing ability.

4.6 ZEIN COMPOSITES—APPLICATIONS

Zein composites and their characteristics are shown in Table 4.6.

TABLE 4.6 Properties of Zein Composites.

Zein–metal filler composites	Zein ceramic composites	Zein clay composites
• Zein and its biopolymer blends are considered as inorganic materials include metal to form biocomposites. • Incorporation of inorganic bioactive phase causes the bonding between bone tissues and zein matrix composites fast	• Bioactive ceramic fillers in the composite increase mechanical properties and hydrophilicity to the material • Degradability can be boosted by mixing zein with synthetic polymers. • Increase in the quantity of ceramic increases the toughness of the composites • Zein–silver composites exhibit resorbability of scaffolds. It helps to increase the properties of zein-based biomaterials.	• It helps to boost the different properties of zein-based biomaterial. • Zein clay material is effectively used in bone tissue engineering • Increase in clay content in the composite helps to increase the production of bead formation in the nanofiber • Zein clay composites/films/foams have various applications in the areas of biomedical engineering • Composite used in medical therapies in dermatology • Composites exhibit less adsorption of water.

Due to biodegradability and biocompatibility, zein got important application in food and pharmaceutical industry. Zein has been added with many inorganic compounds of less mechanical properties, which exhibited composite materials with improved mechanical properties. Zein is an amorphous polymer, which exhibits plasticizing, viscoelasticity, and glass transition temperature. Zein films are produced by the wet or dry process via solubilization and exhibit thermoplastic properties. The mineralization process will help zein fibers to provide a platform to fabricate new biomimetic scaffolding. Nanoparticles from zein are used to produce edible capsulated films, several drugs, and bioactive compounds. It includes livermectin, coumarin, and 5-flourouracil. Zein nanocarriers are proved to show peak cytotoxicity and cellular uptake against MCF-7 breast cancer cells. Zein nanocarriers can evoke an efficient tumor-targeted delivery system for breast cancer.

For polyurethane-cellulose acetate–zein drugs composite exhibited blood-clotting property than pristine PU nanofibers, Cellulose acetate in the composite enhanced its hydrophilicity and permeability to air and moisture, CA also enhanced a suitable environment for the world which promoted healing of the wound easy. PEG and zein together will form thermoplastic material that is very effective for bone regeneration.[3] Zein–PCL composites will increase the hydrophilicity of the composite and degrade faster than neat PCL scaffold. Zein as a protein is degradable in enzymatic media. It cannot be degraded in the absence of enzymes like pepsin, thermolysin, and trypsin. Cheriyan et al. used different enzymes to produce water-soluble zein. Zein can act as a delivery vehicle due to its efficient interaction with hydrophobic and hydrophilic drugs.[15] Zein-PLA spray is also used for root canal infection. Patel et al. developed curcumine filled zein colloidal nanoparticles. It is very efficient as an oral drug. It was found that encapsulation efficiency increased when decrease in curcumine proportion added. Columns of colloidal particles were also studied in another work by changing the ratios of curcumine in it. In the same way, Paris et al. synthesized zein by wet and dry bioprocess to develop deoiled corn using different proteases.[28]

Kemppainen et al.[35] made 3D-designed PGs. They changed scaffold modulus, which helps new cartilage degeneration. In the same field, another researcher Chen et al. designed PGs elastomer that is used to enhance mechanical property especially myocardinal tissue that can enhance cardiac function. Another scholar Dashdorg et al. in his study developed zein nanofiberous mats by electrospinning. They fabricated Zein/Ag nanocomposite mats by using electrospinning. They fabricated zein silver nanocomposite mats and found effective in wood dressing.

Rai et al. have synthesized bioactive zein-PGS composites and zein-PCL composites for cardiac patch treatments. The bioactive polymer matrix/ scaffold protects the attachment and growth of seeded myogenic and vasculogenic cell limits. Zein can be used effectively as a protective, strong, impermeable, and coating agent for food packaging.[50]

Plant proteins are highly efficient in biomedical applications including soft tissue engineering. Zein, a prolamine-based corn protein, has excellent biomedical properties. The main disadvantage of zein is its low mechanical properties and aqueous stability. To compensate that it is added with polymer scaffold and less toxic solvents. Biocompatibility, antimicrobial characteristics, and mechanical properties make the zein biocomposites a potential biomaterial for biomedical applications.

TABLE 4.7 Characteristics and Morphology of Polymer/Zein Composites.

Material	Morphology	Enhanced Property	References
Zein	Film	Structural	[68]
Zein	Sheet	Tensile	[69]
Zein/PEG	Film	Mechanical	[64]
Zein/PVP	Film	Degradation	[57]
Zein-wax	Film	Controlled release of lysozyme	[3]
Zein-PP	Film	Water barrier	[15]
Zein	Membrane	Electrostatic deposition	[28]
Zein-glycerol	Film	Oxygen permeability	[35]
Zein	Film	Mechanical and physical	[50]
Zein/PCL/HDI	Film	Mechanical and elastic	[71]
Zein/PCL	Coating	Compressive strength and mechanical	[21]

4.6.1 APPLICATIONS OF ZEIN-BASED COMPOSITES

Zein-Based Biomaterials: Zein and its composites based are used for biomedical applications like in drug delivery and tissue engineering. Zein application as in the form of film, which is glossy, tough, and grease proof properties has lots of applications. Zein-based compounds exhibit properties like low water permeability, which helps to resist microbial attacks. Zein rate is calculated which helps to analyze the mechanical properties of zein sheets. Zein can be combined with polymer to improve the properties of the materials.

Zein Filler Composites: Inorganic material is another promising material, which is used to make zein filler composites. The most commonly used inorganic fillers are bioceramics, which are shown below Zein/HA, Zein/

calcium phosphate, zein/bioactive gases, zein/natural clay montmorillonite, and bentonite

(a) *Zein/Ha Composites*: Bioactive ceramics are important in bone tissue engineering due to its natural mineral phase of bones. It is divided into bioinert ceramics, biodegradable ceramics, and bioactive ceramics

Bioinert ceramics	Bioactive ceramics	Biodegradable and bioactive ceramics
Al_2O_3, ZrO_2, C, vitreous C, SiN etc	HA, bioactive glasses-ceramics, calcium carbonate	Phosphate glasses, calcium aluminate, calcium sulfate etc

These bioactive ceramic fillers in polymer will increase the mechanical and hydrophilicity of the material. From the reported studies it was proved that the addition of HA increases the properties of zein-based composites. It helps in biodegradability and mechanical strength. These scaffolds give better cell adhesion and proliferation with good mechanical strength, as shown in Table 4.8.

TABLE 4.8 Characteristics and Morphology of Zein Composites.

Composites	Enhanced properties	References
HA/Zein	Scaffold with good interconnectivity	[52]
Zein/PCL/HA	Scaffold with good mechanical and physical property	[60]
Nano HA/Zein	Mechanical property	[72]
PLGA/Zein/HA	Bone cartilage interface regeneration. Bioactivity and hydrophobicity	[73]
Zein/HACC	Antibacterial activity	[74]

***Zein/Clay Composites*:** Inorganic fillers synthesized from natural clay are highly efficient in the field of food packaging. It helps to increase physical and mechanical properties of zein materials. Agenda is to synthesize zein-based compounds with optimal mechanical, biological, and chemical characteristics, which can replace petroleum-based polymers. Zein/clay materials can be used for biomedical applications. Clay fillers like montmorillonite, mica, kaolinite, zeolite, etc., can be added with zein, which can be organomodified or unmodified. It was found that not only the concentration of clay but the type of clay also has a strong effect on changing the properties of zein fibers. The increase in the clay concentration increases the probability of the formation of the beads. Except unmodified mica, the remaining clay forms exhibit uniform distribution. Clay are natural and nontoxic material with good mechanical, physical, and biological properties. Clay-based compounds can be used as biosensors, scaffold for tissue engineering, and drug delivery.[49,52,59] Montmorillonite clay is also added with zein to enhance the mechanical properties and water uptake reduction. These biocomposites exhibited good mechanical properties, as the filler increases young modulus also increases and the composites become stiffer in nature.

***Other Composites*:** Zein composites were recently synthesized with silver nanoparticles through silver nitrogen coordinate-covalent bond. Composites displayed properties like antimicrobial activities and hemocompatibility. Zein/CNT composites are also good in tissue engineering applications. Zein/PU composites are used in wound-healing applications.

4.7 CONCLUSIONS

Zein production is a low-cost process due to the extraction of zeins from renewable sources. It has good applications in different areas. It was a wrong conception that chemical modification of zein does not influence the biodegradability. Need of cytotoxicity and plasticization on zeins biocompatibility can be studied. Zein with inorganic nanoparticles as fillers have high mechanical strength and more protein-based compounds can be studied. Zein is a biopolymer that possesses qualities like good aspect ratio, biocompatibility, renewability, and easy way of processing with good biodegradability. Compared to other biopolymers originated from plants, zein is the most attracted product for the synthesis of fillers, coating agents, films, sheets, paper, microspheres, NPs, and nanofibers. To enhance the physical, mechanical, and biological properties of zein, different surface modifications have to be used and have to add cross-linking agents, plasticizers, inorganic materials, or biopolymers with zein-based composites. In this review, the performance of zeins and zein-based composites was explained. From the literature, it was clear that due to smart behavior, easy synthesis, recycling, low cost, environment friendly, biocompatibility, etc. make zeins more efficient. By reading this chapter, the researchers could find new approaches, which help them in designing a new zein-based system for different applications.

KEYWORDS

- **zein**
- **barrier properties**
- **biomedical field**
- **adhesives**
- **MALDI-TOFF spectra**

REFERENCES

1. Anderson, T. J.; Lamsal, B. P. Zein Extraction from Corn, Corn Products, and Coproducts and Modifications for Various Applications: A Review. *Cereal Chem.* **2011,** *88*(2), 159–173.
2. Anderson, T. J.; Ilankovan, P.; Lamsal, B. P. Two-Fraction Extraction of A-Zein from DDGS and its Characterization. *Ind. Crops Prod.* **2012,** *37*(1), 466–472.
3. Arcan, I.; Yemenicioğlu, A. Development of Flexible Zein–Wax Composite and Zein–Fatty Acid Blend Films for Controlled Release of Lysozyme. *Int. Food Res. J.* **2013,** *51*(1), 208–216.
4. Argos, P.; Pedersen, K.; Marks, M. D.; Larkins, B. A. A Structural Model for Maize Zein Proteins. *J. Biol. Chem.* **1982,** *257*(17), 9984–90.
5. Arulmozhi, V.; Pandian, K.; Mirunalini, S. Ellagic Acid Encapsulated Chitosan Nanoparticles for Drug Delivery System in Human Oral Cancer Cell Line (KB). *Colloids Surf. B* **2013,** *1*(110), 313–20.
6. Assadpour, E.; Mahdi Jafari, S. A Systematic Review on Nanoencapsulation of Food Bioactive Ingredients and Nutraceuticals by Various Nanocarriers. *Crit. Rev. Food Sci. Nutr.* **2018,** *59*(19), 3129–51.
7. Aswathy, R. G.; Sivakumar, B.; Brahatheeswaran, D.; Fukuda, T.; Yoshida, Y.; Maekawa, T.; Kumar, D. S. Biocompatible Fluorescent Zein Nanoparticles for Simultaneous Bioimaging and Drug Delivery Application. *Adv. Nat. Sci.: Nanosci. Nanotechnol.* **2012,** *3*(2), 025006.
8. Bothast, R. J.; Schlicher, M. A. Biotechnological Processes for Conversion of Corn in to Ethanol. *Appl. Microbiol. Biotechnol.* **2005,** *67*(1), 19–25.
9. Boundy, J. A.; Turner, J. E.; Wall, J. S.; Dimler, R. J. Influence of Commercial Processing on Composition and Properties of Corn Zein. *Cereal Chem.* **1967,** *44*(3), 281–287.
10. Breiteneder, H.; Mills, E. C. Plant Food Allergens—Structural and Functional Aspects of Allergenicity. *Biotechnol. Adv.* **2005,** *23*(6), 395–399.
11. Carter, R.; Reck, D. R. U.S. Patent No. 3,535,305, Washington, DC: U.S. Patent and Trademark Office, 1970.
12. Carter, R.; Reck, D. R. Low Temperature Solvent Extraction Process for Producing High Purity Zein. US Patent 3,535,305, 1970.
13. Cheryan, M. U.S. Patent No. 6,433,146. Washington, DC: U.S. Patent and Trademark Office, 2002.

14. Deng, L.; Zhang, X.; Li, Y.; Que, F.; Kang, X.; Liu, Y.; Feng, F.; Zhang, H. Characterization of Gelatin/Zein Nanofibers by Hybrid Electrospinning. *Food Hydrocoll.* **2019,** 75, 72–80.
15. Doğan Atik, İ.; Özen, B.; Tıhmınlıoğlu, F. Water Vapour Barrier Performance of Corn-Zein Coated Polypropylene (Pp) Packaging Films. *J. Therm. Anal. Calorim.* **2008,** *94*(3), 687–693.
16. Elzoghby, A. O.; Samy, W. M.; Elgindy, N. A. Protein-based Nanocarriers as Promising Drug and Gene Delivery Systems. *J. Control Release* **2012,** *161*(1), 38–49.
17. Esen, A. Separation of Alcohol-Soluble Proteins (Zeins) from Maize in to Three Fractions by Differential Solubility. *Plant Physiol.* **1986,** *80*(3), 623–627.
18. Evans, C. D.; Manley, R. H. Stabilizing Zein Dispersions Against Gelation. *Ind. Eng. Chem. Res.* **1943,** *35*(2), 230–232.
19. Farrell, A. E.; Plevin, R. J.; Turner, B. T.; Jones, A. D.; O'hare, M.; Kammen, D. M. Ethanol can Contribute to Energy and Environmental Goals. *Science* **2006,** *311*(5760), 506–508.
20. Fox, G.; Manley, M. Hardness Methods for Testing Maize Kernels. *J. Agric. Food Chem.* **2009,** *57*(13), 5647–57.
21. Frenot, A.; Chronakis, I. S. Polymer Nanofibers Assembled by Electrospinning. *Curr. Opin. Colloid Interface Sci.* **2003,** *8*(1), 64–75.
22. Jing, L.; Wang, X.; Liu, H.; Lu, Y.; Bian, J.; Sun, J.; Huang, D. Zein Increases the Cytoaffinity and Biodegradability of Scaffolds 3d-Printed with Zein and Poly (E-Caprolactone) Composite Ink. *ACS Appl. Mater. Interfaces* **2018,** *10*(22), 18551–18559.
23. Kasaai, M. R. Zein and Zein-Based Nano-Materials for Food and Nutrition Applications: A Review. *Trends Food Sci Technol.* **2018,** *79*, 184–97.
24. Katayama, H.; Kanke, M. Drug Release from Directly Compressed Tablets Containing Zein. *Drug Dev. Ind. Pharm.* **1992,** *18*(20), 2173–2184.
25. Kemppainen, J. M.; Hollister, S. J. Tailoring the Mechanical Properties of 3d-Designed Poly (Glycerol Sebacate) Scaffolds for Cartilage Applications. *J. Biomed. Mater. Res.* **2010,** *94*(1), 9–18.
26. Khan, I.; Khan, M.; Umar, M. N.; Oh, D. H. Nanobiotechnology and Its Applications in Drug Delivery System: A Review. *IET Nanobiotechnol.* **2015,** *9*(6), 396–400.
27. Kim, S.; Sessa, D. J.; Lawton, J. W. Characterization of Zein Modified with A Mild Cross-Linking Agent. *Ind. Crops Prod.* **2004,** *20*(3), 291–300.
28. Klinkesorn, U.; Sophanodora, P.; Chinachoti, P.; Decker, E. A.; McClements, D. J. Encapsulation of Emulsified Tuna Oil in Two-Layered Interfacial Membranes Prepared Using Electrostatic Layer-By-Layer Deposition. *Food Hydrocoll.* **2005,** *19*(6), 1044–1053.
29. Labib, G. Overview on Zein Protein: A Promising Pharmaceutical Excipient in Drug Delivery Systems and Tissue Engineering. *Expert Opin. Drug Deliv.* **2018,** *15*(1), 65–75.
30. Landry, J.; Guyon, P. Zein of Maize Grain: I Isolation by Gel Filtration and Characterization of Monomeric and Dimeric Species. *Biochimie* **1984,** *66*(6), 451–460.
31. Larkins, B. A.; Pedersen, K.; Marks, M. D.; Wilson, D. R. The Zein Proteins of Maize Endosperm. *Trends Biochem. Sci.* **1984,** *9*(7), 306–308.
32. Lawton, J. W. Zein: A History of Processing and Use. *Cereal Chem.* **2002,** *79*(1), 1–18.
33. Lawton, J. W. Plasticizers for zein: Their Effect on Tensile Properties and Water Absorption of Zein Films. *Cereal Chem.* **2004,** *81*(1), 1–5.

34. Leo, M.; Logan, W. A. U.S. Patent No. 2,882,265. Washington, DC: U.S. Patent and Trademark Office, 1959.
35. Liang, J.; Xia, Q.; Wang, S.; Li, J.; Huang, Q.; Ludescher, R. D. Influence of Glycerol on the Molecular Mobility, Oxygen Permeability and Microstructure of Amorphous Zein Films. *Food Hydrocoll.* **2015,** *44*, 94–100.
36. Luo, Y.; Pan, K.; Zhong, Q. Casein/Pectin Nanocomplexes as Potential Oral Delivery Vehicles. *Int. J. Pharm.* **2015,** *486*(1–2), 59–68.
37. Luo, Y.; Teng, Z.; Wang, T. T.; Wang, Q. Cellular Uptake and Transport of Zein Nanoparticles: Effects of Sodium Caseinate. *J. Agric. Food Chem.* **2013,** *61*(31), 7621–7629.
38. Luo. Y.; Wang, Q. Zein-Based Micro-and Nano-Particles for Drug and Nutrient Delivery: A Review. *J. Appl. Polym. Sci.* **2014,** *131*(16).
39. Manley, R. H.; Evans, C. D. Inventor Process for Extracting Prolamines Patent 2354393, 1944.
40. Manley, R. H.; Evans, C. D. The Critical Peptization Temperatures of Zein in Concentrated Ethyl Alcohol. *J. Biol. Chem.* **1942,** *143*(3), 701–702.
41. Manley, R. H.; Evans, C. D. Binary Solvnts for Zein. *Ind Eng Chem.* **1943,** *35*, 661–665.
42. Mannheim, A.; Cheryan, M. Water-Soluble Zein by Enzymatic Modification in Organic Solvents. *Cereal Chem.* **1993,** *70*, 115–115.
43. Matsushima, N.; Danno, G. I.; Takezawa, H.; Izumi, Y. Three-Dimensional Structure of Maize A-Zein Proteins Studied by Small-Angle X-Ray Scattering. *Biochim. Biophys.* **1997,** *1339*(1), 14–22.
44. Momany, F. A.; Sessa, D. J.; Lawton, J. W.; Selling, G. W.; Hamaker, S. A.; Willett, J. L. Structural Characterization of A-Zein. *J. Agric. Food Chem.* **2006,** *54*(2), 543–547.
45. Morawsky, N.; Martino, G. T.; Guth, J.; Tsai, J. Jeffcoat, R. U.S. Patent No. 5,518,717. Washington, DC: U.S. Patent and Trademark Office, 1996.
46. Oehlke, K.; Adamiuk, M.; Behsnilian, D.; Gräf, V.; Mayer-Miebach, E.; Walz, E. Greiner, R. Potential Bioavailability Enhancement of Bioactive Compounds using Food-Grade Engineered Nanomaterials: A Review of The Existing Evidence. *Food Funct.* **2014,** *5*(7), 1341–59.
47. Oncley, J. L.; Jensen, C. C.; Gross Jr, P. M. Dielectric Constant Studies of Zein Solutions. *J. Phys. Chem.* **1949,** *53*(1), 162–174.
48. Osborne, T. B.; Cornelison, R. W. U.S. Patent No. 691,966. Washington, DC: U.S. Patent and Trademark Office, 1902.
49. Paliwal, R.; Palakurthi, S. Zein in Controlled Drug Delivery and Tissue Engineering. *J. Control Release* **2014,** *189*, 108–22.
50. Panchapakesan, C.; Sozer, N.; Dogan, H.; Huang, Q.; Kokini, J. L. Effect of Different Fractions of Zein on the Mechanical and Phase Properties of Zein Films at nano-scale. *J. Cereal Sci.* **2012,** *55*(2), 174–182.
51. Parris, N.; Coffin, D. R. Composition Factors Affecting the Water Vapor Permeability and Tensile Properties of Hydrophilic Zein Films. *J. Agric. Food Chem.* **1997,** *45*(5), 1596–1599.
52. Qu. N.; Dickey, L. C. Extraction and Solubility Characteristics of Zein Proteins from Dry-Milled Corn. *J. Agric. Food Chem.* **2001,** *49*(8), 3757–3760.
53. Payne, R. A.; Tyrpin, H. T. 1990. Method of Producing an Aqueous Zein Solution. *Patent EP, 383428.*

54. Rai, R.; Tallawi, M.; Grigore, A.; Boccaccini, A. R. Synthesis, Properties and Biomedical Applications of Poly (Glycerol Sebacate) (PGS): A Review. *Prog. Polym. Sci.* **2012,** *37*(8), 1051–1078.
55. Saito, H.; Shinmi, O.; Watanabe, Y.; Nishimura, K. Aso. K. Papain-Catalyzed Hydrolysis of Zein in an Aqueous Organic System. *Agric Biol Chem.* **1998,** *52*(3), 855–856.
56. Müller, V.; Piai, J. F.; Fajardo, A.R.; Fávaro, S.L.; Rubira, A.F. Muniz, E.C. Preparation and Characterization of Zein and Zein-Chitosan Microspheres with Great Prospective of Application in Controlled Drug Release. *J. Nanomater.* **2011**.
57. Sadat, A.; Joye, I. J. Peak Fitting Applied to Fourier Transform Infrared and Raman Spectroscopic Analysis of Proteins. *Appl. Sci.* **2020,** *10*(17), 5918.
58. Selling, G. W.; Biswas, A.; Patel, A.; Walls, D. J.; Dunlap, C.; Wei, Y. Impact of Solvent on Electrospinning of Zein and Analysis of Resulting Fibers. *Macromol. Chem. Phys.* **2008,** *208*(9), 1002–1010.
59. Serizawa, T.; Iida, K.; Matsuno, H.; Kurita, K. Prolonged Degradation of End-Capped Polyelectrolyte Multilayer Films. *Polym. Bull.* **2006,** *57*(3), 407–413.
60. Sessa, D. J.; Mohamed, A.; Byars, J. A. Chemistry and Physical Properties of Melt-Processed and Solution-Cross-Linked Corn Zein. *J. Agric. Food Chem.* **2008,** *56*(16), 7067–7075.
61. Shao, P.; Liu, Y.; Ritzoulis, C.; Niu, B. Preparation of Zein Nanofibers with Cinnamaldehyde Encapsulated in Surfactants at Critical Micelle Concentration for Active Food Packaging. *Food Packag. Shelf Life* **2019,** *22*, 100385.
62. Sousa, F. F. O. D. Luzardo-Álvarez, A.; Blanco-Méndez, J.; Otero-Espinar, F. J.; Martín-Pastor, M.; Macho, I. S. Use of 1H NMR STD, WaterLOGSY, and Langmuir Monolayer Techniques for Characterization of Drug–zein Protein Complexes. *Eur. J. Pharm. Biopharm.* **2013,** *85*(3), 790–798.
63. Sousa, F. F. O.; Luzardo-Álvarez, A.; Blanco-Méndez, J.; Martín-Pastor, M.; Nmr Techniques in Drug Delivery: Application to Zein Protein Complexes. *Int. J. Pharm.* **2012,** *439*(1–2), 41–48.
64. Takagi, K. Teshima, R.; Okunuki, H.; Sawada, J. Comparative Study of In Vitro Digestibility of Food Proteins and Effect of Preheating on the Digestion. *Biol. Pharm. Bull.* **2003,** *26*(7), 969–973.
65. Takahashi, H.; Yanai, N. U.S. Patent No. 5,342,923. Washington, DC: U.S. Patent and Trademark Office, 1994.
66. Tillekeratne, M.; Easteal, A. J. Modification of Zein Films by Incorporation of Poly (Ethylene Glycol) S. *Polym. Int.* **2000,** *49*(1), 127–134.
67. Torres-Giner, S.; Gimenez, E.; Lagaron, J. M. Characterization of the Morphology and Thermal Properties of Zein Prolamine Nanostructures Obtained by Electrospinning. *Food Hydrocoll.* **2008,** *22*(4), 601–614.
68. Trezza, T. A.; Vergano, P. J. Grease Resistance of Corn Zein Coated Paper. *J. Food Sci.* **1945,** *9*(4), 912–915.
69. Ullah, S.; Hashmi, M.; Khan, M. Q.; Kharaghani, D.; Saito, Y.; Yamamoto, T. Kim is Silver Sulfadiazine Loaded Zein Nanofiber Mats as a Novel Wound Dressing. *RSC Adv.* **2019,** *9*(1), 268–277.
70. Wang, Y.; Padua, G. W. Tensile Properties of Extruded Zein Sheets and Extrusion Blown Films. *Macromol. Mater. Eng.* **2003,** *288*(11), 886–893.

71. Wang, Y.; Filho, F. L.; Geil, P.; Padua, G. W. Effects of Processing on the Structure of Zein/Oleic Acid Films Investigated by X-Ray Diffraction. *Macromol. Biosci.* **2005,** *5*(12), 1200–1208.
72. Wani, T. A.; Shah, A. G.; Wani, S. M.; Wani, I. A.; Masoodi, F. A.; Nissar, N.; Shagoo, M. A. Suitability of Different Food Grade Materials for the Encapsulation of Some Functional Foods Well Reported for their Advantages and Susceptibility. *Crit. Rev. Food Sci. Nutr.* **2016,** *56*(15), 2431–54.
73. Wu, S.; Myers, D. J.; Johnson, L. A Factors Affecting Yield and Composition of Zein Extracted from Commercial Corn Gluten Meal. *Cereal Chem.* **1997,** *74*(3), 258–263.
74. Yang, S. B.; Rabbani, M. M.; Ji, B. C.; Han, D. W.; Lee, J. S.; Kim, J. W. Yeum, J. H. Optimum Conditions for the Fabrication of Zein/Ag Composite Nanoparticles from Ethanol/H2O Co-Solvents using Electrospinning. *Nanomaterials* **2016,** *6*(12), 230.
75. Zhang, Y.; Cui, L.; Che, X.; Zhang, H.; Shi, N.; Li, C.; Chen, Y.; Kong, W. Zein-Based Films and Their Usage for Controlled Delivery: Origin, Classes and Current Landscape. *J. Control Release* **2015,** *206*, 206–19.
76. Zhong, Q.; Jin, M. Nanoscalar Structures of Spray-Dried Zein Microcapsules and In Vitro Release Kinetics of the Encapsulated Lysozyme as Affected by Formulations. *J. Agric. Food Chem.* **2009,** *57*(9), 3886–3894.
77. Zhong, Q.; Jin, M. Nanoscalar Structures of Spray-dried Zein Microcapsules and In vitro Release Kinetics of the Encapsulated Lysozyme as Affected by Formulations. *J. Agric. Food Chem.* **2000,** *57*(9), 3886–3894.

CHAPTER 5

Nanocellulose Extracted from Nutshells as a Potential Filler for Polymers: A Green Approach

JIJA THOMAS[1] and RANIMOL STEPHEN[2]

[1]*Department of Chemistry, St Mary's College, Sulthan Bathery, Wayanad, Kerala, India*

[2]*Department of Chemistry, St. Joseph's College (Autonomous), Devagiri, Calicut, Kerala, India*

ABSTRACT

Today, the world focuses on environmental management and sustainable development through eco-friendly materials. Nanocellulose is one such green material, which is abundantly available in nature. High mechanical strength, low density, high crystallinity, renewability, nontoxicity, and biodegradability are among the unique qualities of nanocellulose. Nanocellulose can be broadly classified as cellulose nanocrystals (CNCs) and nanocellulose fibers. There are a lot of works reported on the isolation of nanocellulose from agricultural and municipal wastes. This chapter draws attention to the extraction, properties and application of nanocellulose from nutshells. This gives a comprehensive idea about the structure, history, sources and methods of isolation of both cellulose nanofibers (CFs) and CNCs. The surface hydroxyl group makes nanocellulose hydrophilic and hence it is incompatible with hydrophobic polymer moiety. This can be addressed by suitable surface modifications. Nanocellulose has potential applications in various fields, such as functional papers, optics, packaging, hygiene products,

Foundation and Growth of Macromolecular Science: Advances in Research for Sustainable Development. Meegle S. Mathew, PhD, Józef T. Haponiuk, PhD & Sabu Thomas, PhD, DSc (Eds.)

energy storage, catalysis, adsorbents, skin care, biosensors, and in polymer nanocomposites.

5.1 INTRODUCTION

Considering the rapid depletion of petroleum products and growing environmental concerns, there is a rising demand for development of environmentally compatible materials and cellulose is one among them. Cellulose is the world's most abundant organic polymer with an estimated production of 1.5 trillion tons annually. Cellulose can be derived from plants, amoebae, fungi, bacteria, and sea animals.[1] Tunicate is considered one of the best sources of nanocellulose owing to its length and high crystallinity, but high cost of harvesting and limited availability creates hurdle.[2]

Reuse and revalorization of agricultural and industrial waste for the production of nanocellulose are gaining more relevance from economic and environmental points of view. There are about 2000 species of valuable fiber plants in different parts of our world. Plant-derived cellulose is found to be a mixture of hemicellulose, lignin, pectin, and other substances. Cellulose is the reinforcing material holding hemicellulose and lignin as a matrix in wood. Hemicellulose is a cell wall polysaccharide that binds strongly with cellulose. It is a branched polysaccharide with several sugar monomers such as pentoses (β-D-xylose, α-L-arabinose), hexoses (β-D-mannose, β-D-glucose, α-D-galactose), and uronic acids (β-D-glucuronic acid, α-D-4-*O*-methylglucuronic acid, and α-D-galacturonic acid). They are amorphous, low molecular weight, and more reactive compared with cellulose which makes them more susceptible to hydrolysis. Lignin is a complex cross-linked polymer of phenyl-propane, methoxy groups, and noncarbohydrate polyphenolic substance which acts as a glue filling the voids in between cellulose and hemicellulose. Cellulose content varies in a range of 25–60% depending on the plant source. Biocompatibility and renewability in addition to their abundance make plant-based cellulose a focus of attention. High crystallinity and amorphous transition temperature are some unique properties which differentiate it from starch. Abundance of surface hydroxy groups gives the benefit of tailoring and hence has numerous applications.[3] A simple molecular structure of cellulose is represented in Figure 5.1.

Cellulose is a linear homopolymer of D-glucopyranoside units linked by 1,4β-glycosidic linkages. It is a tasteless, odorless, hydrophilic polymer with the molecular formula $(C_6H_{10}O_5)_n$, where n is the degree of polymerization.

Cellulose is insoluble in water and is the most organic solvent. It is chiral and biodegradable with a density of about 1.50 g/cm.[3] It melts at 467°C and can be broken down chemically into glucose units by treating it with concentrated mineral acids at high temperature.[4] Each monomer has two free secondary OH groups at C-2 and C-3 positions and one free primary OH group at C-6 position.[5] Several hundreds to 10 million glucose units condense together to form a straight chain polysaccharide unit referred to as CFs, which has a diameter of 3–15 nm and an aspect ratio in the range of 20–200. The free hydroxyl group of one such polysaccharide thread hydrogen bonds with another thread via intermolecular hydrogen bonding to form microfibers with diameter ranging from 2 to 20 nm. Depending on their origin, the microfibril diameters may vary. These OH groups also function as the active sites for chemical modifications.[6] Cellulose microfibrils on acid hydrolysis undergo transverse cleavage along the amorphous regions and further sonication results in a rod-like material with a relatively low aspect ratio, which is referred to as cellulose whiskers, nanorods or rod-like cellulose crystals which re-aggregate via hydrogen bonds leading to another cellulose structure called microcrystalline cellulose (MCC). MCC is a commercially available material which finds application as a rheology control agent and as a binder in the pharmaceutical industry.[6–7]

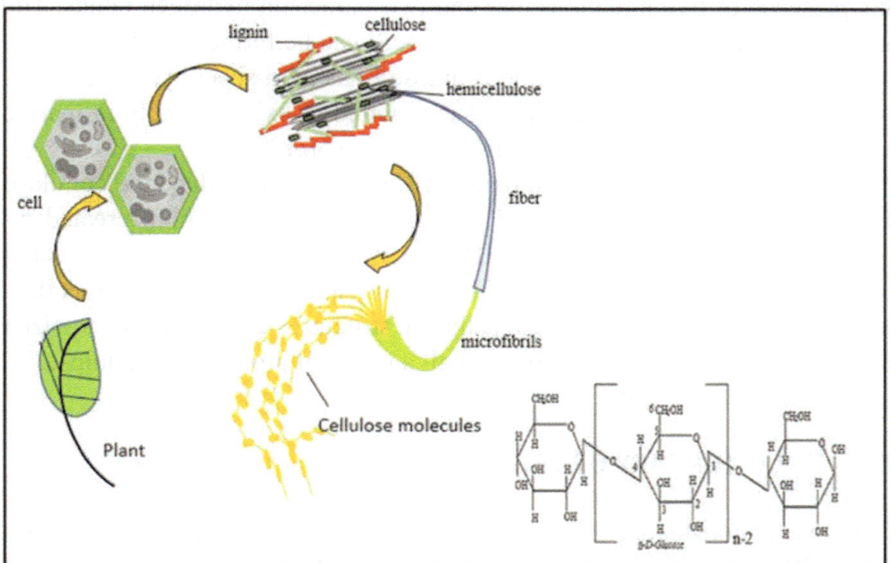

FIGURE 5.1 Hierarchial structure of plant cell.

Cellulose can exist in six different polymorphic forms known as cellulose I, II, III, IV, V, and VI. Polymorphs are solids having the same chemical composition and different crystal structures which results in variations in density, solubility, melting point, and tensile strength. The natural form of cellulose is cellulose I which is partly crystalline with structures I_α and I_β. ^{13}C NMR spectroscopic studies confirmed that natural cellulose exists in two distinct polymorphic forms cellulose I_α and cellulose I_β. Cellulose produced by bacteria and algae is enriched in I_α, whereas cellulose of higher plants consists mainly of I_β. Cellulose I_α has a single-chain triclinic structure and I_β has two-chain monoclinic structures. The remarkable mechanical properties of nanocellulose can be attributed to its special arrangements. Cellulose I_α is thermodynamically less stable compared with cellulose I_β, as it can irreversibly change to I_β at high temperatures. Cellulose I_α and cellulose I_β both have parallel unit cells, bound together through hydrogen bonding of different patterns and hence they differ in their crystalline structure as shown in Figure 5.2. Cellulose II has a folded chain structure with antiparallel unit cells bound to itself which makes it less polar. Cellulose I is metastable, which when dispersed in water or chemically treated can irreversibly change into cellulose II. Cellulose II has a monoclinic structure similar to cellulose I_β. Cellulose II is formed mainly by mercerization or regeneration. The greater stability of cellulose II in comparison with other polymorphs is due to the extensive hydrogen bonding. It is characterized by the presence of five intermolecular and two intrachain hydrogen bonds. The intermolecular hydrogen bonds in cellulose II can be further divided as intra-sheet and inter-sheet bonds, while cellulose I has intra-sheet hydrogen bonds as shown in Figure 5.3. Cellulose II is reported to exist naturally in marine algae Halicystis[8] and in gram-positive bacterium *Sarcina*.[9]

With various chemical treatments, it is possible to produce the cellulose III and cellulose IV structures. Cellulose III and IV are both created by reversible reactions with cellulose III being hexagonal and cellulose IV being orthogonal. Most of the research is done on either cellulose I or II as they are the most common.[10] All these cellulose structures constitute a nano-sized crystalline region surrounded by a small amount of amorphous cellulose.

5.2 BRIEF HISTORY OF CELLULOSE

Cellulose is considered as an interesting material due to its amazing properties. With the advent of nanotechnology, a lot of works are reported on the extraction of nanocellulose from plant sources and their utilization in a variety of applications.

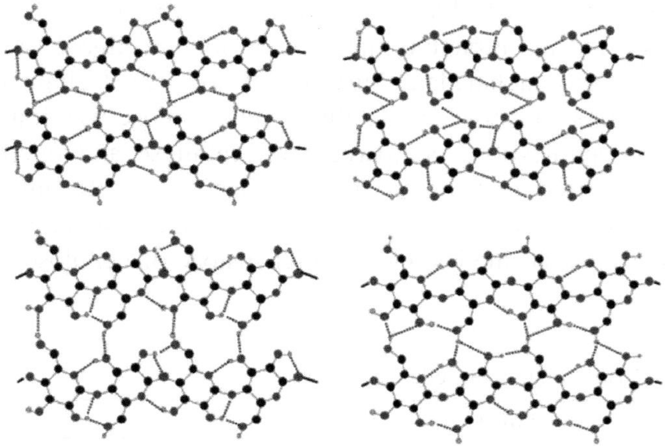

FIGURE 5.2 Hydrogen bonding patterns in cellulose I_α and I_β.[11]

Source: Reprinted with permission from Ref. [11]. Copyright © 2004 American Chemical Society

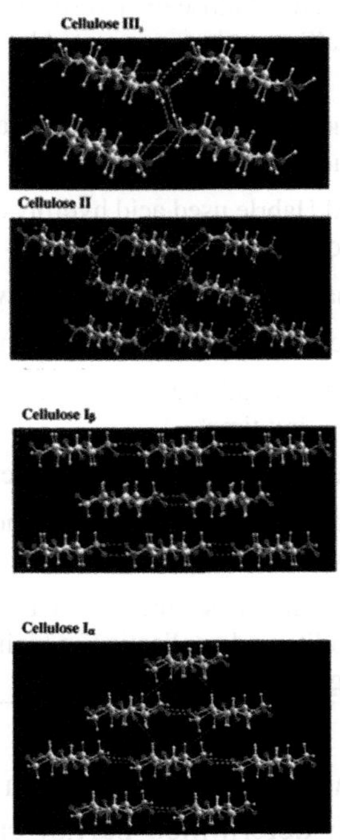

FIGURE 5.3 Projections of the crystal structures of cellulose III, cellulose II, cellulose I_α, and cellulose I_β down the chain axes directions. C, O, and H atoms are represented as gray, red, and white balls, respectively. Covalent and hydrogen bonds are represented as full and dashed sticks, respectively.[12]

Source: Reprinted with permission from Ref. [12]. Copyright © 2004 American Chemical Society

Cellulose being renewable and strong compared with steel, glass fiber, etc. is considered an attractive material for solving global issues.[13–14] Its attractive properties include large surface area, low density, high mechanical strength, stiffness, and biodegradability. Nanocellulose was found to have many intense applications compared with cellulose and microcellulose. Nanocellulose-based composites find applications in the medical field, electrical devices, packaging materials, barrier films, membranes, optical appliances, paints, cosmetics, etc. A brief history of cellulose is given in Table 5.1.

TABLE 5.1 A Brief History of Cellulose.

Year	Major works	Ref
1838	Anselme discovered and isolated cellulose	15
1858	Nageli discovered partial crystallinity of cellulose using polarized light microscopy	16
1914	Nishikawa found cellulose microstructure to have rod-like crystalline regions connected by amorphous regions	18
1922	Cellulose structure identified as used today by Staudinger and Fritschi	27
1940s	Nickerson and Habrle used acid hydrolysis to isolate cellulose nanocrystals	19
1950s	First SEM image using palladium shadowing technique	26
1983	CNF prepared for the first time using homogenizer	23
1995	CNC derived from tunicate was used as a reinforcement in latex for the first time	20
2004	Pure cellulose nanofiber prepared by electrospinning	24
2011	First pilot plant for the production of nanocellulose established by Innventia in Sweden	25
After 2012	Reliable and economic sources, extraction techniques, reinforcing ability and application in high performance material is highly explored	28

Anselme discovered and isolated cellulose in 1838.[15] After 20 years of cellulose discovery, Nageli determined partial crystallinity of cellulose in 1858 by using polarized light microscopy.[16] Later, during 1910 to 1930, crystalline structure was confirmed and characterized using powder X-ray diffraction.[17] In 1914, Nishikawa explained cellulose microstructure to have rod-like crystalline regions connected by amorphous regions.[18] Nickerson and Habrle and many others in 1940s developed a method for acid hydrolysis of cellulose using hydrochloric acid and ferric chloride.[19] This procedure for the isolation of cellulose nanocrystals (CNCs) remains nearly the same even today. The rod-like structure of CNC was first confirmed using SEM by Morehead. He used palladium-shadowing technique to show higher contrast. Favier et al.[20] were the first to demonstrate the use of CNC derived from tunicate as reinforcement in latex with increased storage modulus.[20] CNC is reported as a reinforcement in polymers, such as polyethylene, polyurethane, polyester, and many more.[21–22] After 2012, cellulose is identified as a new generation material capable of replacing synthetic nanomaterials in many fields of application. The revalorization of agricultural wastes and green methods for the extraction of nanocellulose are gaining great relevance with a number of scientific publications in this area.

5.3 TYPES OF NANOCELLULOSE

Cellulosic materials with at least one of its dimensions less than 100 nm can be termed nanocellulose. Two major types of nanocellulose are cellulose nanofibrils (CNFs) and CNCs, which vary appreciably in their properties as given in Table 5.2. Bacterial nanocellulose (BNC) is another category, which is successfully synthesized by certain strains of bacteria under favorable conditions.

TABLE 5.2 Types of Nanocellulose.

	Other names	**Dimensions**
Micro/nano fibrillated cellulose (NFC)	Cellulose nanofibrils (CNF), cellulose microfibrils (CF), nanofibrillated cellulose (NFC), microfibrillated cellulose (MFC)	Composed of crystalline and amorphous regions with width length greater than 1 μm, of 5–100 nm and aspect ratio of 10–100, are often produced through mechanical treatment.
Cellulose nanocrystals (CNC)	Nanocrystalline cellulose (NCC), whiskers, nanowhiskers, rod-like cellulose	Width of 5–20 nm, lengths of 50–350 nm and aspect ratio of 5–30.

Purified cellulose can be physically separated to produce CNFs containing both crystalline and amorphous regions of the cellulose fiber or chemically separated through acid hydrolysis to produce CNCs containing only the crystal domains of the cellulose fibers.[29–30] Different structures of nanocellulose result in different properties and uses. CNCs with no amorphous region have high crystallinity and hence a higher Young's modulus. The aspect ratio of CNC and CNF vary appreciably as 8 and 50, respectively.[30] CNFs have a higher length compared with CNC, resulting in the availability of more bonding sites to interact with a polymer matrix.

Both types of nanocellulose do increase the toughness of the composite. CNFs can increase the viscosity of water by several orders of magnitude to create stable gels with less than 1 wt% added. Like CNCs, the gels created by CNFs are shear thinning, but because they lack the liquid crystal property of the CNCs, there is no shear thickening region and the increase in viscosity is not as high. CNFs can also be used to stabilize emulsions. Droplet size in emulsions will be large while using CNFs because of the large size of fibers compared with CNCs.[31–32]

Recently, CNCs are getting tremendous importance in the scientific world owing to the combination of many attractive properties, such as biodegradability, lightweight, nontoxicity, stiffness, renewability, adaptable surface chemistry, gas impermeable, optical transparency, low thermal expansion, and improved mechanical properties.

Environmental benefits together with improved physiochemical properties make CNCs a better reinforcement material compared with Kevlar, Boron nanowhiskers, carbon nanotube, and carbon fibers, as given in Table 5.3. There is a substantial enhancement in mechanical properties due to greater interfacial interaction. CNCs can be used as effective reinforcement with low density in composites with significant increases in elastic modulus, glass transition temperature while retaining the optical properties of the host polymer. Nanocellulose is a green alternative for many nonbiodegradable nanoscale materials and a great volume of research is going on in this area.

TABLE 5.3 Mechanical Properties of CNC and other Reinforcing Materials.

Material	Tensile strength (GPa)	Elastic modulus (GPa)	Density(g cm^{-1})	Ref
CNC	7.5–7.7	110–220	1.6	34
Kevlar	3.8	124–130	1.4	35
Steel wire	4.1	210	7.8	36
Boron nanowhisker	2–8	250–950	–	33
Carbon fiber	1.5–5.5	150–500	1.8	36

5.4 SOURCES OF NANOCELLULOSE

Nanocellulose can be isolated from a variety of natural sources which can be broadly classified as (1) lignocellulosic source; (2) animal, algae, and bacterial sources.

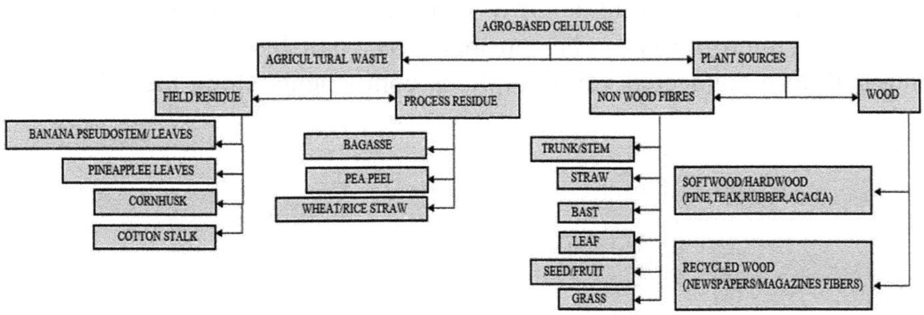

FIGURE 5.4 Classification of lignocellulosic sources.

A major source of cellulose available at low cost includes higher to lower plants which get constantly replenished by photosynthesis. There are about 2000 species of valuable fiber plants in different parts of our world. Such lignocellulosic sources include wood (hard and soft wood), seed (cotton), bast (flax, hemp), cane (bamboo, bagasse), leaf (pineapple, sisal), straw (rice, wheat), fruits, fruit peels, and nutshells as shown in Figure 5.4. Biocompatibility and renewability in addition to their abundance make plant-based cellulose a focus of attention for numerous new applications. There is a large amount of waste produced in the agricultural sector and is unavoidable. Valorization of these solid wastes is of great significance. Since the last few decades, such agricultural wastes are identified as potential sources of cellulose. This is considered as an excellent solution to reduce their accumulation. Literature shows a large variety of cellulose sources, such as corn stalk, cotton flax, hemp, jute, ramie, cereal straw, banana peels, sisal, rice husk, sugarcane bagasse, wood,[37–47] etc. Still, there is a constant search for new and unexplored sources of cellulose.

5.5 NUTSHELLS AS SOURCE OF NANOCELLULOSE

Recently, revalorization of agricultural waste is a relevant topic of research. Nutshells are the least explored source of cellulose among various agricultural wastes. A few such nutshells are shown in Figure 5.5. All these are verified as efficient sources of cellulose with reported yields between 30 and 60%. Lignocellulosic materials contain cellulose along with various cementing materials, such as hemicellulose and lignin. The strength and stiffness of nutshells are primarily due to the presence of cellulose microfibrils. Intermolecular and intramolecular bonding causes the agglomeration of these microfibrils into a crystalline array with their width ranging from 5 to 30 nm.[37] Each microfibrils can be considered as a hair-like strand with crystalline cellulose connected via amorphous components along the microfibril axis.

FIGURE 5.5 Images of various nutshells.

Low economic and energy cost together with the advantage of waste disposal makes agricultural wastes a notable source of cellulose. Complicated structure of cellulose is considered as an important challenge in the cost-effective extraction of cellulose. Integrated production of nanocellulose and biofuel is a proposed solution to make this economical process. Sugarcane bagasse, a residue of the alcohol industry was effectively utilized for the production of nanocellulose by Mandal.[48] Shells of peanut,[49] Pistachio,[50] walnut,[51] coconut,[52] plum seed,[53] and fruit shells shells of camellia oleifera Abel,[54] almond[55] are among recently explored for the isolation of cellulose. Plant-derived cellulose exists as a mixture of hemicellulose, lignin, pectin, and other substances. Cellulose content varies in a range of 25–60% depending on the plant source. The chemical composition of various nutshells and other plant-based cellulose sources is summarized in Table 5.4.

TABLE 5.4 Chemical Composition of Various Cellulose Sources.

Materials	Cellulose	Hemicellulose	Lignin
Almond	29.9 ± 0.7	25.1 ± 0.7	30.1 ± 0.5
Peanut shell	35.7	30.2	18.7
Camellia oleifera Abel	17.8	19.5	15.8
Sago seed shell	36.5	22.5	23.6
Soy hull	56.4	12.5	18.0
Pistachio	43.08 ± 0.19	25.30 ± 0.46	16.33 ± 0.41
Walnut	36.38 ± 0.05	27.85 ± 0.31	43.70 ± 0.57
Chestnut	44.12 ± 0.23	16.28 ± 0.35	36.58 ± 0.26
Poplar	44.12 ± 0.23	30.21 ± 0.11	21.24 ± 0.31
Coconut shell	38.47 ± 0.39	22.36 ± 1.47	28.04 ± 0.57
Wheat straw fiber	43.2 ± 0.15	34.1 ± 1.2	22.0 ± 3.1
Ramie fiber	69.83	9.63	3.98
Areca nut husk fiber	34.18	2.83	31.60
Pineapple leaf fiber	81.25 ± 2.45	12.31 ± 1.35	3.36 ± 0.58
Sugarcane bagasse	43.6	27.24	27.7
Rubber wood	45 ± 3	20 ± 2	29 ± 2
Kenaf bast	63.5 ± 0.5	17.6 ± 1.4	12.7 ± 1.5
Sengkang leaves	37.3 ± 0.6	33.4 ± 0.2	24 ± 0.8
Palm rachis	44.0	28.0	14

Crystallinity is a key factor, as it determines the mechanical strength and reinforcing ability of nanocellulose. Native cellulose exists as cellulose I which may change to a more stable cellulose II structure on chemical treatment. Percentage of crystallinity increased with each step of chemical treatments and reached the maximum value after acid hydrolysis. This was described as, due to the removal of noncellulosic components on pretreatments and removal of amorphous cellulose on acid hydrolysis leaving only crystalline cellulose in the suspension. The range of crystallinity in literature is from 50 to 95%. Crystallite size depends on the strength of acid, reaction time, and temperature, and increase in these parameters will result in shortening of the cellulose chain. Literature reports a crystallite size in the range 1–10 nm. The percentage of crystallinity observed for nanocellulose from various nutshells and other plant-based cellulosic materials are tabulated in Table 5.5.

TABLE 5.5 % Crystallinity of Various Cellulose Sources.

Raw materials	% Crystallinity
Ground nut shell	74
Sago seed shell	Cellulose I 69
	Cellulose II 72
Plum seed shell	54
Camellia oleifera Abel shell	72
Pistachio shell	67
Peanut shell	95
Soy hull	77.8
Walnut shell	72
Almond shell	67.50
Citrus waste	77
Lotus leaf stalk	55
Wheat straw fiber	57.5
Ramie fiber	55.48
Areca nut husk fiber	37
Pineapple leaf fiber	35.97

Native cellulose exists as cellulose I. Allotropic form in which cellulose is obtained in most studies is cellulose I. Presence of cellulose I is confirmed by XRD peaks at $2\theta = 14.5$ (101), 17.5 (110), 22 (200), and 34.6 (004), [56–57] whereas cellulose II is confirmed by characteristic peaks at $2\theta = 12.0$ (110), 20.0 (210) and 22.0 (200). A mixture of polymorphs cellulose I and cellulose II was reported in some studies. Alkaline and acid hydrolysis may result in a change in crystalline arrangement from cellulose I to cellulose II.[57] Zeta potential is a measure of stability of the CNC suspension. Zeta potential value within the range –15 to 15 mV will result in the agglomeration of nanocellulose since it does not have enough charge to repel each other. For a suspension of nanocellulose to be homogeneous and stable, zeta potential value should be less than –30 mV and greater than 25 mV. Sulfuric acid hydrolysis will result in the grafting of negatively charged sulfate group on nanocellulose and hence a greater value of zeta potential is expected. Low values of zeta potential indicate weak electrostatic repulsion between CNCs. Negative zeta potential value below –30 mV indicates the stability of nano-suspension together with high resistance to agglomeration. The presence of sulfate group on the surface of CNC increases the suspension stability due to electrostatic repulsion, at the same time decreases the thermal stability. Thermogravimetric analysis (TGA) reveal the least thermal stability for CNCs compared with the raw sample and α-cellulose.

Morphology of cellulose at various stages was analyzed using SEM, TEM, and AFM in various studies. Dimensions of nanocellulose will strongly depend on the source, processing techniques, and hydrolysis conditions. TEM images of CNC isolated from various sources, such as soy hull and pistachio nutshell are shown in Figure 5.6. This confirms that the final shape, size, length, and width are highly influenced by the source and extraction techniques adopted for the preparation.

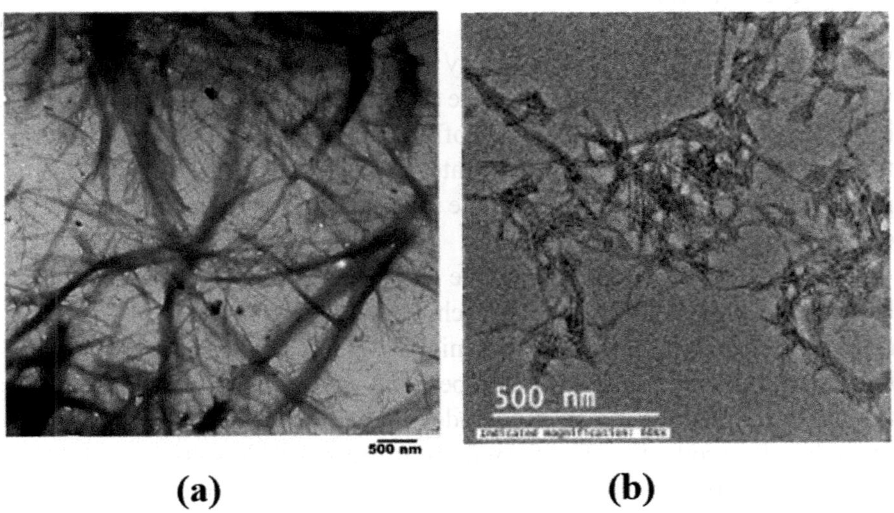

FIGURE 5.6 TEM images of CNC isolated from various sources. (a) Soy hull CNC[58] [Reproduced with permission from Elsevier], (b) pistachio nutshell.[59]

Source: a- Reprinted with permission from Ref. [58]. Copyright © 2007 Elsevier Ltd.; b- Reprinted with permission from Ref. [59]. Copyright © 2017 Elsevier Ltd.

5.6 METHODS FOR ISOLATION OF NANOCELLULOSE FROM AGROWASTES

Nanocellulose can be isolated from lignocellulosic biomass through a process involving two stages: (1) pretreatment to remove noncellulosic components, such as lignin, hemicellulose, wax, and oil to isolate MCC, (2) mechanical, chemical or enzymatic treatments, chemomechanical or bacterial methods to generate nanocellulose. Different chemical and mechanical methods or combination of methods used for the preparation of nanocellulose is summarized in Figure 5.7.

Isolation techniques have a profound influence on the morphology and properties of nanocellulose. CNCs with an average diameter of 5–35 nm and length of a few hundred nms or CNF with an average diameter of 3–100 nm and length in the micrometer range can be isolated from cellulose and can be used as a reinforcement in polymer matrices. Isolation of spherical CNC from sago seed shells with 10–15 nm size and from groundnutshells with length 67–172 nm and width 5–18 nm using sulfuric acid hydrolysis are reported in literature.[60–61]

5.6.1 PRETREATMENT

Pretreatments involve most commonly alkali treatment and bleaching. Alkali treatment will depolymerize the native cellulose structure and hence the defibrillation of external cellulose microfibrils. This is followed by bleaching to fully eliminate the cementing material from the fiber. Even the same procedures on different natural source materials may produce nanocellulose of varying dimensions and yield.

The pretreatment steps to produce high-purity cellulose, such as sulfur-based pulping and chlorite-based bleaching involve the emission of poisonous and toxic gases which are cancerogenic. Delignification using chlorine-free bleaching agents, such as hydrogen peroxide is considered an environmentally friendly technology, as it degrades into water and oxygen leaving no secondary by-products.[62] The formation of irrecoverable salts, consumption of large quantities of water for its neutralization, and incorporation of salts into the biomass matrix are among the disadvantages of sodium hydroxide or ammonium hydroxide catalyzed alkaline hydrolysis. Further, the disposal of this salt and alkaline wastewater results in environmental pollution. Also, the presence of NaOH may cause recrystallization or partial degradation of cellulose resulting in a significant reduction of cellulose crystallinity index.[63]

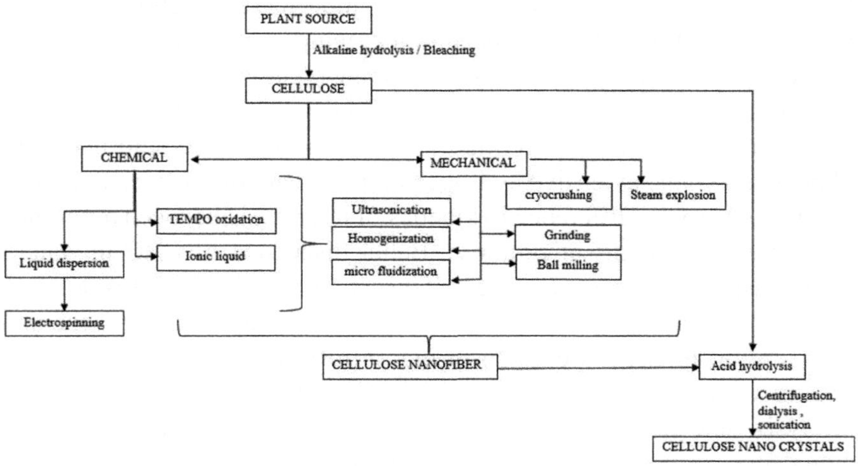

FIGURE 5.7 Possible routes for the preparation of nanocellulose.

Steam explosion is considered to be another effective pretreatment method. This process involves milling of biomass followed by subjecting it to high pressure of 14–16 bar at a temperature 200–270°C. Further pressure inside the digestor is dropped down quickly causing a steam explosion. This method helps to remove hemicellulose and lignin as water-soluble moiety leaving behind cellulose in the sample. Steam explosion is more effective for hardwoods than softwoods.

STAGE II

5.6.2 EXTRACTION OF CELLULOSE NANOCRYSTALS

Acid hydrolysis is the most common method to extract nanocellulose from lignocellulosic starting material. Acid hydrolysis can remove amorphous region to produce crystalline cellulose. Depending on the source and reaction conditions, such as acid used and its concentration, time and temperature, the yield and quality of nanocellulose may vary. In order to maximize the yield and preserve the morphology of nanocellulose, reaction conditions are to be optimized. Acid hydrolysis will be followed by repeated centrifugation, dialysis against deionized water, and finally ultrasonication to get homogeneous dispersion of crystalline nanocellulose.[64]

Use of different inorganic acids, such as sulfuric (H_2SO_4), hydrochloric (HCl), tungstophosphoric acid, and rarely nitric (HNO_3), phosphoric

(H_3PO_4), and hydrobromic acid (HBr)[65] or their mixtures under controlled hydrolysis conditions of acid concentration, temperature, time and agitation are reported in various studies. Recently, some organic acids, such as oxalic, p-toluene sulfonic, maleic, and benzene sulfonic acids are also being used to prepare nanocellulose.[66] Hydronium ions generated during acid hydrolysis will break down the extensive hydrogen bonding network between the cellulose chain due to the difference in kinetics of hydrolysis between amorphous and crystalline domains in cellulose. Better dispersion of CNC in water is obtained using sulfuric acid as the sulfate ester creates an electrostatic layer in between and thus improving the stability of the suspension. A schematic representation of acid hydrolysis using sulfuric acid is shown in Figure 5.8. Elazzouzi-Hafraoui et al. investigated the influence of temperature and hydrolysis time on the sulfuric acid hydrolysis of cotton and found that length decreases with increase in temperature.[67]

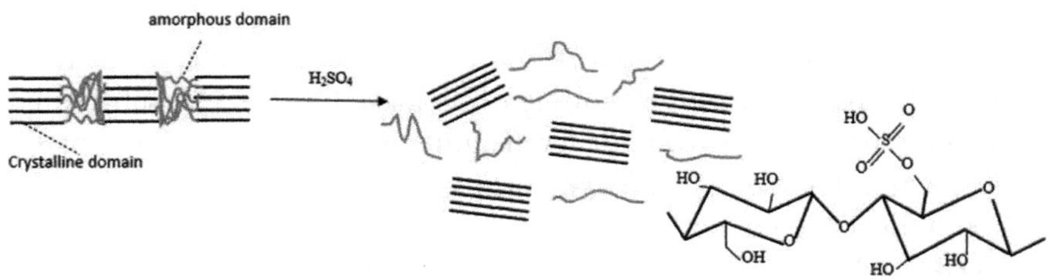

FIGURE 5.8 Acid hydrolysis of cellulose using sulfuric acid.

Valorization of waste liquor containing many undesirable products and sugar by over degradation of cellulose and residual sulfuric acid creates environmental concerns. Low thermal stability of resultant nanocellulose leads to limitations in processing techniques at relatively high temperature. Yield, degree of polymerization, and crystallinity index are found to decrease, with an increase in temperature and acid treatment time.[68] Preparation of nanocellulose from the lignocellulosic biomass by using a co-catalyst along with sulfuric acid could be a green approach to overcome vigorous conditions, such as high temperature and long hours of hydrolysis. Metal salt catalysts could possibly break the hydrogen bonding and cause the degradation of cellulose.[69–71] This is considered to be a less corrosive and environmentally friendly method compared with inorganic acids. The catalytic ability of metal salts including group 1 (alkali metals), group 2 (alkaline earth metals), group 13 metal and transition metals, were widely investigated by Peng and his co-workers.[72] Recently, Chen et al. utilized a one-pot oxidative hydrolysis process in acidic condition using $H_2O_2/Cr\ (NO_3)_3$ solution as a bleaching agent and hydrolysis medium to isolate nanocellulose from selected municipal waste. Resultant nanocellulose showed a high crystallinity index of 62.6–83.6% and fiber length ranging from 15.6 to 46.2 nm.[73]

Ionic liquids and deep eutectic solvents are gaining tremendous attention due to their environmental benefits. Ionic liquid, IL can solubilize cellulose and act as a catalyst in hydrolysis of cellulose. IL possesses lots of advantages like excellent electrical conductivity, ionic mobility, chemical, and thermal stability, along with the selective dissolution of many organic and inorganic substances and recyclability.[74] Man et al. carried out MCC hydrolysis in 1-butyl-3-methylimidazolium hydrogen sulfate. The resultant nanocrystalline cellulose (NCC) showed high crystallinity but reduced thermal stability. As cellulose is soluble in ionic liquids, it can also be easily regenerated from ionic liquids.[75] Myllymaki and Aksela could successfully delignify lignocellulosic materials for pulping process.[76] They directly used microwave irradiation or pressure and further regenerated with the same degree of polymerization. But there was a significant change in its morphology with the addition of water, ethanol, or acetone leaving lignin and other extractives in the solution. This method resulted in higher yield compared with steam explosion and chemical pretreatments. Zhang and his coworkers prepared cellulose acetate with a high degree of substitution in 1-allyl-3-methylimidazolium chloride (AMIMC1) without any catalyst. Handa and his group used a Zn-based ionic liquid for the efficient O-acetylation of cellulose. Acylation and carbonylation of cellulose in 1-butyl-3-methylimidazolium chloride (BMIMCl) in the absence of any catalysts was done under mild conditions using a slight excess of reagent by Heinze and Barthel in a short reaction time.[77] Novo et al. reported a supercritical water hydrolysis method for the extraction of CNCs. This method is a green alternative for nanocellulose extraction. They carried out supercritical water hydrolysis at 12°C and 20.3 MPa for 60 min and obtained highly crystalline rod-shaped CNCs (79%). Low yield (21.9%) and need for complex reactors make this substandard to conventional mineral acid hydrolysis.[78] Surface carboxylated NCC was prepared with high yield by ultrasonic-assisted oxidation using 2,2,6,6-tetramethylpiperidine-1-oxyl (TEMPO) by Mishra et al.[79] This method resulted in thinner nanocellulose with higher degree of fibrillation.[79] Highly crystalline and uniform c-NCC was prepared by Leung et al.[80] using a simple, sustainable, and low-cost method. This novel one-step procedure used ammonium persulfate (APS) oxidation at 60°C for 16 h. Reaction time was substrate-dependent with only 3 h for bacterial cellulose. APS is a novel method, as it is nontoxic. This method requires no chemical pretreatment for isolating nanocellulose from biomass, which is a major advantage compared with acid hydrolysis.[80] A novel enzyme-assisted catalysis method was suggested by Satyamurthy and Vigneswaran for the controlled hydrolysis of MCC obtained from cotton fiber, to generate nanocellulose in an anaerobic microbial consortium. The major advantages of this method were biocompatible, less energy consumption, and no surface sulfation with preservation of the chemical structure of cellulose.[81] A few such resources, their dimensions, and techniques used for the isolation of nanocellulose are reviewed in Table 5.6 as well.

TABLE 5.6 Dimensions of Nano Cellulose with Techniques Used for the Isolation of Nanocellulose from Various Sources.

Sources	Dimensions	Technique used (chemical/ mechanical)	Ref
Pea nut shell	Nanocellulose with an average particle size of 343 nm	Acid hydrolysis using sulfuric acid	48
Pistachio nut shell	CNF with diameter 20–30 nm	Acid hydrolysis using sulfuric acid	49
Walnut shell cellulose	_	Alkaline hydrolysis and bleaching	50
Plum seed shell	CNC with length of 100–800 nm and height less than 14 nm	Acid hydrolysis using sulfuric acid	52
Camellia oleifera Abel fruit shells	CNC with average diameter of 6 ± 2 nm and length 500 ± 100 nm	Acid hydrolysis using sulfuric acid	53
Almond nut shell	_	dissolution of bleached almond shell in ionic liquid, coagulation of cellulose-ionic liquid solution in water, and freeze-drying	55
Sago seed shells	Spherical CNC with 10–15 nm size	Acid hydrolysis using sulfuric acid	60

TABLE 5.6 *(Continued)*

Sources	Dimensions	Technique used (chemical/ mechanical)	Ref
Ground nut shells	CNC with length 67–172 nm and width 5–18 nm	Acid hydrolysis using sulfuric acid	61
Shea nut shells	–	Alkaline hydrolysis	82
Soy hull	CNF with diameter 50–100 nm	Acid hydrolysis using sulfuric acid	83
Sugarcane bagasse	CNC of 20–60 nm diameter and 250–480 nm length	Acid hydrolysis using sulfuric acid	84
Bleached potato pulp	CNF of approximately 5 nm width	Disintegration in a blender followed by homogenization by 15 passes in a homogenizer at 500 bars and 90–95°C	85
Soybean stock	CNF, 50–10 nm in width and several micrometer lengths	Cryocrushing and subsequent defibrillation at 500–1000 Pa	86

Pretreatment using alkalis, such as sodium hydroxide, potassium hydroxide, calcium hydroxide, hydrazine or ammonium hydroxide will remove hemicellulose and lignin. Swelling of cellulose in alkali may increase its surface area and decrease its crystallinity.[28] Organic solvents, such as methanol, ethanol, acetone, and ethylene glycol can also be of the same function.

5.6.3 EXTRACTION OF CELLULOSE NANOFIBERS

Mechanical treatment involves high shear force which cleaves the cellulose fiber.

(1) REFINING AND HOMOGENIZATION

Homogenization is a technique used widely for the preparation of nanofibrillar cellulose. Refining is done in most case, prior to homogenization. High-pressure homogenization may break inter and intramolecular hydrogen bonding and release nanocellulose of 10–20 nm width. In homogenization technique, the cellulose suspension will be passed through a vessel placed between two valve seats, and high pressure is applied resulting in fibrillation of cellulose. To improve the degree of fibrillation, multiple homogenization steps can be done. One of the drawbacks is high-energy consumption which can be reduced by various pretreatment methods. Another major issue is the clogging of the system, which can be minimized by reducing the fiber size prior to homogenization.[87]

(2) MICROFLUIDIZATION

This technique was used for the first time by Zimmermann.[88] In this method sample slurry will be passed through a Z- or Y-shaped chamber having a channel size of 200–400 μm, followed by the application of high pressure. The resulting shear force causes the collision of suspension with channel walls, resulting in delamination of the fibers. High-energy consumption and clogging, can be dealt with pretreatment and reverse flow through the chamber respectively as it has no in-line moving parts.

(3) GRINDING

In this method, sample slurry is passed through a super mass collider grinder with a static and rotating stone rotor. The gap between the rotors can be changed according to the type of raw materials. The resulting shear force can split up the cell wall structure and cellulosic bonds producing NFC. Even though the process has many advantages such as high efficiency, low energy consumption, large capacity, and less clogging, the high mechanical strength employed may damage the fiber resulting in low crystallinity, thermal stability, and physical strength.[89]

(4) BALL MILLING

This technique employs a hollow cylindrical jar partly filled with balls (metal, ceramic, or zirconia) and the sample is added to it. As the container rotates the sample, balls and walls of the container rotate causing fibrillation. Ball-to-cellulose mass ratio, milling time, and ball size are important factors determining the properties of CNF. Pretreatment using alkali can weaken the hydrogen bonds and fibrillation is achieved easier.[90]

(5) CRYOCRUSHING

Cryocrushing is another method used to prepare CNF. In this method, cellulose sample is frozen in liquid nitrogen and mechanically crushed. The high shearing force causes the release of exerted pressure on the ice crystals on the cell walls and subsequently breaks down the cell walls forming CNFs.[91]

(6) ULTRASONICATION

Sample in this method is exposed to ultrasonic waves. This method employs alternative production of high and low-frequency waves to create, expand, and collide the gas bubbles in an aqueous solution. These hydrodynamic forces weaken the hydrogen bonding and release CFs. The source of cellulose, sonication time, and power output are found to affect the morphology of CNFs.[92]

(7) STEAM EXPLOSION

Steam explosion, a thermomechanical process employs pressurized steam followed by a sudden release of pressure to break down the glycosidic and hydrogen bonds and isolate CNFs. The advantages of this method are minimum energy consumption, environmental impact, equipment corrosion, and the use of less hazardous chemicals. Nonuniformity is the major drawback of this method. Further, combination of steam explosion with high-speed blending and ultrasonication is effective in this regard.[93]

(8) EXTRUSION

In an extruder, the sample is introduced into the feeding zone. The shear temperature and pressure are maintained with the kneading and heating zones. Screw speed, screw configuration, screw length-to-diameter ratio (L/D), temperature, feed rate, and die shape/size may be varied to achieve desired morphology and structure. At higher pass, more defibrillation is achieved, but thermal stability and crystallinity decreases. This is due to the fiber damage at higher passes.[94]

5.7 POLYMER COMPOSITES WITH NANOCELLULOSE FROM NUTSHELLS

A composite consists of multiple components resulting in better combination of properties. Properties vary depending on the particle size, shape, distribution, and interfacial interaction between the constituents. Nanofiller is expected to show enhanced properties due to large surface area and hence better interaction. Literature shows that nanocellulose is suitable to be used as reinforcement in hydrophilic and hydrophobic matrices either biodegradable or nonbiodegradable. Hydrophobic surface modification can improve the dispersion of nanocellulose in hydrophobic matrices and hence enhance the properties.

Bano et al.,[61] successfully isolated CNCs from groundnut shells with a total yield of 12%. The isolation technique includes chemical pretreatment and acid hydrolysis resulting in rod-shaped nanocellulose.[61] They fabricated a polymer nanocomposite membrane of sulfonated poly (ether ether ketone) SPEEK with ethylene glycol using CNC as reinforcement. This membrane can be used as a polymer electrolyte membrane with applications in fuel cells. CNC as a reinforcement could improve the mechanical and thermal stability, reduce water uptake and resist oxidation while retaining the proton conductivity. Surface hydroxyl and sulfonic acid groups could effectively interact with ionic moieties of polymer matrix and generate a pathway within the membrane for the conduction of protons.

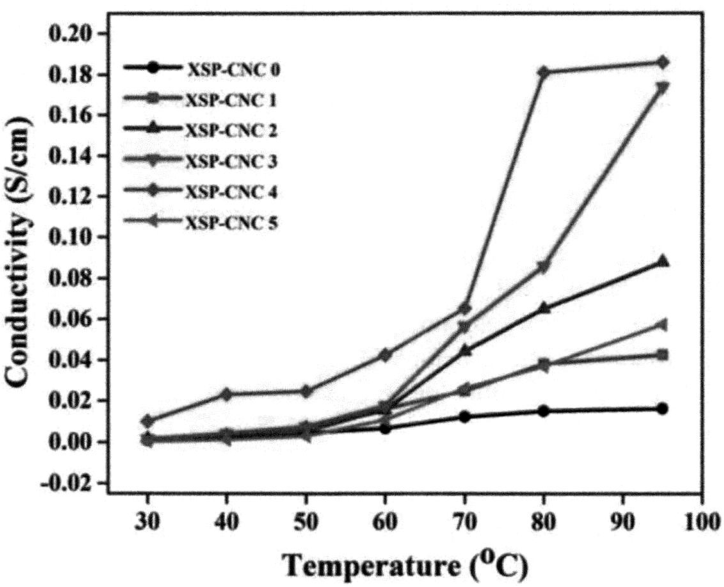

FIGURE 5.9 Proton conductivity of XSPEEK/CNC membranes at different temperatures and 95% relative humidity.[61]

Source: Reprinted with permission from Ref. [61]. Copyright © 2018 Elsevier Ltd

The proton conductivity of SPEEK membranes with varying quantities of CNC was studied. With increasing temperature, conductivity also increased for all membranes as shown in Figure 5.9. Incorporation of CNC resulted in greater conductivity compared with neat SPEEK membrane and maximum is shown by 4 wt% CNC. Shaniba et al.,[49] isolated both MCC and NC from peanut shells via alkaline hydrolysis, bleaching, and acid hydrolysis.[49] FTIR spectra of MCC prove the complete removal of hemicellulose and lignin. Dynamic light scattering (DLS) measurement of freeze-dried sample shows that 99% of particles have a particle size of 343 nm. Silane modification of the prepared peanut shell powder was done using triethoxyvinylsilane and dicumyl peroxide. Biocomposites were prepared by two roll mixing mills with sulfur as a vulcanizing agent. Surface modification helped to improve the interfacial interaction between polymer and filler. Composites were prepared via compression molding. Styrene butadiene rubber (SBR) reinforced with silane-modified peanut shell powder at filler loading 10Phr showed enhanced thermal and mechanical properties. Gaonkar isolated MCC from coconut shell and compared physiochemical properties, such as moisture, pH of 2% solution, starch content, cellulose, total ash, copper number, average particle size, degree of polymerization, intrinsic viscosity, flow rate, angle of repose, and compressibility with commercially available MCC.[52] Most of the properties were comparable while particle size, molecular mass, and degree of polymerization were slightly less compared with commercial MCC. Nanocellulose was isolated by Frone from plum seed shells via two different methods, one alkali hydrolysis and subsequent sonication (cellulose nanofiber, CF) and acid hydrolysis (cellulose nano crystals, CN).[53] The prepared CN was having a length between 100 and 800 nm and crystallinity increased from 34% for raw PS to 51 and 54% depending on the method of preparation. polylactic acid/poly(3-hydroxybutyrate) (PLA/PHB) blend composite reinforced with nanocellulose was prepared via solution casting. Nanocellulose was found to enhance the thermal stability and crystallinity of the membrane. Cellulose nano rystals resulted in better Young's modulus and storage modulus compared with CFs emphasizing its better reinforcing ability. The effect of CN and CF on the mechanical properties of nanocomposite is shown in Figure 5.10.

Addition of CN resulted in an increase in both tensile strength (12 MPa to 15 MPa) and Young's modulus (1705 MPa to 2791 MPa) while CF resulted in small effect on the mechanical properties. The strong intermolecular interaction between the polymer matrix and CN nanofiber is favored by the surface sulfate group. Better interaction along with its higher crystallinity resulted in good mechanical properties. The drop in elongation at break with the addition of CN is due to the reduced chain mobility. Figure 5.11

shows mainly three mass loss regions. PLA/PHB/CN nanocomposite shows higher peak temperatures corresponding to the main degradation peak. This is attributed to better dispersion and presence of sulfate groups on the surface resulting in greater interaction with the matrix. Due to the thermal decomposition of cellulose, higher residue was determined for both PLA/PHB/CN and PLA/PHB/CF in an inert atmosphere at 710°C and under oxygen at 720°C compared with neat PLA/PHB matrix.

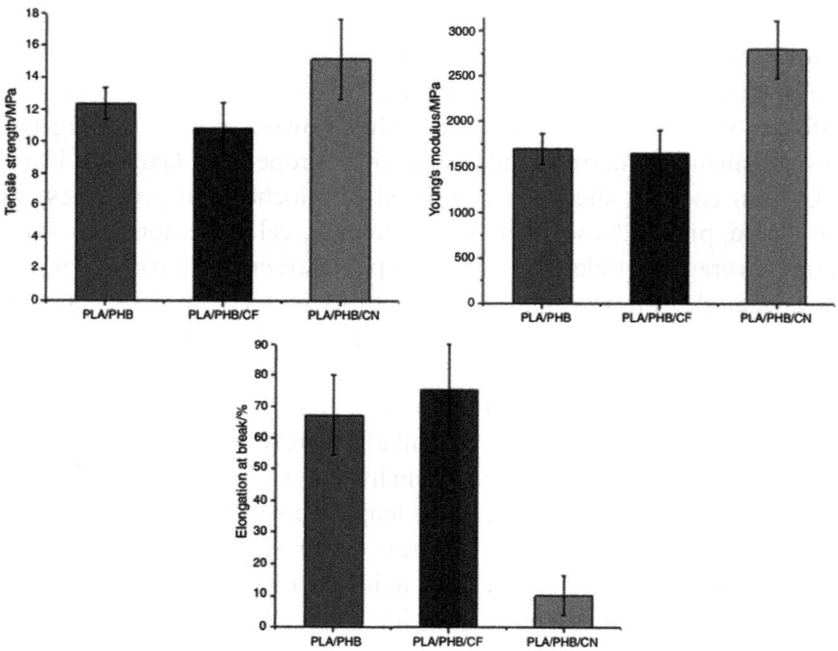

FIGURE 5.10 Young's modulus, tensile strength, and elongation at break for neat PLA/PHB and biocomposite film containing CF.[53]

Source: Reprinted with permission from Ref. [53]. Copyright © 2019, Akadémiai Kiadó, Springer Nature

Fruit shells of *Camellia oleifera* Abel contain about 17.5% of cellulose. Yao et al. isolated CNC from fruit shells of *Camellia oleifera* Abel via alkali treatment, H_2O_2 bleaching, and sulfuric acid hydrolysis and studied the structural and optical properties.[54] The prepared CNC had a needle-like structure, with 72% crystallinity, thermal stability above 230°C, and high aspect ratio. Nanopaper prepared from CNC suspension using vacuum filtration showed a high transmittance above 90% for high visible light. Modification of CNC with butyric anhydride could improve its dispersion in organic solvents. They fabricated biodegradable polylactic acid nanocomposites with both CNC and modified CNC. Compared with CNC, butyrated CNC showed enhanced mechanical properties due to better interfacial interaction.

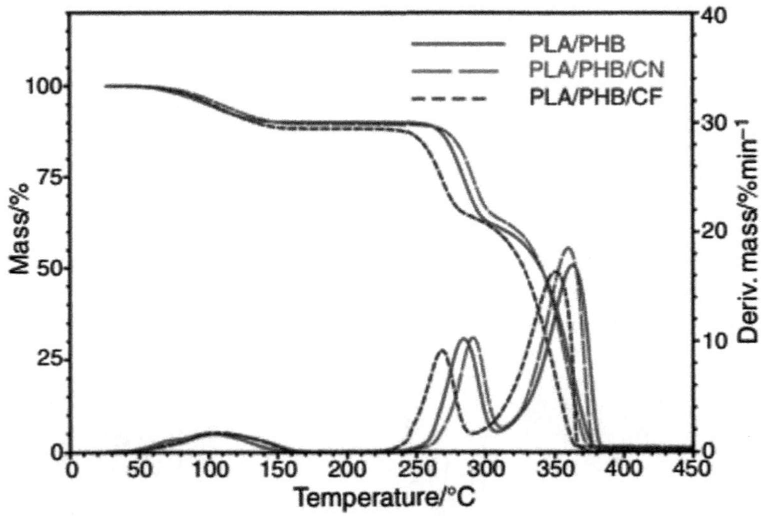

FIGURE 5.11 Thermogravimetric curves (TG) and their derivatives (DTG) for neat PLA/PHB and biocomposite film containing CF.[53]

Source: Reprinted with permission from Ref. [53]. Copyright © 2019, Akadémiai Kiadó, Springer Nature

Taghizadeh et al. used pistachio hull extract as a reducing and stabilizing agent for the biosynthesis of copper nanoparticles from copper (II) acetate monohydrate.[95] Cu nanoparticles were immobilized on pistachio shell powder and copper/pistachio shell nanocomposites were fabricated for rapid catalytic applications as shown in Figure 5.12. FE-SEM image of copper/pistachio shell nanocomposites in Figure 5.13 shows Cu nanoparticles as bright dots well dispersed and immobilized on pistachio shell powder. TEM image in Figure 5.14 shows that Cu NPs are roughly spherical with size ranging between 15 and 60 nm. Copper as a good electrical conductor can easily transfer electrons from one adsorbed species to another and hence acts as a catalyst for reductive elimination of organic pollutants such as 4-nitrophenol(4-NP), methyl orange (MO), rhodamine B (RhB) and methylene blue (MB) individually or simultaneously using $NaBH_4$ at ambient temperature from industrial wastewater.

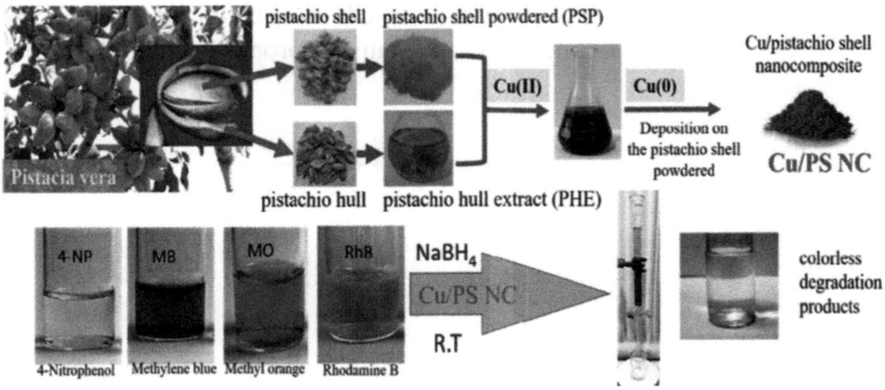

FIGURE 5.12 Green synthesis and evaluation of catalytic activity of Cu/pistachio shell nanocomposite (Cu/PS NC).[95]

Source: Reprinted with permission from Ref. [95]. Copyright © 2018 Elsevier Ltd.

FIGURE 5.13 The FE-SEM images of pistachio shell (a) and Cu/pistachio shell nanocomposite (b and c).[95]

Source: Reprinted with permission from Ref. [95]. Copyright © 2018 Elsevier Ltd.

Movva et al.[50] isolated nanocellulose crystals from pistachio nutshells using alkali hydrolysis with NaOH, H_2O_2 bleaching followed by acid hydrolysis with sulfuric acid.[50] The prepared CNCs showed 95% crystallinity and enhanced thermal degradation. Nanocomposite of polyester resin with different weight percentages of CNCs was prepared. It showed an increase in tensile and flexural strength due to strong adhesion. Agwuncha et al.[82] utilized shea nutshell, an agricultural waste from shea butter production industry for the production of nanocellulose.[82] They investigated the optimum conditions, such as temperature, time, and alkali concentration required for the isolation of nanocellulose from shea nutshell. The isolated cellulose showed the excellent thermal property.

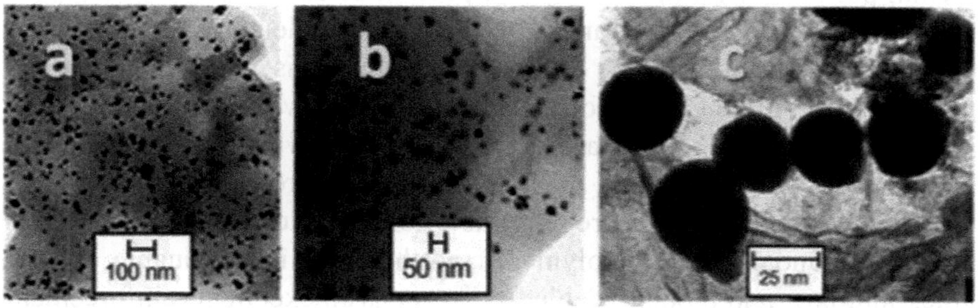

FIGURE 5.14 The TEM images of Cu/pistachio shell nanocomposite at various magnifications.[95]

Source: Reprinted with permission from Ref. [95]. Copyright © 2018 Elsevier Ltd.

Maaloul et al.[55] extracted CF from almond shell.[55] The chemical modification with sodium trimetaphosphate (STMP) resulted in a 3D microporous biomaterial with a large number of adsorption sites. This was proved to be a low-cost biosorbent bead for the adsorption of Cu (II) ions with good adsorption capacity of about 128.24 mg/g. Harini et al. isolated and characterized cashew nutshell starch (CNS) and walnut shell cellulose (WNC) and prepared cellulose-reinforced starch film.[51] Alkaline hydrolysis using NaOH and sodium hypochlorite bleaching was performed, subsequently filtered and oven dried at 60°C for 16 hrs to prepare cellulose from walnut shell powder. This study identified walnut shell cellulose as a good reinforcement for CNS film. The resultant CNS-WNC starch film showed high crystallinity, good oxygen transfer rate (OTR), mechanical and physical properties. Thermal degradation temperature of the film was found to be between 298 and 302°C. The surface roughness of the film increased with the concentration of cellulose. Two percent walnut shell cellulose-reinforced starch film incorporated with hydrophilic active compounds from pomegranate peel extract showed active packaging properties. Economic and environmental benefits together with desirable properties make nutshells a promising sustainable material.

5.8 CONCLUSIONS

This chapter emphasizes the utilization of agricultural by-products, especially nutshells for the preparation of nanocellulose. As agricultural production is increasing to meet the present-day requirements, agro-based wastes are also unavoidable. Most of these are burnt off creating environmental pollution with the emission of poisonous gases. These agricultural field and product residues are proved to be excellent resources for the production of nanocellulose and hence recycling is the demand of the time. Considering the issues of energy crisis and environmental pollution, materials of natural origin are gaining more relevance. Nutshells with almost same weights as that of nuts are usually thrown away as waste. They are cost-effective and are renewable sources of cellulose. Cellulose is the most abundant organic polymer on earth, which is sustainable, renewable, nontoxic, and biodegradable. It can act as an excellent reinforcement material in the polymer matrix. Nanocellulose-based materials find application in packaging, flame-retardants, biomedicine, optics, adsorbents, and catalysis. It has excellent mechanical properties such as high tensile strength and elastic modulus. Hydrophobic surface modifications, such as silylation, carboxylation, phosphorylation, sulfonation, and esterification can improve its compatibility in polymer matrices. This review explores nutshells as an alternative raw material for the production of nanocellulose.

KEYWORDS

- **nanocellulose**
- **nutshells**
- **agrowaste**
- **nanocomposites**
- **plant fiber**
- **extraction**

REFERENCES

1. Moon, R. J.; Martini, A.; Nairn, J.; Simonsen, J.; Youngblood, J. Cellulose Nanomaterials Review: Structure, Properties and Nanocomposites. *Chem. Soc. Rev.* **2011,** *40*(7), 3941–3994.
2. Zhao, Y.; Zhang, Y.; Lindström, M. E.; Li, J. Tunicate Cellulose Nanocrystals: Preparation, Neat Films and Nanocomposite Films with Glucomannans. *Carbohydr. Polym.* **2015,** *117*, 286–296.
3. Li, Y. L.; Wang, W. X.; Zhou, J.; Chen, H. S.; Zhao, J. C.; Wang, B. D. Fatigue Crack Growth and Fracture of 30 Wt% B4c/6061al Composites. *Fatigue Fract. Eng. Mater. Struct.* **2017,** *40*(9), 1378–1388.

4. Matsuzaki, R.; Ueda, M.; Namiki, M.; Jeong, T. K.; Asahara, H.; Horiguchi, K.; Hirano, Y. Three-Dimensional Printing of Continuous-Fiber Composites by in-Nozzle Impregnation. *Sci. Rep.* **2016,** *6*(1), 1–7.
5. Stelte, W.; Sanadi, A. R. Preparation and Characterization of Cellulose Nanofibers from Two Commercial Hardwood and Softwood Pulps. *Ind. Eng. Chem. Res.* **2009,** *48*(24), 11211–11219.
6. John, M. J.; Thomas, S. Biofibres and Biocomposites. *Carbohydr. Polym.* **2008,** *71*(3), 343–364.
7. Orts, W. J.; Shey, J.; Imam, S. H.; Glenn, G. M.; Guttman, M. E.; Revol, J. F. Application of Cellulose Microfibrils in Polymer Nanocomposites. *J. Polym. Environ.* **2005,** *13*(4), 301–306.
8. Sisson, W. A. Some X-Ray Observations Regarding the Membrane Structure of Halicystis. *Contrib. Boyce Thompson Inst.* **1941,** *12*, 31–44.
9. Canale-Parola, E. Biology of the Sugar-Fermenting Sarcinae. *Bacteriol. Rev.* **1970,** *34*(1), 82–97.
10. Kukle, S.; Grāvītis, J.; Putniņa, A.; Stikute, A. The Effect of Steam Explosion Treatment on Technical Hemp Fibres. In *Environment. Technologies. Resources. Proceedings of the International Scientific and Practical Conference,* **2011,** *1*, 230–237).
11. Šturcová, A.; His, I.; Apperley, D. C.; Sugiyama, J.; Jarvis, M. C. Structural Details of Crystalline Cellulose from Higher Plants. *Biomacromolecules* **2004,** *5*(4), 1333–1339.
12. Wada, M.; Chanzy, H.; Nishiyama, Y.; Langan, P. Cellulose Ii Crystal Structure and Hydrogen Bonding by Synchrotron X-Ray and Neutron Fibre Diffraction. *Macromolecules* **2004,** *37*, 8548–8555.
13. Moon, R. J.; Martini, A.; Nairn, J.; Simonsen, J.; Youngblood, J. Cellulose Nanomaterials Review: Structure, Properties and Nanocomposites. *Chem. Soc. Rev.* **2011,** *40*(7), 3941–3994.
14. Panaitescu, D. M.; Donescu, D.; Bercu, C.; Vuluga, D. M.; Iorga, M.; Ghiurea, M. Polymer Composites with Cellulose Microfibrils. *Polym. Eng. Sci.* **2007,** *47*(8), 1228–1234.
15. Payen, A. Mémoire Sur La Composition Du Tissu Propre Des Plantes Et Du Ligneux. *Comptes Rendus* **1838,** *7*, 1052–1056.
16. Nägeli, C. Die Stärkekörner, Pflanzenphysiologische Untersuchungen Von Nägeli Und Cramer.1858.
17. Singh, G. Biodegradation of Nanocellulose and Microbial Community Response: Effect of Surface Modification and Morphology, Doctoral Dissertation, Virginia Polytechnic Institute and State University, 2015
18. Chawla, K. K. *Composite Materials: Science and Engineering*. Springer Science and Business Media.2012.
19. Nickerson, R. F.; Habrle, J. A. Cellulose Intercrystalline Structure. *Ind. Eng. Chem. Res.* **1947,** *39*(11), 1507–1512.
20. Favier, V.; Canova, G. R.; Cavaillé, J. Y.; Chanzy, H.; Dufresne, A.; Gauthier, C. Nanocomposite Materials from Latex and Cellulose Whiskers. *Polym. Adv. Technol.* **1995,** *6*(5), 351–355.
21. Kargarzadeh, H.; Sheltami, R. M.; Ahmad, I.; Abdullah, I.; Dufresne, A. Cellulose Nanocrystal: A Promising Toughening Agent for Unsaturated Polyester Nanocomposite. *Polymer* **2015,** *56*, 346–357.

22. Pei, A.; Malho, J. M.; Ruokolainen, J.; Zhou, Q.; Berglund, L. A. Strong Nanocomposite Reinforcement Effects in Polyurethane Elastomer with Low Volume Fraction of Cellulose Nanocrystals. *Macromolecules* **2011**, *44*(11), 4422–4427.
23. Herrick, F. W.; Casebier, R. L.; Hamilton, J. K.; Sandberg, K. R. In *Microfibrillated Cellulose: Morphology and Accessibility*. J Appl Polym Sci Appl Polym Symp, vol CONF8205234-Vol 2. ITT Rayonier Inc.; Shelton, WA, 1983.
24. Viswanathan, G.; Murugesan, S.; Pushparaj, V.; Nalamasu, O.; Ajayan, P. M.; Linhardt, R. J. Preparation of Biopolymer Fibres by Electrospinning from Room Temperature Ionic Liquids. *Biomacromolecules* **2006,** *7*(2):415–418
25. Crotogino, R. NanoCellulose. In *International Symposium on Assessing the Economic Impact of Nanotechnology*, 28, 2012.
26. Morehead, F. F. Ultrasonic Disintegration of Cellulose Fibers Before and After Acid Hydrolysis. *Text. Res. J.* **1950,** *20*(8), 549–553.
27. Staudinger, H.; Fritschi, J. Über Isopren und Kautschuk. 5. Mitteilung. Über die Hydrierung des Kautschuks und über seine Konstitution. *Helv. Chim. Acta.* **1922,** *5*(5), 785–806.
28. Rajinipriya, M.; Nagalakshmaiah, M.; Robert, M.; Elkoun, S. Importance of Agricultural and Industrial Waste in the Field of Nanocellulose and Recent Industrial Developments of Wood Based Nanocellulose: A Review. *ACS Sustain. Chem. Eng.* **2018,** *6*(3), 2807–2828.
29. Aitomaki, Y.; Oksman, K. Reinforcing Efficiency of Nanocellulose in Polymers. *React. Funct. Polym.* **2014,** 85, 151–156.
30. Xu, X.; Liu, F.; Jiang, L.; Zhu, J. Y.; Haagenson, D.; Wiesenborn, D. P. Cellulose Nanocrystals vs. Cellulose Nanofibrils: A Comparative Study on their Microstructures and Effects as Polymer Reinforcing Agents. *ACS Appl. Mater. Interfaces* **2013,** *5*(8), 2999–3009.
31. Quennouz, N.; Hashmi, S. M.; Choi, H. S.; Kim, J. W.; Osuji, C. O. Rheology of Cellulose Nanofibrils in the Presence of Surfactants. *Soft Matter.* **2016,** *12*(1), 157–164.
32. Hu, Z.; Ballinger, S.; Pelton, R.; Cranston, E. D. Surfactant-Enhanced Cellulose Nanocrystal Pickering Emulsions. *J. Colloid Interface Sci.* **2015,** *439*, 139–148.
33. Ding, W.; Calabri, L.; Chen, X.; Kohlhaas, K. M.; Ruoff, R. S. Mechanics of Crystalline Boron Nanowires. *Compos. Sci. Technol.* **2006,** *66*(9), 1112–1124.
34. Septevani, A. A.; Annamalai, P. K.; Martin, D. J. In *Synthesis and Characterization of Cellulose Nanocrystals as Reinforcing Agent in Solely Palm Based Polyurethane Foam.* AIP Conference Proceedings. AIP Publishing LLC, **2017,** *1904*(1), 020042.
35. D'Aloia, J.; Newell, J. A.; Del Vecchio, C.; Hill, C.; Santino, D.; Russell, K. Enhancement of The Compressive Strength of Kevlar-29/Epoxy Resin Unidirectional Composites. *High Perform. Polym.* **2008,** *20*(3), 357–364.
36. Borjesson, M.; Westman, G. Crystalline Nanocellulose-Preparation, Modification, and Properties. Cellulose-Fundamental Aspects and Current Trends. *InTech, Rijeka*. https:// doi. org/10.5772/61899, 2015.
37. Mandal, A.; Chakrabarty, D. Isolation of Nanocellulose from Waste Sugarcane Bagasse (Scb) and its Characterization. *Carbohydr. Polym.* **2011,** 86, 1291–1299.
38. Reddy, N.; Yang, Y. Structure and Properties of High Quality Natural Cellulose Fibres from Cornstalks. *Polymer* **2005,** *46*(15), 5494–5500.
39. Yin, C.; Li, J.; Xu, Q.; Peng, Q.; Liu, Y.; Shen, X. Chemical Modification of Cotton Cellulose in Supercritical Carbon Dioxide: Synthesis and Characterization of Cellulose Carbamate. *Carbohydr. Polym.* **2007,** *67*(2), 147–154.

40. Cao, X.; Chen, Y.; Chang, P. R.; Muir, A. D.; Falk, G. Starch-Based Nanocomposites Reinforced with Flax Cellulose Nanocrystals. *Express Polym. Lett.* **2008,** *2*(7), 502–510.
41. Ouajai, S.; Shanks, R. A. Composition, Structure and Thermal Degradation of Hemp Cellulose After Chemical Treatments. *Polym. Degrad. Stab.* **2005,** *89*(2), 327–335.
42. Cao, X.; Ding, B.; Yu, J.; Al-Deyab, S. S. Cellulose Nanowhiskers Extracted from Tempo-Oxidized Jute Fibres. *Carbohydr. Polym.* **2012,** *90*(2), 1075–1080.
43. Nishiyama, Y.; Kim, U. J.; Kim, D. Y.; Katsumata, K. S.; May, R. P.; Langan, P. Periodic Disorder Along Ramie Cellulose Microfibrils. *Biomacromolecules* **2003,** *4*(4), 1013–1017.
44. Pelissari, F. M.; Andrade-Mahecha, M. M.; do Amaral Sobral, P. J.; Menegalli, F. C. Nanocomposites Based on Banana Starch Reinforced with Cellulose Nanofibres Isolated from Banana Peels. *J. Colloid Interface Sci.* **2017,** *505*, 154–167.
45. Almeida, E. V. R.; Frollini, E.; Castellan, A.; Coma, V. Chitosan, Sisal Cellulose, and Biocomposite Chitosan/Sisal Cellulose Films Prepared from Thiourea/Naoh Aqueous Solution. *Carbohydr. Polym.* **2010,** *80*(3), 655–664.
46. Das, A. M.; Ali, A. A.; Hazarika, M. P. Synthesis and Characterization of Cellulose Acetate from Rice Husk: Eco-Friendly Condition. *Carbohydr. Polym.* **2014,** *112*, 342–349.
47. Cerqueira, D. A.; Rodrigues Filho, G.; da Silva Meireles, C. Optimization of Sugarcane Bagasse Cellulose Acetylation. *Carbohydr. Polym.* **2007**, *69*(3), 579–582.
48. Mandal, A.; Chakrabarty, D. Studies on the Mechanical, Thermal, Morphological and Barrier Properties of Nanocomposites Based on Poly (Vinyl Alcohol) and Nanocellulose from Sugarcane Bagasse. *J. Ind. Eng. Chem.* **2014,** *20*(2), 462–473.
49. Shaniba, V.; Sreejith, M. P.; Aparna, K. B.; Jinitha, T. V.; Purushothaman, E. Mechanical and Thermal Behavior of Styrene Butadiene Rubber Composites Reinforced with Silane-Treated Peanut Shell Powder. *Polym. Bull.* **2017,** *74*(10), 3977–3994.
50. Movva, M.; Kommineni, R. Extraction of Cellulose from Pistachio Shell and Physical and Mechanical Characterisation of Cellulose-Based Nanocomposites. *Mater. Res. Express* **2017,** *4*(4), 045014.
51. Harini, K.; Mohan, C. C.; Ramya, K.; Karthikeyan, S.; Sukumar, M. Effect of Punica Granatum Peel Extracts on Antimicrobial Properties in Walnut Shell Cellulose Reinforced Bio-Thermoplastic Starch Films from Cashew Nut Shells. *Carbohydr. Polym.* **2018,** *184*, 231–242.
52. Gaonkar, S. M.; Kulkarni, P. R. Microcrystalline Cellulose from Coconut Shells. *Acta Polym.* **1989,** *40*(4), 292–293.
53. Frone, A. N.; Panaitescu, D. M.; Chiulan, I.; Gabor, A. R.; Nicolae, C. A.; Oprea, M.; Puitel, A. C. Thermal and Mechanical Behaviour of Biodegradable Polyester Films Containing Cellulose Nanofibres. *J. Therm. Anal. Calorim.* **2019,** *138*(4), 2387–2398.
54. Yao, J.; Huang, H.; Mao, L.; Li, Z.; Zhu, H.; Liu, Y. Structural and Optical Properties of Cellulose Nanocrystals Isolated from the Fruit Shell of Camellia Oleifera Abel. *Fibers Polym.* **2017,** *18*(11), 2118–2124.
55. Maaloul, N.; Oulego, P.; Rendueles, M.; Ghorbal, A.; Díaz, M. Enhanced Cu (Ii) Adsorption using Sodium Trimetaphosphate-Modified Cellulose Beads: Equilibrium, Kinetics, Adsorption Mechanisms, and Reusability. *Environ. Sci. Pollut. Res.* **2020,** 1–17.
56. Shahabi-Ghahafarrokhi, I.; Khodaiyan, F.; Mousavi, M.; Yousefi, H. Preparation and Characterization of Nanocellulose from Beer Industrial Residues using Acid Hydrolysis/Ultrasound. *Fibers Polym.* **2015,** *16*(3), 529–536.

57. Peng, Y.; Gardner, D. J.; Han, Y.; Kiziltas, A.; Cai, Z.; Tshabalala, M. A. Influence of Drying Method on the Material Properties of Nanocellulose I: Thermostability and Crystallinity. *Cellulose* **2013,** *20*(5), 2379–2392.
58. Alemdar, A.; Sain, M. Isolation and Characterization of Nanofibres from Agricultural Residues–Wheat Straw and Soy Hulls. *Bioresour. Technol.* **2008,** *99*(6), 1664–1671.
59. Marett, J.; Aning, A.; Foster, E. J. The Isolation of Cellulose Nanocrystals from Pistachio Shells Via Acid Hydrolysis. *Ind. Crops Prod.* **2017,** *109*, 869–874.
60. Naduparambath, S.; Jinitha, T. V.; Shaniba, V.; Sreejith, M. P.; Balan, A. K.; Purushothaman, E. Isolation and Characterisation of Cellulose Nanocrystals from Sago Seed Shells. *Carbohydr. Polym.* **2018,** *180*, 13–20.
61. Bano, S.; Negi, Y. S.; Illathvalappil, R.; Kurungot, S.; Ramya, K. Studies on Nano Composites of Speek/Ethylene Glycol/Cellulose Nanocrystals as Promising Proton Exchange Membranes. *Electrochim. Acta.* **2019,** *293*, 260–272.
62. Abraham, E.; Deepa, B.; Pothan, L. A.; Jacob, M.; Thomas, S.; Cvelbar, U.; Anandjiwala, R. Extraction of Nanocellulose Fibrils from Lignocellulosic Fibres: A Novel Approach. *Carbohydr. Polym.* **2011,** *86*, 1468–1475.
63. El Oudiani, A.; Ben Sghaier, R.; Chaabouni, Y.; Msahli, S.; Sakli, F. Physico-Chemical and Mechanical Characterization of Alkali-Treated Agave Americana, l. Fiber. *J. Text. Inst.* **2012,** *103*(4), 349–355.
64. Lu, Z.; L. Fan, H. Zheng, Q. Lu, Y. Liao, B. Huang Preparation, Characterization and Optimization of Nanocellulose Whiskers by Simultaneously Ultrasonic Wave and Microwave Assisted. *Bioresour. Technol.* **2013,** *146*, 82–88.
65. Xu, Q.; Gao, Y.; Qin, M.; Wu, K.; Fu, Y.; Zhao, J. Nanocrystalline Cellulose from Aspen Kraft Pulp and its Application in Deinked Pulp. *Int. J. Biol. Macromol.* **2013,** *60*, 241–247.
66. Cheng, M.; Qin, Z.; Chen, Y.; Liu, J.; Ren, Z. Facile One-Step Extraction and Oxidative Carboxylation of Cellulose Nanocrystals Through Hydrothermal Reaction by Using Mixed Inorganic Acids. *Cellulose* **2017,** *24*(8), 3243–3254.
67. Elazzouzi-Hafraoui, S.; Nishiyama, Y.; Putaux, J. L.; Heux, L.; Dubreuil, F.; Rochas, C. The Shape and Size Distribution of Crystalline Nanoparticles Prepared by Acid Hydrolysis of Native Cellulose. *Biomacromolecules* **2008,** *9*(1), 57–65.
68. Shahabi-Ghahafarrokhi, I.; F. Khodaiyan, M. Mousavi H. Yousefi Preparation and Characterization of Nanocellulose from Beer Industrial Residues using Acid Hydrolysis/Ultrasound. *Fibers Polym.* **2015,** *16*, 529–536.
69. Alvira, P.; Tomás-Pejó, E.; Ballesteros, M.; Negro, M. J. Pretreatment Technologies for an Efficient Bioethanol Production Process Based on Enzymatic Hydrolysis: A Review. *Bioresour. Technol.* **2010,** *101*(13), 4851–4861.
70. Zhang, Y.; Li, Q.; Su, J.; Lin, Y.; Huang, Z.; Lu, Y.; Zhu, Y. A Green and Efficient Technology for the Degradation of Cellulosic Materials: Structure Changes and Enhanced Enzymatic Hydrolysis of Natural Cellulose Pretreated by Synergistic Interaction of Mechanical Activation and Metal Salt *Bioresour. Technol.* **2015**, *177*, 176–181.
71. Wang, W.; Yuan, T. Q.; Cui, B. K. Fungal Treatment Followed by Fecl 3 Treatment to Enhance Enzymatic Hydrolysis of Poplar Wood for High Sugar Yields. *Biotechnol. Lett.* **2013**, *35*(12), 2061–2067.
72. Peng, L.; Lin, L.; Zhang, J.; Zhuang, J.; Zhang, B.; Gong, Y. Catalytic Conversion of Cellulose to Levulinic Acid by Metal Chlorides. *Molecules* **2010**, *15*(8), 5258–5272.

73. Chen, Y. W.; Lee, H. V. Revalorization of Selected Municipal Solid Wastes as New Precursors of "Green" Nanocellulose Via A Novel One-Pot Isolation System: A Source Perspective. *Biotechnol. Lett.* **2018**, *107*, 78–92.
74. Abd Hamid, S. B.; Al Amin, M.; Ali, M. E. Zeolite Supported Ionic Liquid catalyst for The Synthesis of Nanocellulose from Palm Tree biomass. *Micro Nano Lett.* **2014**, *925*, 52–56.
75. Man, Z.; Muhammad, N.; Sarwono, A.; Bustam, M. A.; Kumar, M. V.; Rafiq, S. Preparation of Cellulose Nanocrystals using an Ionic Liquid. *J. Polym. Environ.* **2011**, *19*,726–731.
76. Myllymäki, V.; Aksela, R. Dissolution and Delignification of Lignocellulosic Materials with Ionic Liquid Solvent under Microwave Irradiation, 2005, WO Pat, 17001.
77. Barthel, S.; Heinze, T. Acylation and Carbanilation of Cellulose in Ionic Liquids. *Green Chem.* **2006**, *8*(3), 301–306.
78. Novo, L. P.; Bras, J.; García, A.; Belgacem, N.; da Silva Curvelo, A. A. A Study of The Production of Cellulose Nanocrystals Through Subcritical Water Hydrolysis. *Ind. Crops Prod.* **2016**, *93*, 88–95.
79. Mishra, S. P.; Thirree, J.; Manent, A. S.; Chabot, B.; Daneault, C. Ultrasound-Catalyzed Tempo-Mediated Oxidation of Native Cellulose for the Production of Nanocellulose: Effect of Process Variables. *Bioresources* **2011**, *6*, 121–143.
80. Leung, A. C.; Hrapovic, S.; Lam, E.; Liu, Y.; Male, K. B.; Mahmoud, K. A.; Luong, J. H. Characteristics and Properties of Carboxylated Cellulose Nanocrystals Prepared from A Novel One-Step Procedure. *Small* **2011**, *7*(3), 302–305.
81. Satyamurthy, P. N. Vigneshwaran A Novel Process for Synthesis of Spherical Nanocellulose by Controlled Hydrolysis of Microcrystalline Cellulose using Anaerobic Microbial Consortium. *Enzyme Microb. Technol.* **2013**, *52*, 20–25.
82. Agwuncha, S. C.; Owonubi, S.; Fapojuwo, D. P.; Abdulkarim, A.; Okonkwo, T. P.; Makhatha, E. M. Evaluation of Mercerization Treatment Conditions on Extracted Cellulose from Shea Nut Shell using Ftir and Thermogravimetric Analysis. *Mater. Today* **2021**, *38*, 958–963.
83. Neto, W. P. F.; Silvério, H. A.; Dantas, N. O.; Pasquini, D. Extraction and Characterization of Cellulose Nanocrystals from Agro-Industrial Residue-Soy Hulls. *Ind. Crops Prod.* **2013**, *42*, 480–488.
84. Raghvendra, K. M.; Sravanthi, L. Fabrication Techniques of Micro/Nano Fibres Based Nonwoven Composites: A Review. *Mod. Chem. Appl.* **2017**, *5*(206), 2.
85. Dufresne, A.; Dupeyre, D.; Vignon, M. R. Cellulose Microfibrils from Potato Tuber Cells: Processing and Characterization of Starch-Cellulose Microfibril Composites. *J. Appl. Polym. Sci.* **2000**, *76*(14), 2080–2092.
86. Wang, B.; Sain, M. Dispersion of Soybean Stock-Based Nanofibre in A Plastic Matrix. *Polym. Int.* **2007**, *56*(4), 538–546.
87. Wei, J.; Zhou Y, Lv, Y.; Wang, J.; Jia, C.; Liu, J.; Zhang, X.; Sun, J.; Shao, Z. C. Carboxymethyl Cellulose Nanofibrils with A Treelike Matrix: Preparation and Behavior of Pickering Emulsions Stabilization. *ACS Sustain. Chem. Eng. 2019*, *7*(15), 12887–12896
88. Zimmermann, T.; Pöhler, E.; Geiger, T. Cellulose Fibrils for Polymer Reinforcement. *Adv. Eng. Mater.* **2004**, *6*(9), 754–761.

89. Siqueira, G.; Oksman, K.; Tadokoro, S. K.; Mathew, A. P. Re-Dispersible Carrot Nanofibres with High Mechanical Properties and Reinforcing Capacity for Use in Composite Materials. *Compos. Sci. Technol.* **2016**, *123*, 49–56.
90. Zhang L, Jia Y, He H, Yin J, Chen R, Zhang C, Shen W, Wang X. Multiple Factor Analysis on Preparation of Cellulose Nanofibre by Ball Milling from Softwood Pulp. *Bio. Resources* **2018**, *13*(2), 2397–2410
91. Dufresne, A. Cellulose Nanomaterials as Green Nanoreinforcements for Polymer Nanocomposites. *Philos. Trans. R Soc. A* **2018**, *376*(2112), 20170040
92. Mokhena, T. C.; Jacobs, N. V.; Luyt, A. S. Nanofibrous Alginate Membrane Coated with Cellulose Nanowhiskers for Water Purification. *Cellulose* **2018a**, *25*(1):417–427
93. Mokhena, T. C.; Sefadi, J. S.; Sadiku, E. R.; John, M. J.; Mochane, M. J.; Mtibe, A. Thermoplastic Processing of Pla/Cellulose Nanomaterials Composites. *Polymers* **2018**, *10*(12), 1363.
94. Deepa, B.; Abraham, E.; Cherian, B. M.; Bismarck, A.; Blaker, J. J.; Pothan, L. A.; Leao, A. L.; De Souza, S. F.; Kottaisamy M. Structure, Morphology and Thermal Characteristics of Banana Nano Fibres Obtained by Steam Explosion. *Bioresour. Technol.* **2011,** *102*(2):1988– 1997.
95. Taghizadeh, A.; Rad-Moghadam, K. Green Fabrication of Cu/Pistachio Shell Nanocomposite using Pistacia Vera l. Hull: An Efficient Catalyst for Expedient Reduction of 4-Nitrophenol and Organic Dyes. *J. Clean. Prod.* **2018**, *198*, 1105–1119.

CHAPTER 6

Synthesis and Characterization of Polymer/Metal Nanocomposite Magnetic Material

M. BANERJEE[1], PREETI SACHDEV[1], ARUNA JOSHI[1], AAKANKSHA CHOUDHARY[1], SWATI NAGAR[1], and G. S. MUKHERJEE[2]

[1]*Nano Science and Technology Lab., School of Physics, Devi Ahilya University, Indore, Madhya Pradesh, India*

[2]*Defence Research & Development Organization, DRDO Metcalfe House DESIDOC Complex, Delhi, India*

ABSTRACT

Organic polymer-material-based magnets may have several attributes, such as low density, flexibility, processability at low temperatures, property modulation, solubility, biocompatibility, and semiconducting or insulating abilities. Magnetic materials are the backbone of modern digital technologies; of late, stretchable magnetoelectronics have gained substantial interest over the last few years because of their scope for exciting new applications offered by arbitrary surface geometries possible after fabrication. There are reports about the generation of organic molecular magnets through synthetic routes. Although much attention has been paid by researchers on the development of organic molecular magnets through chemical synthesis, but such synthetic methods are quite difficult to realize organic magnets. In this backdrop, magnetic materials have been prepared by embedding magnetic nanoparticles of transition metals, such as cobalt, iron, and nickel on different polymer matrices, such as polyvinyl alcohol (PVA), polymethylmethacrylate (PMMA) using ion beam sputtering (IBS) technique. Characterization of

Foundation and Growth of Macromolecular Science: Advances in Research for Sustainable Development. Meegle S. Mathew, PhD, Józef T. Haponiuk, PhD & Sabu Thomas, PhD, DSc (Eds.)

these magnetic materials has been made using instruments, such as grazing incidence X-ray diffraction (GIXRD), magneto optical Kerr effect (MOKE), and X-ray photoelectron spectroscopy (XPS) to understand the microstructural properties, chemical state of the nanoparticles and magnetic properties.

6.1 INTRODUCTION

Magnetic materials are the backbone of modern digital information technologies which require efficient hard-disk storage system. Transition metals and rare earth elements are known to exhibit magnetic properties because of their atoms with unfilled shell and often the outer shell starts to fill before the inner shell are filled up. One such transition metal is cobalt and its nanostructures have potential applications in the area of high-density magnetic recording system.[1] One of the attractive features of transition metals like cobalt (Co) is that they can display ferromagnetic properties even if the thickness of their film goes down to the ultrathin level. However, Co nanoparticle clusters have the disadvantage that they have a tendency to get oxidized when exposed to the atmosphere of oxygen environment,[2–6] such adverse effect of oxidation on cobalt nanoparticle is absolutely undesirable. This undesired oxidation toward cobalt nanoparticles gets all the more aggravated with the reduction of the particle size, particularly in the nanoregime. In order to protect such nanoparticle structures, there are certain reports to apply overcoating[4–7] on the deposited magnetic nanoparticles; such overcoating is generally done by depositing inert metals or some nonmetals like carbon particles over the substrates, such as Au, Ge, glass, Si, and so on.[8,9] There are several candidates materials used for overcoating, for example, copper (Cu), silver (Ag), gold (Au)[9]; however, the carbon (C) overcoat has certain advantages, for example, the carbon nanoparticle (CNP) overcoat is noncorrosive and gives protection against chemical, physical, and magnetic degradation.[10–13]

Our major interest is to evaluate and compare the magnetic properties of such nanocomposite films as PVA/Co/CNP, PVA/Co in which cobalt nanoparticles are embedded in a given polymer matrix system like PVA film where carbon nanoparticles (CNP) were applied as overcoat[4,5] overlaying the cobalt nanostructure in the nanocomposites viz., PVA/Co/CNP films; on the other hand, no overcoat was applied in PVA/Co nanocomposite film.[6] Carbon is nonmetallic as well as a nonmagnetic substance. However, we also presented here some preliminary experimental results of nanocomposite systems using PMMA as polymer matrix film and other metal nanoparticles (Fe, Ni) in a few selective experiments in our current ongoing research.

6.2 EXPERIMENTAL PART

6.2.1 SAMPLE PREPARATION

Float glass substrates were cleaned ultrasonically and films of polymer were deposited by solution casting method. Different thicknesses of Co (and Fe, Ni for a few preliminary study) nanoparticles were deposited on polymer underlayer using IBS (ion beam sputtering) technique using Kaufman gun for all cases of sample preparation.

The deposition was carried out at a base pressure of 2.0×10^{-7} torr. The rate of deposition was 25 Å/min for nanoparticles. The time of deposition was adjusted to get cobalt deposition of different thicknesses (40 nm)[5] A protective overlayer of carbon was deposited also by the ion beam sputtering method.[4,5] The schematic representation of the layered structure of a typical sample is shown in Figure 6.1, where the glass substrate remains as the lowest layer, followed by layers consisting of polymer underlayer (30 μ), metal nanoparticle embedded nanocomposite layer, and carbon nanoparticle (CNP, 2 nm) or without any overcoat as the case may be.

6.2.2 CHARACTERIZATION

The structural measurements were done using GIXRD (Grazing incidence X-ray diffraction) technique. The GIXRD patterns of all nanocomposite samples were recorded by Siemens-5000 Diffractometer using Cu K_α radiation (λ = 1.54 Å) and a grazing incidence of 0.57°. X-ray photoelectron spectroscopic (XPS) measurements of as-deposited films of PVA/Co/CNP and PVA/CNP were carried out on VSW (UK.) instrument using Al-K_α radiation (λ = 8.3 Å) with a pass energy of 40 eV and energy resolution of 0.9 eV. The XPS peaks were fitted using standard XPS Peak 4.1 software program. Magnetization measurements were carried out using the MOKE (magneto optical Kerr effect) technique in which polarized light from He–Ne laser of wavelength 6328 Å was reflected from the sample of interest. Field used in MOKE setup was nearly 1.8 kOe. For all the MOKE measurements, the magnetic field was applied parallel to the sample surface.

6.3 RESULTS AND DISCUSSION

Polymer/metal/overcoat (e.g., PVA/Co/CNP) films were prepared by ion beam sputtering of nanoparticles onto PVA film matrix, and a typical structural presentation is shown in Figure 6.1.

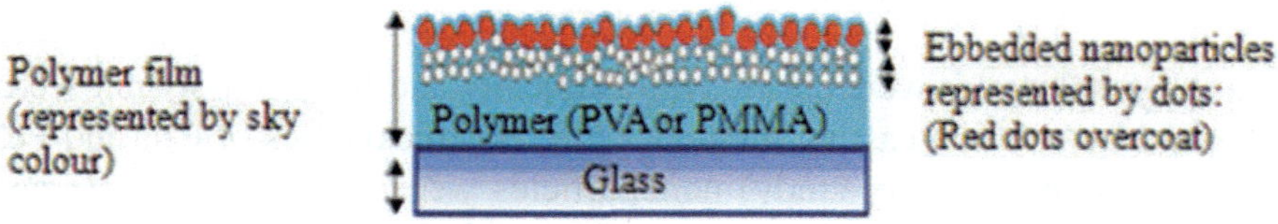

FIGURE 6.1 Schematic diagram of the sample of Polymer/Metal/Overcoat (e.g., PVA/Co/CNP) film prepared by sputtering Co (shown as lower dots) followed by CNP (shown as upper dots as overcoat) into PVA. In PVA/Co film there is no overcoat (i.e., no upper dots).

In PVA/Co/CNP film, the thickness of the cobalt nanocomposite layer has been controlled to get samples with a cobalt layer thickness of 5 nm, 10 nm, and 40 nm, and the samples named hereafter PC5, PC10, and PC40, respectively. The GIXRD patterns of as-deposited sample show the small peaks appearing between 2θ range of 35°–80° are due to cobalt. Such small peaks are characteristic of cobalt with very small thickness.[14] Illustratively, only a single peak of Co (100) is visible for PC5. The PC5 sample shows a broad peak at 41° corresponding to the (100) plane of *hcp* cobalt. The earlier reports on cobalt deposited on glass and silicon do not show any diffraction pattern at such a small thickness of the cobalt layer.[14,15] Further, two peaks due to Co (l00) and Co (101) planes can be seen in the GIXRD pattern of the PC10 sample; whereas the diffraction pattern for PC40 exhibits three peaks for Co (100), Co (101), and Co (110) planes. All three samples show a prominent peak at 19.6° due to PVA, which matches with the reported results in the literature.[16,17] Interestingly, the structure of crystalline zone of PVA remains undisturbed even after nanoparticles having been embedded in PVA film. Because of the presence of polar hydroxyl (OH) groups in PVA, it has a tendency to weakly interact with metals.[4,19,20] Mechanistically, because of such a very weak interacting tendency of PVA, preferably through its amorphous zone with metals, presumably, PVA acted as a facilitator in seeding the nanostructure of cobalt to grow. The interplanar distance of Co (100) d_{100} plane as obtained from Bragg's equation is 0.22 nm for both samples of PC5 and PC10, whereas for PC40, it is 0.216 nm which is slightly less than the values for the other two samples; with the growth of the size of the particles, the exfoliated structure gets more consolidated as a result of compaction owing to more maturing of crystallization, as revealed from the sharpness of the peak in the GIXRD pattern and the reduction in the interplanar distance from 0.22 nm to 0.216 nm as the sample shifts from PC5 to PC40. However, Co (002) plane is not seen in any of the samples, and Co (100) peak intensity increases with the level of cobalt deposition. This is indeed favorable for the development of in-plane magnetization.[2,3] The crystallite size of cobalt particles in the nanocomposite samples as estimated from the Scherer formula ($D = 0.9\lambda/B \cos \theta$)[20–22] is found to be 3.9 nm, 4.62 nm, and 7.07 nm for the samples PC5, PC10, and PC40, respectively.

XPS is a useful technique for investigating the chemical state of an element. XPS is used for depth profiling of the film to know the information about the vertical location of metal particles embedded in the polymer matrix[23] to further confirm the chemical state of cobalt. The sample was etched by Ar+ ion sputtering technique while carrying out the depth profiling experiment.

The details of the survey scan of PC5 as-deposited film, the peaks belonging to C1s peaks dominate the spectrum. These are due to the contribution of carbon from the overcoat layer as well as the carbon from the PVA. The O1s peak is also visible near 530.3 eV due to the contribution from the PVA.

A closer examination of XPS can show a very tiny peak at 779 eV due to Co 2*p* level. The Co 2*p* peak can be clearly seen for the first time in this spectrum after 80 min of "etching" and the Co 2*p* spectrum is found to contain a doublet due to spin–orbit coupling resulting in Co $2p_{3/2}$ and Co $2p_{1/2}$ peaks at 779.0 eV and 794.0 eV, respectively. The energy values of the peaks, the difference of 15.0 eV between the two peaks, a long tail on the high-energy side, and the absence of satellite peaks are consistent with the spectra of metallic cobalt.[24,25] The Co $2p_{3/2}$ peak fitting can be fitted well with one peak (FWHM = 2 eV), indicating the presence of only one kind of cobalt which is nothing but metallic cobalt. This implies that the carbon overcoat and the PVA have protected the cobalt metal nanoparticles from getting oxidized.

C1s peak was recorded to follow the behavior of carbon in the nanocomposite. The C1s spectrum of as-deposited nanocomposite is seen to contain two peaks – one at 284.6 eV and the other as a small peak at 287.7 eV. The peak at 284.6 eV is due to pure carbon and is attributed to overcoat carbon. As the sputtering ("etching") time is increased, the peak at 284.6 decreases, that is, the contribution from overcoat carbon decreased and the contribution of carbon from PVA is increased as given in Table 6.1. The C1s spectra after 80 min of sputtering ("etching") can be fitted with three curves with binding energy (BE) at 284.6 eV is attributed to overcoat carbon, whereas the other two peaks at higher energy, that is, at 285.7 and 287.2 eV is due to two different kinds of carbon atoms (prim. and sec.) of PVA.[19]

TABLE 6.1 % Concentration of Carbon from PVA and Carbon from CNP Overcoat As a Function of Etching/Sputtering Time for PVA/Co/CNP Nanocomposite (PC5).

Sputtering time (min.)	% Concentration of carbon	
	Overcoat carbon, C-1s	Polymer carbon, C-1s
As-deposited	94	6
50	47	53
80	30.5	69.5

The magnetic properties of the as-deposited nanocomposites were investigated using MOKE method. In-plane magnetic property measurements of the samples exhibit closed hysteresis loops indicating ferromagnetic behavior. The values of coercivity parallel to the film surface ($Hc_{\|}$), the saturation magnetization (Hs), and the squareness of the loops given by Mr/Ms for PC5, PC10, and PC40 are presented in Table 6.2. The $Hc_{\|}$ is obtained as 7.7 Oe, 65.3 Oe, and 102.6 Oe for the samples PC5, PC10, and PC40, respectively. A comparison of the $Hc_{\|}$ value of cobalt films on other substrates reported earlier that the as-deposited 40 nm Co film on glass is only 26 Oe[25] and the as-deposited Co film on glass as well as silicon, with thickness from 50 to 120 nm, is reported to be less than 10 Oe.[14] This indicates that PVA/Co/CNP nanocomposites display better magnetic properties than those reported for corresponding inorganic-based samples; the higher value of Hc may be due to the prominence of Co (100) and *hcp* structure of the cobalt nanoparticles embedded in PVA favoring the development of in-plane magnetization. The increment of Hc value has resulted from the increase of the particle size and increase in the intensity of Co (100) plane.[3] The squareness of magnetic Mr/Ms and Hs values also increases in the same manner.

TABLE 6.2 Magnetization Parameters of PVA/Cobalt/CNP Nanocomposites As a Function of Thickness of Co Layer.

Sample	$Hc_{\|}$ Coercivity {Oe}	H_S Saturation field (H,) {Oe}	Mr/Ms, Squareness
PC5	7.7	50	0.482
PC10	65.3	200	0.81
PC40	102.6	250	0.91

6.3.1 POLYMER/METAL NANOPARTICLES WITHOUT ANY OVERCOAT

6.3.1.1 PVA/CO MAGNETIC FILM

Like PVA/Co/CNP films, in the same way, the PVA/Co nanocomposite film has been prepared by embedding Cobalt (Co) nanoparticles in a PVA film matrix in which Co nanoparticle film of 5 nm was deposited without any overcoat. Generation of nanocrystalline Co with the hcp phase is revealed from the GIXRD pattern of the film and it also indicates that there is apparently no change in the crystalline structure of PVA even after sputtering of the cobalt nanoparticles. The average particle size of Co nanoparticles in the PVA/Co film is found to be 2.1 nm using Scherer formula.[19,21] The chemical state of the nanoparticles of Co was examined for PVA/Co film using XPS spectroscopic analysis, and it confirms that Co exists in the metallic state only. Thus, the chemical state of cobalt nanoparticles in PVA/Co nanocomposite without any overcoating retains its metallic characteristics.

XPS survey scan of the as-deposited (unetched) sample exhibited C1s and O1s peaks. As the etching/sputtering time was increased, the intensity of Co peak increased while the intensity of C1s and O1s peaks decreased. A narrow scan of Co peaks revealed that in the post-etched sample spectra, Co is present in the metallic phase. Co 2p peak is present in the XPS spectra recorded after 80 mins of sputtering. The Co 2p spectrum could be fitted well with one peak indicating the presence of only one kind of cobalt, thus indicating the metallic cobalt. The binding energy of Co 2p state as determined from the XPS study is found to be 15.09 eV which matches well with the reported value of cobalt metal[25,28]

The experimental data of MOKE for the measurement of in-plane magnetic properties show well-defined hysteresis loop with coercivity $Hc_{\|}$ of 42.8 Oe indicating soft ferromagnetic behavior. A comparison of the $Hc_{\|}$ value of cobalt films of 40 nm on substrates like the glass is reported to be 26 Oe;[27–34] whereas the corresponding result for the same on glass and on Si substrates reveals that $Hc_{\|}$ value is to be less than 10 Oe.[28] It is worthy to note that it can give a higher value of the magnetic property ($Hc_{\|}$) in the PVA/Co nanocomposites system compared with similar systems with inorganic substrates. More importantly, the property like $Hc_{\|}$ for 5 nm layer of Co nanoparticles embedded in the PVA matrix of PVA/Co/CNP nanocomposite having an overlayer of 2 nm of carbon[2] showed a $Hc_{\|}$ value of 7 Oe (Table 6.3). Clearly, this study showed that nanocomposites based on Co and PVA without the application of any overcoat in the nanocomposite film system showed better $Hc_{\parallel}$ magnetic property value. PVA/Co nanocomposite appears to exhibit better magnetic properties compared with that of PVA/Co/CNP; thus, it seems that there is no need for overcoat in such a system of magnetic materials unlike in other systems of materials.

TABLE 6.3 Magnetic Property ($Hc_{\parallel}$) Evaluated by MOKE Method of PVA/Co/CNP and PVA/Co Magnetic Films.

Type of co-embedded PVA polymer magnetic film	Overcoat thickness	Magnetic property ($Hc_{\parallel}$)	Remarks
PVA/Co/CNP magnetic film	2 nm thickness of CNP overcoat	7.70 Oe	If there is no overcoat in cobalt-embedded PVA polymer nanocomposite, the value of magnetic property ($Hc_{\parallel}$) is higher than those of similar nanocomposite systems with inorganic matrix system
PVA/Co magnetic film	0 nm thickness (i.e., no overcoat)	42.80 Oe	

This indicates that there is apparently no change in the crystalline structure of PVA even after the sputtering of the cobalt nanoparticles. The average particle size of Co nanoparticles in the PVA/Co film is found to be 2.1 nm using the Scherer formula.[19,21]

6.3.1.2 NANOPARTICLES AND MATRIX MATERIAL

Thus, from the above observation, it reveals that an overcoat is redundant for nanoparticle embedded polymer nanocomposite systems without overcoating, the coercivity magnetic property is more pronounced than its counterpart having overcoating. Therefore, for further studies, similar samples have been prepared *without using any overcoat* in nanoparticle embedded polymer nanocomposite systems. In order to examine if there is any influence of polymer matrix in the magnetic properties, polymethylmethacrylate (PMMA) is an optical material and is used here as matrix material.[29] To examine the influencing role of polymer matrix on the magnetic property of the nanocomposites, samples have also been prepared by embedding magnetic nanoparticles (Fe) on the PMMA as well as PVA matrix separately till the embedded thickness of 5 nm is achieved in both nanocomposites without any overcoat.

GIXRD pattern of PMMA/Fe and PVA/Fe films nanocomposite reveals the presence of nanoparticles with *bcc* phase (Figs. 6.2 and 6.3). This phase is desirable as it is mainly responsible for the magnetic property in the PMMA/Fe and PVA/Fe films.

The magnetic measurement of the prepared PMMA/Fe and PVA/Fe nanocomposite film was carried out by using the MOKE technique. From the record of MOKE measurement, well-defined hysteresis loop has been observed in each of the samples. The value of the coercively ($Hc_{\parallel}$) is found to be respectively 24.39 Oe and 50.34 Oe for PMMA/Fe and PVA/Fe nanoparticle embedded polymer films. It may be noted that the magnetic property in the PVA/Fe is found to be higher than that of PMMA/Fe.

The earlier work reported on Fe deposited on glass by the IBS technique shows the $Hc_{\parallel}$ value of 2.5–5 Oe for Fe layer with a thickness of 300–350 nm[32,34]; thus the polymer is proved to be a good substrate host for the guest magnetic nanoparticles to facilitate the growth of magnetic nanostructures to display better magnetic property.

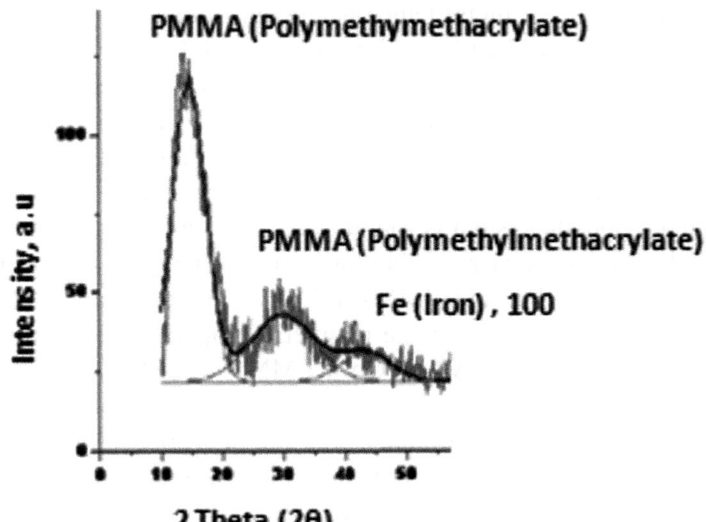

FIGURE 6.2 GIXRD spectral pattern of as-deposited PMMA/Fe film (5 nm thickness).

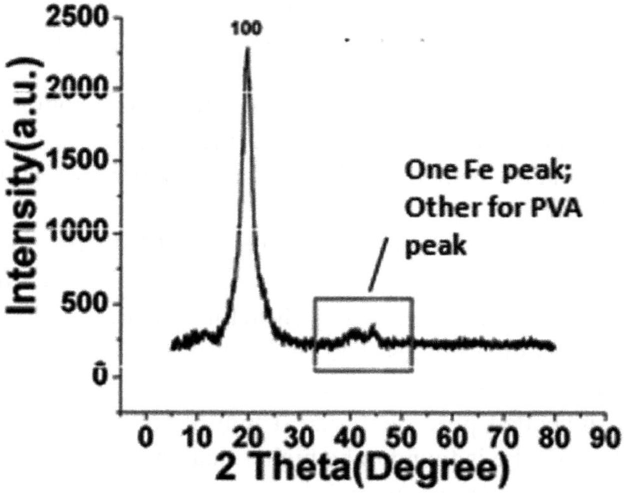

FIGURE 6.3 GIXRD pattern of as-deposited PVA/Fe nanocomposite film (5 nm thickness).

Recently, we developed interest to examine the influence of the type of metal nanoparticle on the magnetic property of the nanoparticle embedded polymer nanocomposite film (polymer/M; M=Fe, Co, Ni). For this, nanocomposite of metal nanoparticle embedded layer of 10 nm in a given polymer matrix like PMMA[28] was prepared. It may be noted that we report here the result corresponds to the nanoparticle embedded layer of 10 nm instead of 5 nm; this is due to the fact that PMMA/Ni nanocomposite corresponding to 5 nm did not show any measurable magnetic property. The particle size of the nanoparticles of Fe, Co, and Ni in the PMMA/metal nanocomposite system is found to be 1.69 nm, 1.4 nm, and 2.06 nm respectively. And we observed the magnetic property of such samples as PMMA/Co, PMMA/Fe, and PMMA/Ni nanocomposite are 32.6 Oe, 22 Oe, and 6.59 Oe respectively; the magnetic property follows the pattern in the given condition as PMMA/Co > PMMA/Fe > PMMA/Ni. Thus, it appears that the nature of metal nanoparticle, as well as the type of matrix play an important role to influence the magnetic property of the metal nanoparticle embedded polymer nanocomposite material system. Thus, from the preliminary studies (Polymer/M; M= Fe, Co, Ni), it implicates that the size and type of metal nanoparticle may influence the magnetic properties and allied attributes.

6.4 CONCLUSIONS

Magnetic transition metal nanoparticles can be embedded by ion beam sputtering technique to generate magnetic materials which hold promise for possible application in hi-tech areas of magnetoelectronics and for the storage of data for information technology. The polymer matrices are biocompatible materials, thus may be explored for some biomedical applications.

KEYWORDS

- **polymer**
- **nanoparticles**
- **carbon**
- **magnetic materials**
- **ion beam sputtering**

REFERENCES

1. Wang, H.; Wong, S. P.; Cheung, W. Y.; Ke, N.; Chiah, M. F.; Liu, H.; Zhang, X. X. Microstructure Evolution, Magnetic Domain Structures, and Magnetic Properties of Co–C Nanocomposite Films Prepared by Pulsed-filtered Vacuum Arc Deposition. *J. Appl. Phys.* **2000,** *88*, 2063.
2. Grundy, P. J. Thin Film Magnetic Recording Media. *J. Phys. D: Appl. Phys.* **1998,** *31*, 2975.
3. Park, I.-W.; Yoon, M.; Kim, Y.-M.; Yoon, H.; Song, H. J.; Volkov, V. I.; Avilov, A.; Park, Y. Magnetic Properties and Icrostructure of Cobalt Nanoparticles in a Polymer Film; Solid State Communications, **2003,** *126* (7), 385–389, doi: 10.1016/S0038-1098(03)00189-3
4. Banerjee, M.; Sachdev, P.; Mukherjee, G. S. Studies on Magnetic Nanocomposites of Carbon Cobalt Vinyl-Polymer Prepared by Ion Beam Sputtering Technique. *J. Sci. Conf.* **2009,** *1*, 86–92.
5. Banerjee, M.; Sachdeva, P.; Mukherjee, G. S. Preparation of PVA/Co/Ag Film and Evaluation of Its Magnetic and Microstructural Properties. *J. Appl. Phys.* **2012,** *111*, 094302; https://aip.scitation.org/journal/jap
6. Sachdeva, P.; Banerjee, M.; Mukherjee, G.S., Magnetic and Microstructural studies on PVA/Co nanocomposite prepared by Ion Beam Sputtering Technique. *Def. Sci. J.* **2014,** *64* (3), 290–294.

7. Ishak, I. M.; Quintela, A.; James, S. W.; Ashwell, G. J.; Lopez-Higuera, J. M.; Tatam, R. P. Modification of the Refractive Index Response of Long Period Gratings Using Thin Film Overlays. *Sens. Actuators B Chem.* **2005,** *107* (2), 738–741.
8. Sukumar, B.; Hazra, S. K. Graphene–Noble Metal Nano-Composites and Applications for Hydrogen Sensors. *J. Carbon Res. C* **2017,** *3*, 29; doi:10.3390/c30400299.
9. Sharma, A.; Tripathy, J.; Tripathy, S.; Ugochukwu, K. C. Investigation of Magnetic and Structural Properties of Au Capped Fe Thin Films. *Phys. B: Condens. Matter.* **2019,** *560*, 81–84.
10. Chadeu, E.; Oulahal, N.; Dubost, L.; Favergeon, F.; Degraeve, P. Anti-listeria Innocua Activity of Silver Functionalized Textile Prepared with Plasma Technology. *Food Control* **2010,** *21*, 505.
11. Veres, T.; Cai, M.; Cochrane, R. W.; Roorda, S. Ion Beam Modification of Co/Ag Multilayers II. Variation of Structural and Magnetic Properties with Co Layer Thickness. *J. Appl. Phys.* **2000,** *87*, 8513.
12. Kundu, S. Interface Modification of Ag/Co System: X-ray Reflectivity Study. *Nucl. Instrum. Methods Phys. Res. B: Beam Interact. Mater. At.* **2003,** *212*, 489.
13. Viegas, A. D. C.; Geshev, J.; Schelp, L. F.; Schmdt, J. E. Correlation Between Giant Magnetoresistance and Magnetic Interactions in a CoAg Multilayered/Granular System. *J. Appl. Phys.* **1997,** *82*, 2466.
14. Yu, M.; Liu, Y.; Sellmyer, D. J. Structural and Magnetic Properties of Nanocomposite Co:C Films *J. Appl. Phys.* **1999,** *85*, 4319.
15. Kharmouche, A.; Cherif, S.-M.; Bourzami, A.; Layadi, A.; Schmerber, G. Structural and Magnetic Properties of Evaporated Co/Si(100) and Co/glass Thin Films. *J. Appl. Phys.* **2004,** *37*, 2583.
16. Mukherjee, G. S. Modification of Poly(vinyl alcohol) for improvement of Mechanical Strength and Moisture Resistance. *J. Mater. Sci.* **2005,** *40*, 3017.
17. Mukherjee, G. S.; Shukla, N.; Singh, R. K.; Mathur, G. N. Studies on the properties of Carboxymethylated Polyvinyl Alcohol. *J. Sci. Ind. Res.* **2004,** *63*, 596–602.
18. Mukherjee, G. S. Calorimetric Characterization Of Membrane Materials Based On Polyvinyl Alcohol. *J. Therm. Anal. Calorim.* **2009,** *96* (1), 21–25.
19. Bian, B.; Bain, A.; Kwon, S.-J.; Laughlin, D. E. High Coercivity Co-alloy Thin films on Polymer Substrates. *IEEE Trails. Magn.* **2001,** *37*, 1640.
20. Banerjee, M.; Jain, A.; Mukherjee, G. S. Microstructural and Optical Properties of Polyvinyl Alcohol/Manganese Chloride Composite Film. *Polym. Compos.* **2019,** *40*, E765–775.
21. Sharma, A.; Tripathi, S.; Brajpuria. R.; Shripathi, T.; Chaudhari, S. M. Thickness Dependent Structural, Magnetic and Transport Properties of Nanostructured Cobalt Thin Films. *J. Nanosci. Nanotechnol.* **2007,** *7*, 2041.
22. Cullity, B. D. *Elements of X-ray Diffraction*, Addison Wesley Publishing Company Inc.: Boston, 1956; p 84, 99.
23. Hofmann, S. Sputter-depth Profiling for Thin-film Analysis. *Philos. Trans. Royal Soc. A Mathematical Physical and Engineering Sciences* **1814,** *362*, 55–75. doi: 10.1098/rsta.2003.1304
24. Mi, W. B.; Guo, L.; Jiang, E. Y.; Li, Z. Q.; Wu, P.; Bai, H. L. Structure and Magnetic Properties of Facing-Target Sputtered Co–C Granular Films. *J. Phys. D: Appl. Phys.* **2003,** *36*, 2393.

25. Yang, H. T.; Shen, C. M.; Wang, Y. G.; Su, Y. K.; Yang, T. Z.; Gao, H. J. Stable Cobalt Nanoparticles Passivated with Oleic Acid and Triphenylphosphine. *Nanotechnology* **2004,** *15*, 70.
26. Sharma, A.; Brajpuriya, R.; Tripathi, S.; Chaudhari, S. M. Study of Annealed Co Thin Films Deposited by Ion Beam Sputtering. *J. Vac. Sci. Technol. A* **2006,** *24*, 74.
27. Al-Kuhaili, M. F. Characterization of Thin Films Produced by the Thermal Evaporation of Silver Oxide. *J. Phys. D: Appl. Phys.* **2007,** *40*, 2847.
28. Gautam, A.; Tripathy, P.; Ram, S. X-ray Diffraction in Ag-metal Reinforced Polymer of Polyvinyl Alcohol of Thin Laminates. *J. Mater. Sci.* **2006,** *41*, 3007.
29. Ghosh, P.; Mukherjee, G. S. Photopolymers (I): Photoinitiating Role of Monochloroacetic Acid in the Synthesis of Poly(methyl methacrylate). *Polym. Adv. Technol.* **1999,** *10*, 687-694.
30. Kharmouche, A.; Cherif, S. M.; Bourzami, A.; Layadi, A.; Schmerber, G. Structural and Magnetic Properties of Evaporated Co/Si(100) and Co/glass Thin Films. *J. Phys. D: Appl. Phys.* **2004,** *37*, 2583.
31. Kundu, S. The Angular Distribution of Sputtered Silver Atoms. *Nucl. Instrum. Methods Phys. Res. B: Beam Interact. Mater. At.* **2003,** *212*, 489.
32. Veres, T.; Cai, M.; Cochrane, R. W.; Roorda, S. Ion-beam Modification of Co/Ag Multilayers II: Variation of Structural and Magnetic Properties with Co Layer Thickness. *J. Appl. Phys.* **2000,** *87*, 8504.
33. Iwatsubo, S.; Takahashi, T.; Naoe, M. Magnetic and Structural Properties of Fe Films Deposited by Ion-beam Sputtering with a High-energy Assisted Process. *Thin Solid Films* **1996,** *281–282*, 484–487.
34. Kharmouche, A.; Cherif, S. M.; Bourzami, A.; Layadi, A.; Schmerber, G. Structural and Magnetic Properties of Evaporated Co/Si(100) and Co/glass Thin Films. *J. Phys. D: Appl. Phys.* **2004,** *37*, 2583–2587.

CHAPTER 7

Polymeric Materials to Improve Durability and Sustainability of Cement Concrete: A Brief Overview

MAINAK GHOSAL[1,2]

[1]*Coal Ash Institute of India, Dum Dum, Kolkata, West Bengal, India*

[2]*School of Advanced Materials, Green Energy and Sensor Systems, Indian Institute of Engineering Science and Technology, Shibpur, Howrah, West Bengal, India*

ABSTRACT

Polymer usage in the construction industry is more than 100 years old. Polymers are used in this sector under a new terminology named "admixture," which means something which is added to the cementitious mix or concrete from outside. Polymers or admixtures in concrete have multifold functions, namely water-reducing, increasing workability, and increasing mechanical properties like compressive strength, tensile strength, flexural strength, shear strength, torsional strength, elasticity, etc. In the construction sector, they are hydrophilic and form solvent with water having 35–40% active chemicals, and the amount added is generally a minuscule amount by weight of cement. In the market, they are as follows: water reducers; set-controlling admixtures; air entertainers; specialty admixtures; shrinkage-reducing admixtures; SBR latexes, and others.

But the polymers in the construction industry have evolved with time, be it thermosets or thermoplastics, from first-generation lignosulfonate polymers to second-generation and third-generation polymers. This chapter discusses

Foundation and Growth of Macromolecular Science: Advances in Research for Sustainable Development. Meegle S. Mathew, PhD, Józef T. Haponiuk, PhD & Sabu Thomas, PhD, DSc (Eds.)

the whole trajectory along with the literature surveys of how nanomaterial additions have helped in addressing unique property changes of the cement concrete.

7.1 INTRODUCTION

7.1.1 BASICS OF POLYMERS

Most polymers are organic, and organic materials generally are made of hydrogen and carbon linked through intramolecular covalent bonds called "hydrocarbons." The carbon atom having four valence electrons participates in the covalent bonding. Thus, in methane (CH_4) a single contributory covalent bond is formed where the two bonding atoms of carbon and hydrogen contribute one electron each, while in ethylene and acetylene the carbon–carbon double and triple bonds between two atoms result from the sharing of two or three pairs of electrons. The term "*polymer*" comes from the Greek word "*polus*," *meaning* "many" or "much" and "*meros*" or "*mers*," *meaning* "part," that refers to large molecules or macromolecules.[34] The molecules in polymers are huge in weight when compared to the hydrocarbon molecules, thus they are referred to as macromolecules. In each molecule, the atoms of macromolecules are bonded together by strong covalent atomic bonds and generally, each carbon atom singly bonds forming the backbone of a macromolecule. Not only size, but the polymers also have different properties when compared to their hydrocarbons, as the hydrocarbon ethylene (C_2H_2) is a gas while its polymer polyethylene (PE) is a solid polymeric liquid.

7.1.1.1 POLYMER STRUCTURES

As shown in Figure 7.1 for example, the repeating units of polyethylene (PE) and polyvinyl chloride (PVC) are H–C–H and H–C–Cl, respectively. The PE chain structure can also be represented as $(–CH_2–CH_2–)_n$, where the repeat units are marked by dashes, and the subscript "n" indicates the number of repeating units. So, with the addition of many ethylene monomer units, we get PE molecule. Considering the case of vinyl chloride monomer ($H_2C=CHCl$) unlike that of ethylene monomer, where one Cl atom replaces one of the four H atoms. Its polymerization formula is represented as $(H_2C=CHCl)_n$.[35] Other polymer structures having different chemistries are also possible like tetrafluoroethylene monomer ($F_2C=CF_2$) polymerizes to form polytetrafluoroethylene (PTFE) having the trade name Teflon™. Polymers also may be depicted using the widely applicable form like $(H_2C{=}CHR)_n$, where "R" represents either an atom like H or Cl for PE or PVC, respectively or an organic group such as CH_3, C_2H_5, and C_6H_5 which are termed as methyl, ethyl, and phenyl groups, respectively.[35] The angle between the singly bonded carbon atoms is not strictly 180° as shown in Figure 7.1, but close to 109° while the C–C bond length comes to be 0.154 nm. Along with the structure of each chain, there may be 1000–100,000 covalent bonded carbon atoms. Thus, the polymers are formed or joined together by covalent carbon–carbon atoms bonding, and as shown in Figure 7.2 the "*mer*" repeating entities may be linear, branched, cross-linked, or networked with the direction of the arrow representing the increasing direction of strength. It may be pointed out that the more the secondary bonding, the more the cross-links and more is the ability of the polymers to carry strength like the arrow shows.

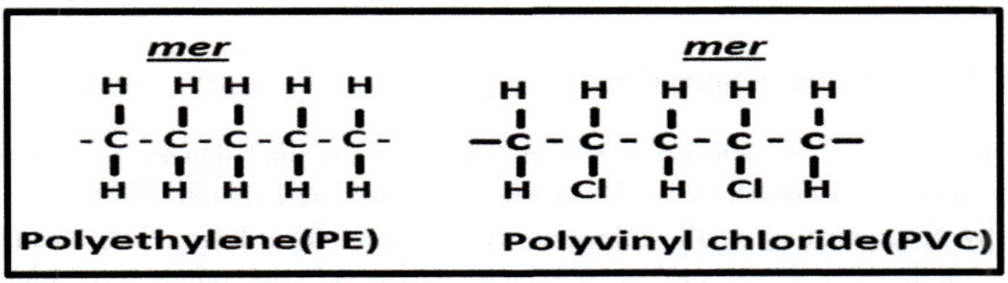

FIGURE 7.1 Basic "*mer*" structure of polymers.

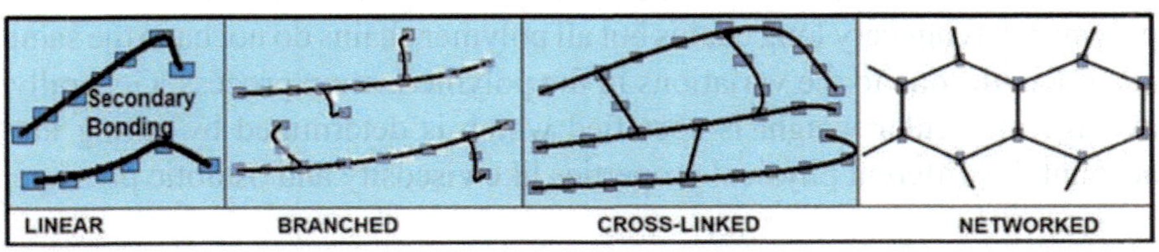

FIGURE 7.2 Schematic relation between covalent carbon–carbon atoms chain configurations with strengths.

We may mathematically express polymers as polymers = many *meros* or many parts or repeating parts/units. From the molecular structural point of view, there are several polymer structures like linear, branched, cross-linked, and networked. Linear polymers have units repeatedly joined together in an end-to-end fashion in single chains. They may have van der Waals interaction or hydrogen bonding between the chains, for example, PE, PVC, polystyrene, nylon, and poly (methyl methacrylate). For branched polymers, the side-branch chains connected to the main chains are believed to have resulted from the side reactions that occur during the polymer synthesis. High-density polyethylene (HDPE) is a branched linear polymer while low-density polyethylene (LDPE) contains short branch chains. Cross-linked polymers have adjacent linear chains joined to one another at various points by covalent bonding like rubbers. The networked polymers are those having three or more active covalent bonds. The distinguishing mechanical and thermal properties of multifunctional polymers can be found in epoxies, polyurethanes, or phenol formaldehyde. From the structural point of view, polymers can be typed as (i) Homopolymer having a configuration as $(\text{-A-A-A-A-A-A-})_n$ or (ii) Copolymers can be further subdivided into: (a) Random type having a configuration as $(\text{-A-A-B-A-B-A-A-A-B-})_n$, (b) Alternating type having a configuration as $(\text{-A-B-A-B-A-B-A-B-})_n$, and (c) Block type having a configuration as $(\text{-A-A-B-B-A-A-B-B-})_n$.[35]

7.1.1.2 POLYMER BEHAVIORS

The physical characteristics of a polymeric compound depend upon the molecular weight, shape, and structure of its polymeric chains. It is generally observed that small molecular weight compounds have small tensile strength while larger molecular weight polymeric compounds have high tensile strength, as shown in Figure 7.3. This may be attributed to the fact that longer chains are anchored better. High molecular weights are found in polymers having very long chains but all polymer chains do not have the same chain lengths due to the variations in the polymerization process. Generally, average molecular weight is specified which is determined by taking into account the different physical properties like viscosity and osmotic pressure.

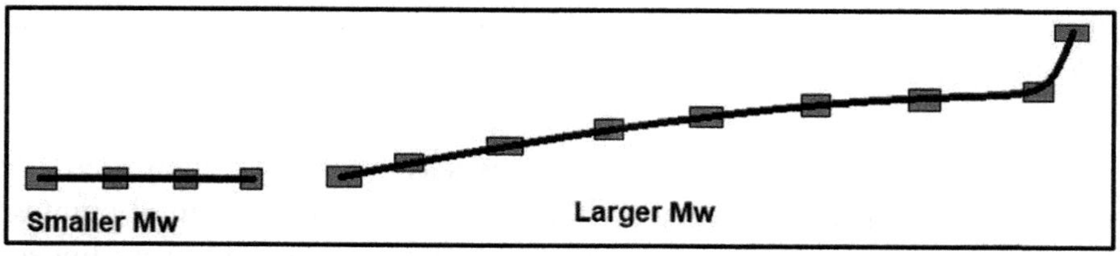

FIGURE 7.3 Schematic representation of linear polymeric chain compounds having different (smaller/larger) molecular weights (M_w).

There are various ways of defining molecular weight and then determining the number of a fraction of chains within each size range. Alternatively, the degree of polymerization (DP), which represents the average number of repeat units in a chain, is expressed by eq 7.1,

$$DP = Mn/m \tag{7.1}$$

where M_n is the average molecular weight of the polymer and m is the molecular weight of the repeating unit or monomer.

A number of properties of polymers are affected by the polymer chains lengths, such as the melting points (MP) or softening temperatures increase with increasing molecular weight (M_w ~100,000 g/mol). At room temperature, polymers with short chains and low molecular weights (M_w ~100 g/mol) behave as liquids or gases while those with medium molecular weights (M_w ~1000 g/mol) are waxy solids or soft resins. Solid polymers, also called as high polymers are those having molecular weights ranging from M_w ~10,000 g/mol to several million g/mol.[36]

7.1.1.3 POLYMER CRYSTALLINITY

Molecular substances having small molecules like water or methane normally exist as totally crystalline (solids) or totally amorphous (liquids). This crystalline state may exist in polymeric materials as shown in Figure 7.4, and it calls for molecules instead of atoms or ions with the atomic arrangement being more complex as with metals or ceramics.

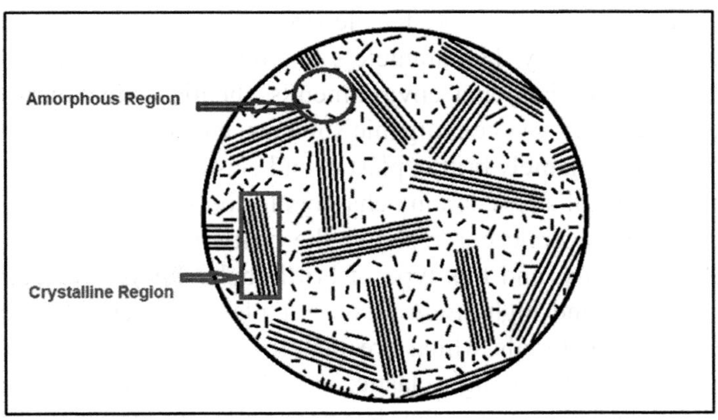

FIGURE 7.4 Schematic representation of crystallinity in a spherulitic polymeric molecule.

In contrast with the metals which are almost 100% crystalline or ceramics which are either totally crystalline or noncrystalline, the polymer's crystallinity degree may vary from totally amorphous to almost (95%) crystalline. As shown in Figure 7.4, the density of a crystalline polymer will be more than the amorphous counterpart of the same material and same weight because the chains are more compactly packed than the haphazard amorphous regions. The crystallinity degree depends upon the rate of cooling during solidification and the chain arrangement. Random chains, which are intertwined in the viscous fluid during crystallization upon cooling, must get proper time to properly align themselves in a straight-line fashion. However, for polymers composed of complex repeat units (e.g., polyisoprene or polybutadiene), crystallization is not favored. The molecular chemistry suggests that for linear polymers, polymerization is easily satisfied as there are no side branches to interfere with. So, the inference is there is lesser crystallization for bulkier or larger side-chained group of atoms.[37] The polymers which are crystalline are normally much stronger and more averse to dissolution and heat softening. A polymer that is in semicrystalline form consists of small crystalline regions (crystallites) interspersed with randomly oriented, coiled, and kinked amorphous regions as shown in Figure 7.4. Extensive literature surveys prove that these crystals are regularly shaped and thin-layered platelets approximately 10–20-nm thick and 10-μm long. It has also been found that the tensile strength and elasticity increase with crystallinity, and annealing causes crystalline regions in a polymeric molecule to grow.

7.1.1.4 MECHANICAL CHARACTERISTICS

The mechanical characteristics of a polymer are expressed in the same parameters as that of metals like tensile strength and modulus of elasticity, which are influenced by the parameters like deformation (strain rate) rate, temperature differentials, and environment's chemical characteristics like the presence of water, oxygen, organic solvents, etc. There are mainly three types of responses, namely brittle, plastic, and elastomeric and thus there are three typical types of stress–strain behaviors for polymeric materials.

A typical stress–strain curve shows the brittle polymer fractures while deforming elastically, and this type of brittle behavior is very common at low temperatures and high strain rates. For plastic material, the behavior resembles many materials which are metallic in nature where the initial distortion/deformation, though elastic, is followed by resulting in plastic distortion/deformation. Lastly, the deformation by the polymers known as elastomers is totally elastic. When a semicrystalline polymer is stretched in tension, the coiled amorphous regions expand in the applied stress direction, and the molecules of the crystallite regions undergo both stretching and bending resulting in a slight increase in layer thickness. Also, the spherulite shape may get completely altered for moderate deformation while complete destruction may occur for large deformations. However, the predeformed spherulitic structure may be restored partially by annealing at elevated temperatures below the polymer's melting point (MP). As previously discussed, the mechanical behavior of a polymer gets impacted by the deformation rate and temperature conditions, like increasing the temperature and/or diminishing the strain rate, the tensile strength, and the modulus decrease with an increase of ductility. Modulus of elasticity or elastic moduli of macromolecules are determined similarly as determined for metals but they vary in range. For example, this modulus for a highly elastic polymer usually lies in the range of 7 MPa to 4 Gpa but for metals, the range is 48–410 GPa.

7.1.1.5 BACKGROUND

Polymers may be derived from flora and fauna as have been there in the pages of history. But do you know that human blood and milk are also forms of natural polymers? This is due to the presence of an adhesive-like material called casein also known as polyamides, which is one of the oldest natural polymers.[1] There are pieces of evidence that suggest that both blood and milk have been used in construction by the Romans and the Indians. Synthetic polymers are man-made polymers, often derived from various sources like wood (or lignin), petroleum oil, etc. Since World War II, the field of material science in general and the macromolecular world, in particular, has been revolutionized by the advent of synthetic polymers.[40]

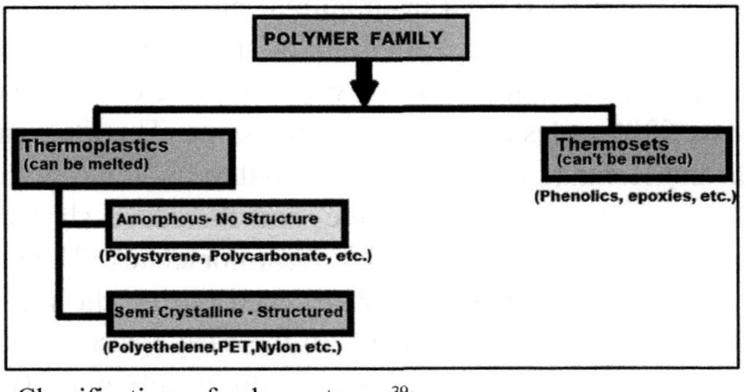

FIGURE 7.5 Classifications of polymer types.[39]

Source: Adapted from Ref. [39].

From the functional perspective, polymers can be differentiated into three main categories: thermosetting, thermoplastics, and elastomers. As the name suggests, the polymeric behaviors to mechanical forces at elevated temperatures can be differentiated into thermoplastic polymers and thermosetting polymers. Thermoplastics soften when heated and can be melted, while thermosets cannot be melted as they become permanently hard during their formation (see Fig. 7.5).[39] The following Table 7.1 gives the differences between the thermosets and thermoplastics.

TABLE 7.1 Differences Between the Thermoset Polymers and Thermoplastic Polymers.

Thermosets	Thermoplastics
Cannot be melted	Can be melted
Large cross-linking (10–50% repeat units)	Linear polymers with little branches
Brittle and hard	Ductile and soft
Do not soften when heating	Soften when heating
Better dimensional stability	Flexible stability
Examples: Vulcanized rubbers, epoxies, phenolics, and some polyester resins	Examples: PE, polystyrene, PVC, and polyethylene terephthalate

7.1.2 BRIEF DESCRIPTION

7.1.2.1 POLYMER-MODIFIED CEMENTITIOUS SYSTEMS

The Romans were credited to have refined the concrete mixture as we know it today. Portland cement was not in vogue and not even invented in those days. The Romans used lime and gypsum as an ancient primary mix for their structures. Roman concrete used even animal blood in mixes composed of pieces of gravel and coarse-sized sand mixed in an aqueous solution of hot lime. The Romans were famous for the wide building of concrete roads and they built some 5300 miles of roads using the same. The literature evidence suggests that the Romans used pozzolanic materials, milk, animal fat, and blood as additives for building concrete, and to reduce shrinkage cracks, they were known to have even used hair from horseback. The natural polymers like eggs, boiled bananas, etc. were used during the middle ages in Europe, China, Mesoamerica, and Peru. In the early years, polymer-related materials could be found in nature just as natural polymers. But advancement in polymer science led to products known as "admixtures," which are added nowadays to the concrete mix from outside, and after hardening, the polymers/admixtures could not be touched or felt in the hardened mix. The polymer can be applied to concrete in three main forms. In the first case, the polymer is directly applied to replace Portland cement in binding aggregates together and is called polymer concrete. In the second, the polymer is applied as an inseminating agent to penetrate an already existing structure mainly for repair purposes and is called polymer-impregnated concrete. And in the third case, the polymer is added from outside called an "admixture" into the concrete and is known as polymer-modified concrete or latex-modified concrete, as the polymer used in this category is mainly of latex type.[2] Strangely, but as a matter of fact, all admixtures or polymers used in cement concrete construction are water-soluble.

7.1.2.1.1 Lignosulfonates

Lignosulfonate (LS) (Fig. 7.6) is the first-generation admixture, and the raw material for its production is sourced from trees. Their molecular weight ranges from 1000 to 140,000. The first recorded application of LS on a road was in 1916, when it was used as dust palliative on a gravel road in Sweden. It has been consistently and extensively used in the United States since late 1940 as a dust palliative and surface stabilizing agent for unsealed roads. Lignosulfonates are a by-product of the pulp and paper industry, and are obtained from various processes like neutralization, precipitation, and fermentation of the waste obtained during the production of paper-making pulp from raw wood. Lignosulfonates having the lowest cost among all the admixtures has excellent dispersing properties and are utilized as superplasticizers in concrete and cement. The dosage of LS varies from 0.3 to 0.5% by weight cement and it can reduce water up to 5–10%. Calcium LS from molasses is a by-product of the sugar industry, used as a retarding and water-reducing admixture for concrete. Calcium LS increases the setting time and decreases the water content in concrete with an increase in strength.[3] When added to clayey soils, the shear stress, maximum cohesion value, and inter-particle friction increase with a 5% optimized addition of calcium LS.[8] The compressive strength of the concrete increased as the temperature during the curing time increased within the range from 28 to 80°C with LS, and reached a maximum value at the addition of 0.2% lignosulfonate by weight of the cement at various temperatures. More addition of the additive resulted in a decrease in the compressive strength of the cement.[5] One percent optimized LS of the replaced cement weight functioned as a retarder which can enhance the compressive strength of the M25 and M30 cement concrete. Further addition of LS decreases compressive strength. From a sustainability point of view, it is estimated that LSs are made from renewable materials, and 14 kg of CO_2 is saved for every kg of LS used in concrete.[4]

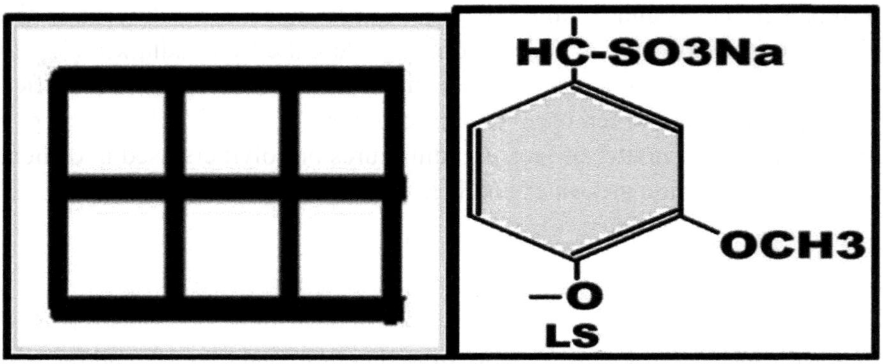

FIGURE 7.6 Schematic networked structure of LS polymer molecule.

Lignosulfonate reduces the cement hydration rate chemically as it is adsorbed on cement grains through the Ca^{+2} bridging, which reduces the calcium ion percentage in the solution. This removal of Ca^{+2} ions will prevent them from entering the setting and hydrating reaction of cement systems, which results in retarding the hydration.[9]

Sodium LS could be used in 0.5% dosages in producing self-curing concrete that decreases the setting issues in sweltering climate cementing by improving the usefulness and water maintenance and enabling the functionality to be expanded without including additional water and the compressive, malleable, and flexural quality tests outcome being positive.[7] Modified LSs and LS superplasticizers, and their compatibility with other admixtures have been of considerable research interest in recent years. With cement mortars and cement pastes, it was found that the retardation of the pastes was in the order of sulfonated naphthalene formaldehyde (SNF) (second-generation admixture) $<$ poly carboxylate ether (PCE) (third-generation admixture) $<$ LS (first-generation admixture) which followed the order of setting times of mortars mixed with admixtures. The loss of workability of the mortars with the LS superplasticizers was similar within the first hour, but less than those with the SNF and PCE superplasticizers. At 28 and 91 days, the porosity of the pastes with the LS superplasticizers at w/c of 0.34 was similar to that with the SNF superplasticizer, but higher than that with the PCE superplasticizer. For modified LS-based superplasticizers, plastic viscosities remain relatively unchanged within the first hour for both w/c ratios (0.34 and 0.4, respectively), whereas mortars with PCE and SNF superplasticizers had some variation in plastic viscosities.[6]

7.1.2.1.2 SULFONATED NAPHTHALENE AND MELAMINE-BASED FORMALDEHYDE (SNF/SMF)

SNF (see Fig. 7.7) and melamine formaldehyde (MF) fall in the group of second-generation admixtures. Their usage as normal plasticizers started in the early 1970s due to the development of high-strength concrete (HSC), which was first applied in skyscrapers in the United States. They are predominantly applied in the textile industry and as concrete superplasticizers in the construction business. With dosage rate @ 0.7–1% by weight of cement, it can reduce water by 10–20% in the concrete mixes. The production of SNF from naphthalene source is by fuming sulfuric acid or oleum ($H_2S_2O_7$); successive reaction with formaldehyde gives rise to polymerization after the sulfonic acid is sodium hydroxide or lime neutralizes it. The molecular weight of SNF/SMF ranges from 4000 to 75,000. The molecular weight effect when investigated by microelectrophoresis and UV-absorption techniques showed that the superplasticizer with the highest molecular weight had the largest negative Zeta Potential, and thus assumed to have a higher dispersing capability. Experiments reveal that for four polymers of MW = 16,000 g/mole, MW = 4000 g/mole, MW = 31,000 g/mole, and MW = 70,000 g/mole, the polymer of MW = 16,000 g/mole is the most adsorbed of the four superplasticizers, whereas the least adsorbed have MW = 70,000 g/mole.[13]

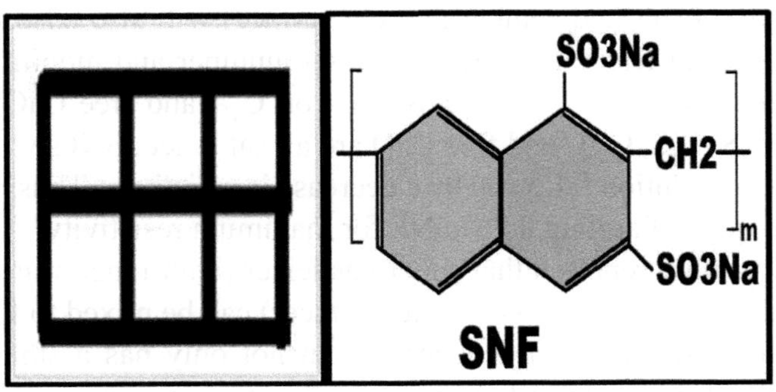

FIGURE 7.7 Schematic networked structure of sulfonated naphthalene formaldehyde polymer molecule.

There exist various discrepancies in the effect of sulfonated melamine formaldehyde (SMF) in cement as presented in various papers due to wrong analytical techniques. Some papers suggest that initial rapid sorption of SMF, probably on an ettringite precursor followed by partial desorption of SMF, but finally, the alkaline nature of the cement predominates. It has been found that the retarding effect of SMF is pronounced at 20°C but diminishes or disappears with increasing temperature up to 55°C. SMF reduces the mean size of $Ca\ (OH)_2$ crystallites and alters their morphology, but promotes the growth of larger ettringite crystals.[17] In one experiment, SNF was added to concrete by two different means using water. It was concluded that water reductions of 22.93% in the concrete mix are possible by the inclusion of SNF in concrete.[14]

Another work deals with the synthesis of a new type of melamine superplasticizer (NMS) with high MW through the reaction between melamine, formaldehyde, and sulfonated glucose. The results showed that NMS could reduce the water content, improve the workability and compressive strength of concrete, and maintain the initial slump of concrete after 1 h.[15]

The electrical resistivity method was used to study the effects of superplasticizers like SNF and LS on cement paste (w/c = 0.4). It was found that SNF improves the fluidity by electrostatic repulsion of the adsorbed surfaces. But after reaching a percentage b.w.c., increasing the dosage of SNF doesn't increase the resistivity but slightly decreases. It has been suggested that when cement particles come in contact with water, its alkali salts and gypsum dissolves, and its reactive minerals like C_3S, C_3A, and free CaO tries to hydrolyze resulting in full of Na^+, K^+, SO_4^{-2}, Ca^{+2}, OH^-, and $Al(OH)_4^-$ ions. Any SNF/LS added will prevent the ions dissolution and increase the resistivity and the higher the SNF used, the larger the surface areas covered and higher the resistivity. But the electrical double layer (formed by adsorption superplasticizer and cement particle surface) will also adsorb the Na^+, K^+, Ca^{+2} ions and leads to a decrease in the number and mobility of these ions resulting in the SNF being adsorbed on C_3A and free CaO minerals. But when C_3S, C_2S, C_3A, and free CaO are all saturated, SNFs will begin to dissolve in the solution followed by a decrease in resistivity. Thus, there is an optimum dosage of around 0.2% SNF for maximum resistivity.[19]

Some literature reported that superplasticizer (high-range water reducer) and normal plasticizer (ordinary water reducer) can be mixed to form a new combined high-range water reducer, which not only has a similar water-reducing effect as the base high-range water reducer but can also reduce the rate of loss of workability of the fresh concrete and achieve a saving in the basic material cost, and also was found to be particularly useful for hot weather concreting.[11] More recently, a new family of products, based on acrylic polymers (AP), has been proposed whose placing characteristics are more effective than those based on SMF or SNF due to its adsorption capacity when compared to electrostatic repulsion effects.[10]

7.1.2.1.3 Polycarboxylate Ether (PCE)

PCEs (see Fig. 7.8) were developed in Asia (mainly Japan) and Europe (mainly Germany) during the 1960s and 1970s, courtesy of the development of Self Compacting Concrete. These third generation polymers/admixtures are extracted from the petrochemicals industry and are chemicals structurally different from normal plasticizers. With a dosage of 0.7–1% by weight of cement, the use of PCEs permits a water reduction to an extent of up to 30% without reducing its workability in w.r.t to a reduction of up to 15% in the case of normal plasticizers. On the other hand, PCEs are more costly than SNF/SMFs or LS.

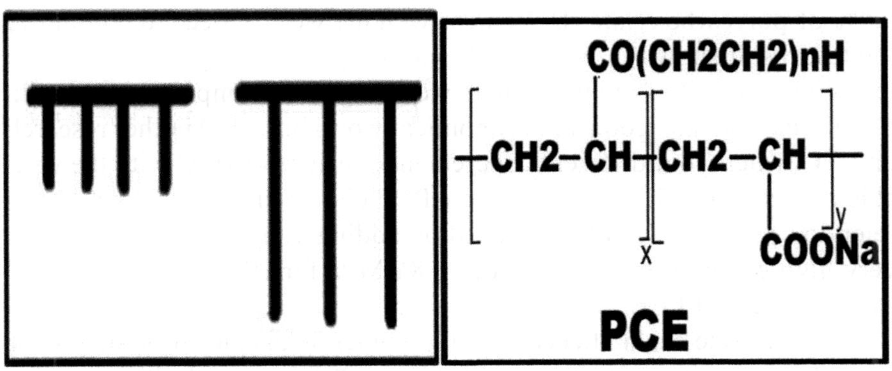

FIGURE 7.8 Schematic comb-like structure of PCE polymer molecule.

In one experiment, three sets of samples were prepared with different amounts of PCE plasticizer (0, 0.5, and 1.0 wt.% of cement). Each pair always contained reference samples (only cement), and 35 wt.% of fine ground recycled concrete was used as a certain substitution for cement in the mixture, due to economic reasons. It was found that adding the PCE increases the bulk density of all mixtures, and the addition of 1.0 wt.% plasticizer improved the compressive strength more favorably than a smaller amount of plasticizer (0.5 wt.%).[16]

Another experiment was conducted to study the effect of three types of superplasticizers (SP)—polycarboxylate-based (PCA) SP, SNF with 0.5% sodium sulfate (LSNF) SP, and SNF with 16.8% sodium sulfate (HSNF) SP, on the properties of the high-performance concrete (HPC). The authors concluded that concrete containing 0.28% PCA SP had the higher slump preservation and mechanical properties, the lower water porosity, carbonation depth, and chloride ion diffusion coefficient than the other SPs.[12]

A paper studies the effect of three different types of superplasticizers namely: SNF (dosage @ 0.5–2% b.w.c.), poly carboxylate ether (PCE) (dosage @ 0.4–1.2% b.w.c), and modified poly carboxylate ether (MPCE) (dosage @ 0.6–1.2% b.w.c) on the workability and mechanical properties of self-compacting concrete (SCC) mixtures. From this comparative study, it was found that MPCE is a better admixture compared to the others in terms of both fresh and hardened properties, and also from an economic point of view.[18]

One study aimed to get the optimized percentage of polycarboxylate to produce HSC of compressive strength of 50 MPa with good workability. The test was conducted on 60 concrete samples (having 15 cm diameter and 30 cm height). Various percentages 0, 0.5, 1.0, 1.5, and 2.0% of polycarboxylate (by cement weight) were added. According to the test result shown in Figure 7.9, optimal polycarboxylate added is in the amount of 1–1.5% to the cement weight, got the compressive strength of 50 MPa, had also good and economical workability.[21] Another research on HPC, where four different concrete mix designs of HPC with the variable of added 0, 0.5, 1, and 2% of the PCE claims that the optimal slump retention of 45 min can be achieved by adding 1 and 2% of PCE, and a very high compressive strength of 53.84 MPa for HPC can be achieved by adding 2% of PCE.[28]

Different side chain effects of PCE polymer in an aqueous system were studied. Results showed that the steric effect of PCE increased with increasing side chain length, which in turn influenced the intermolecular force of PCE polymers as well as the mobility of water molecules.[23]

Different polymeric superplasticizers were added, namely polycarboxylate-ether derivative (PCE), poly naphthalene sulfonate (PNS), and condensate of melamine-formaldehyde sulfonate (SMFC) to improve the injectability performance of the grouts in heritage masonry structures with sodium oleate (a water-repellent agent) to reduce water absorption. Pozzolanic mineral additions like microsilica and metakaolin were used for the enhancement of the strength and setting time. The compatibility between the different admixtures and action mechanisms of the different polymers were studied utilizing zeta potential and adsorption isotherms measurements. The results revealed that the grout composed of air lime, metakaolin, sodium oleate, and PCE was the most effective composition, improving the mechanical strength, injectability, and hydrophobicity.[24]

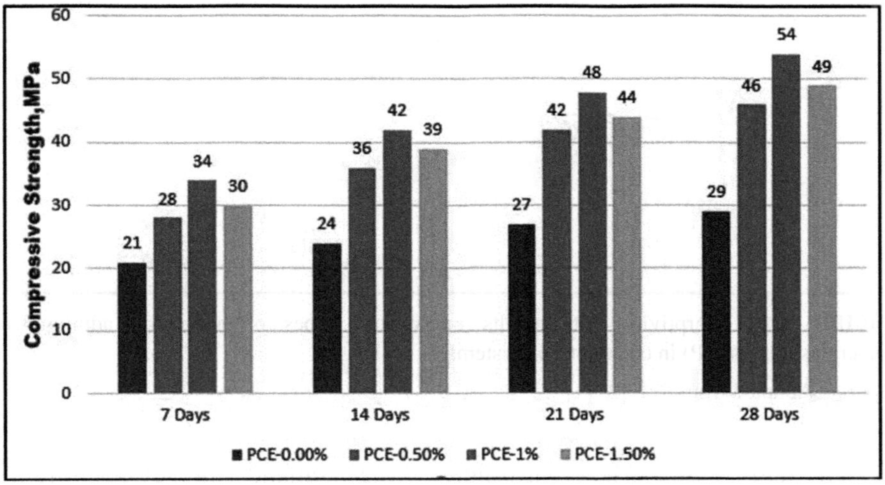

FIGURE 7.9 Strength of PCE polymer of various doses at different testing times.[21]
Source: Adapted from Ref. [21].

For studying polycarboxylate superplasticizers (PCEs) behavior in cement–waste stone powder (SP) paste, two kinds of SPs were measured and the effects of PCEs on their dispersion and adsorption behavior in cement–SP paste were investigated. The results show that the intercalation of PCE with Ledong stone powder (LDSP) was higher than with Haikou stone powder (HKSP) because the intercalation between PCE and muscovite, which is present in LDSP is stronger than the other mineral composites. Thus, a short-side-chain-PCE has good SP tolerance.[25]

The PCE has good compatibility with industrial wastes like ground granulated blast furnace slag (GGBS) and Fly ash. The addition of polycarboxylate gives good workability to the concrete to increase the strength, and the strength is increased more effectively by adding 40% of GGBS instead of cement.[26] In another study, the sorptivity (the tendency to absorb/transmit water) value of the superplasticized mixes is found to be less than that of the control specimen without a superplasticizer (Fig. 7.10).[27]

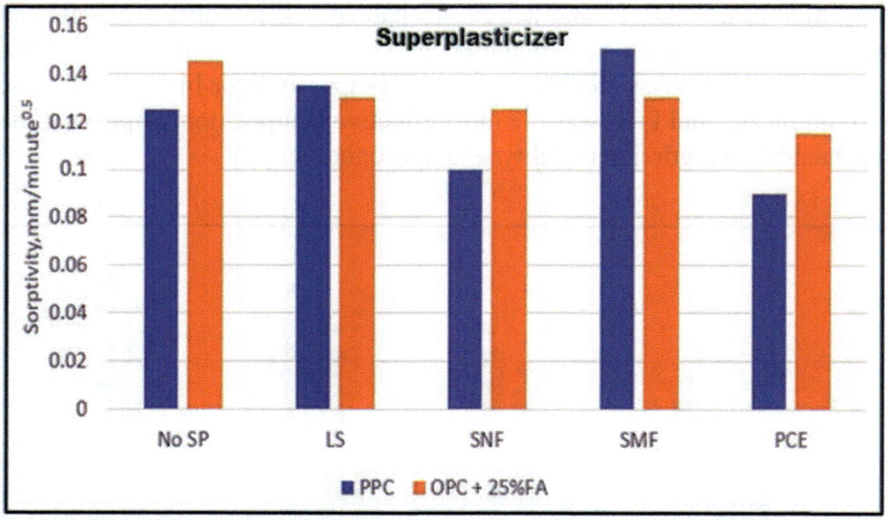

FIGURE 7.10 Sorptivity tests results on various types of polymeric admixtures/superplasticizers (SP) in cementitious systems.[27]

Source: Adapted from Ref. [27]

The temperature effects on fluidity, water demand, and setting time of cement paste with and without superplasticizer have been studied. Four main families of superplasticizers—polycarboxylate (PCE), naphthalene (SNF), melamine (SMF), and lignosulfonate (LS)—were investigated. It was found that—(i) The fluidity decreases with an increase in temperature at lower dosages of superplasticizer; however, the fluidity increases with an increase in temperature at higher dosages of superplasticizer. (ii) For all superplasticizers studied herein, the saturation dosage of superplasticizers obtained from the Marsh cone test increases with an increase in temperature. Thus, the PCE had the least temperature sensitivity, and LS exhibited the most sensitivity.[29] Some results showed that different polymeric structures had different effects on the workability of the mix, and by controlling these copolymer structures we could influence the various work abilities of the cementitious mix.[22]

7.1.2.2 *METHODS OF POLYMER ACTIONS*

The methods of action are different for each of the different polymer families. When water is added to the cementitious mixes, cement grains naturally cluster together to form flocs trapping water inside them. When conventional first- or second- generation polymers are added due to the networked structures (see Fig. 7.11) of LSs and sulfonated naphthalene or melamine formaldehydes, they impart negative charges to all cement grains. Like charges of cement, grains repel each other through a process called electrostatic repulsion, thus releasing the water needed for hydration mechanisms to continue.

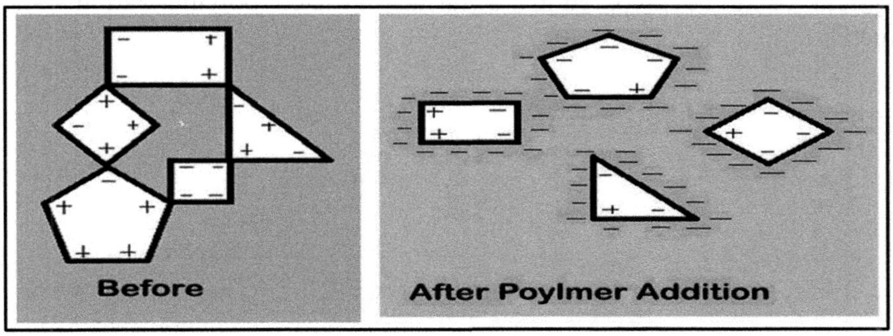

FIGURE 7.11 Charges on cement particles before and after the addition of conventional polymers.

But the third generation polymer PCE acts unconventionally due to the side chains of chain networked (see Fig. 7.8) structure of PCEs. These side chains dig deep into the flocculating structure of the trapped water through a process called steric hindrance, and help in releasing more quantifiable water needed for hydration mechanisms resulting in more hydration products and more strength and workability.[20]

7.1.2.3 POLYMER NANOCOMPOSITES MODIFIED HYBRID CEMENTITIOUS SYSTEMS

Polymers perform in the range of micrometers (10^{-6} m) while nanomaterials act in nanometric (10^{-9} m) levels. Very interesting material properties can be obtained through the behavior of materials in the nanoregime. The properties of materials, especially of nanodimensions in the nanoregime are of heaven and hell different from that of the bulk material. It is seen that with very small additions (1–2%) of nanomaterials, we get much increase in the mechanical strength (30–40%) of the composites. Nanomaterials play the role of nucleating agents apart from other roles of filler, disperser, and supplier of pozzolanic products, and serve the template for the growth of hydration products in a cement concrete polymeric composite. But unlike hydrophilic polymers, these nanomaterials are hydrophobic in nature. So, nanomaterials do not get dissolved in the water naturally. The nanomaterials are either made functionalized or ultrasonic energy is required to tear apart the intermingled nanomaterials as they are found in nature.[30] Other polymeric nanocomposites like Titania/polyaniline (PANI) nanocomposite system could be prepared by a surface-initiated polymerization (SIP) method.[31]

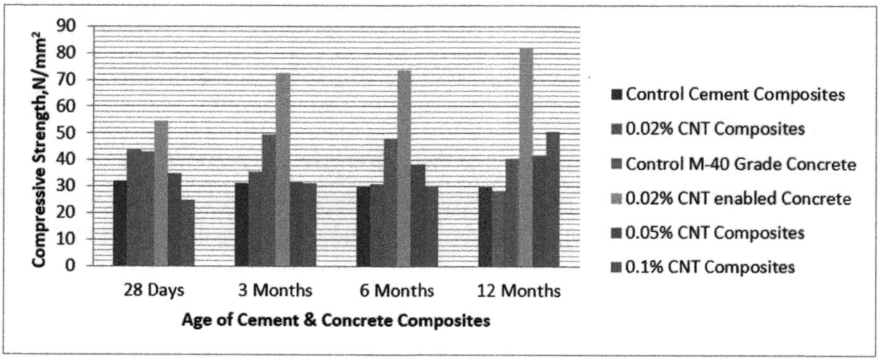

FIGURE 7.12 Variations in the strength of hybrid cementitious systems over the ages.

Source: Reprinted from Ref. [33], author's own work (Ghosal & Chakraborty, 2021)

At present, there exists a chaos regarding the usage of nanomaterials in polymers due to the reasons of toxicity introduced in the environment by these nanosized materials and some byelaws imposed in some foreign countries on nanomaterials usage. Some pieces of literature proposed the materials, multiwalled carbon nanotubes (MWCNT)–epoxy without UV stabilizer as a slow release control. They proposed the present protocol as an interim voluntary standard with a known level of reproducibility and a slightly refined protocol for prenormative validation. Based on the effectiveness of conventional UV stabilizers to reduce polymer aging, it is anticipated that an equally significant reduction of releases by UV-stabilized nanocomposites is intended for outdoor use.[32] Polymer nanocomposites are used as very good performers in cement concrete systems in terms of mechanical characteristics like bulk density and compressive strengths[33] as shown in Table 7.2 and Figure 7.12.

TABLE 7.2 Bulk Density (kg/m^3) and Compressive Strength (N/mm^2) of Ordinary and Hybrid Systems (% Increase w.r.t. 0% Nanomaterial Additions).[33]

% Nano additions in ordinary Portland cement	28-days results		3 months results		6 months results		12 months results	
	Bulk density (% incr.)	Comp. Str. (% incr.)	Bulk density (% incr.)	Comp. Str. (% incr.)	Bulk density (% incr.)	Comp. Str. (% incr.)	Bulk density (% incr.)	Comp. Str. (% incr.)
0% CNT	2315.40	31.89	2290.12	31.20	2309.27	30.01	2229.10	30.01
0.02% CNT (optimized)	2356.24 (1.76%)	43.75 (37.19%)	2337.79 (2.08%)	35.59 (14.07%)	2372.71 (2.75%)	30.89 (2.93%)	2369.02 (6.28%)	28.53 (−4.93%)
0.05% CNT	2189.20 (−5.45%)	34.88 (9.38%)	2306.64 (0.72%)	31.85 (2.08%)	2348.61 (1.70%)	38.55 (28.46%)	2338.68 (4.92%)	41.69 (38.92%)
0.1 % CNT	2267.23 (−2.08%)	24.83 (−22.13%)	2314.70 (1.07%)	31.50 (0.96%)	2353.40 (1.91%)	30.16 (0.50%)	2364.70 (6.08%)	50.78 (69.21%)
M-40 Grade Concrete	2559.34	43.03	2511.28	49.71	2589.38	48.34	2658.00	40.63
0.02% CNT enabled M-40 Concrete	2131.28 (−16.73%)	54.58 (26.84%)	2666.77 (6.19%)	72.37 (45.58%)	2667.45 (3.01%)	73.67 (52.40%)	2680.26 (3.83%)	82.00 (101.8%)

Source: Reprinted from Ref. [33], author's own work (Ghosal & Chakraborty, 2021)

7.2 SUMMARY AND CONCLUSIONS

Construction materials are getting more and more diverse. Take the case of cement which is the main binding material in the cement concrete systems used all over the world. First, it was ordinary Portland cement (OPC), and then came the blended cement like the fly ash or slag cement which used industrial wastes, followed by composite cement. Now, various advanced materials are the new entrants which can reduce CO_2 emissions by a major chunk. Though polymers are popular, there is a general lack of awareness especially their contribution to sustainability. Before using polymer, it is essential to know about the physical and mechanical properties of polymers along with concrete (especially the type of cement used). Also, the new LC3 cement uses additional PCE polymers to the tune of 50% extra, when compared to PCE usage for other cement. As PCEs are very costly when compared to the first and second generation admixtures, more work needs to be carried out. We all know that the first and second generation polymers like LS and SNF/SMF got substituted by the third generation polymers like PCEs, but the later ones are costly. So, researchers tried to modify the SNFs and SMFs to meet the requirements of PCE albeit at a lower cost. Nanomaterials that are generated from waste have added more diversity in the construction material field, but this particular area is also least researched in India though the research results found are very encouraging.

KEYWORDS

- **admixtures**
- **construction**
- **durability**
- **lignosulfonate**
- **melamine**
- **nanomaterials**
- **polymer**

REFERENCES

1. Guo, M.; Wang, G. Milk Protein Polymer and Its Application in Environmentally Safe Adhesives. *Polymers* **2016,** *8*, 324. https://doi.org/10.3390/polym8090324.

2. Li, Z. *Advanced Concrete Technology*; John Wiley & Sons, 2011. ISBN 9780470902431
3. Wang, L. K-P. Effect of Calcium Lignosulfonate on Properties of Concrete at Early Ages. Masters Theses. 5331, 1965. https://scholarsmine.mst.edu/masters theses/5331
4. Sharma, J. et al. Lignosulphonate: An Additive in Concrete. *Int. J. Adv. Res. Sci. Eng.* **2017,** *06* (09), 1514–1524.
5. Satiyawira, B. et al. Effects of Lignosulfonate and Temperature on Compressive Strength of Cement. *Proceedings World Geothermal Congress 2010*, Bali, Indonesia, 25–29 April 2010.
6. Jun Dan, S. Effect of a Newly Developed Lignosulphonate Superplasticizer on properties of Cement Pastes and Mortar. Thesis Paper, University of Singapore, 2008.
7. Priya, E. et al. Comparative Study of Self-Curing Concrete Using Sodium Lignosulphonate and Light Weight Aggregate. *Int. J. Innov. Technol. Explor. Eng. (IJITEE)* **2019,** *8* (6S).
8. Vincent Sam Jebadurai, S. et al. Feasibility Tests of Calcium Lignosulphonate on Clay. *Rasayan J. Chem.* **2017,** *10* (4), 1481–1491.
9. Yousf, M.; Mollah, A.; Polta, P.; Hess, Th.; Vempati, R. R. K.; Cocke, D. L. Chemical and Physical Effect of Sodium Lignosulphonate Superplasticizer on the Hydration of Portland Cement and Solidification/Stabilization Consequences. *Cem. Concrete Res.* **1995,** *25*, 671–682.
10. Collepardi, M. Admixtures Used to Enhance Placing Characteristics of Concrete. *Cem. Concrete Compos.* **1998,** *20* (2–3), 103–112. https://doi.org/10.1016/S0958-9465(98)00071-7.
11. Chang, D. Y.; Chan, S. Y. N.; Zhao, R. P. The Combined Admixture of Calcium Lignosulphonate and Sulphonated Naphthalene Formaldehyde Condensates. *Constr. Build. Mater.* **1995,** *9* (4), 205–209. https://doi.org/10.1016/0950-0618(95)00011-4.
12. Yang, Z.; Hui, Z.; Sun, W. Effect of the Types of Superplasticizers on the Fresh, Mechanical, and Durability Properties of the High-Performance Concrete. *J. Test. Eval.* **2016,** *44*, 20140442. DOI: 10.1520/JTE20140442.
13. Andersen, P. J.; Roy, D. M.; Gaidis, J. M. The Effect of Superplasticizer Molecular Weight on Its Adsorption on, and Dispersion of, Cement. *Cem. Concrete Res.* **1988,** *18* (6), 980–986. https://doi.org/10.1016/0008-8846(88)90035-X.
14. Osuji, S. O. et al. Current Effects of Naphthalene Based Superplasticizer's Addition Process on Water Reduction and Grade C20/25 Concrete's Compressive Strength. *J. Civil Eng. Res.* **2018,** *8* (1), 9–14. DOI: 10.5923/j.jce.20180801.02.
15. Wang, H. et al. Synthesis and the Effects of New Melamine Superplasticizer on the Properties of Concrete. *ISRN Chem. Eng.* **2013,** *2013*, Article ID 708063, 6 pp. http://dx.doi.org/10.1155/2013/708063.
16. Hrůza, J.; Prošek, Z. The Effect of Plasticizer on Mechanical Properties of the Cement Paste with Fine Ground Recycled Concrete. *Acta Polytechnica CTU Proc.* **2017,** *13*, 61. DOI: 10.14311/APP.2017.13.0061.
17. Yilmaz, V. T.; Glasser, F. P. Influence of Sulphonated Melamine Formaldehyde Superplasticizer on Cement Hydration and Microstructure. *Adv. Cem. Res.* **1989,** *2* (7), 111–119. https://doi.org/10.1680/adcr.1989.2.7.111
18. Evangeline, K. et al. Effect of Superplasticizer on Workability and Mechanical Properties of Self-Compacting Concrete. *IOSR Journal of Mechanical and Civil Engineering (IOSR-JMCE), National Conference on Advances in Traffic, Construction Materials and Environmental Engineering (ATCMEE)*, 2015; pp 18–29.

19. Zeng, S. et al. Effects of SNF and LS superplasticizer on Cement Paste Using Electrical Measurement. *Indian J. Eng. Mater. Sci.* **2010,** *17*, 27–33.
20. Ghosal et al. Influence of Cement Behaviour With & Without Polymer Nanocomposites. *Adv. Polym. Mater.* **2018.** River Publishers Series.
21. Sumiati, S.; Mahmuda; Herius, A.; Subrianto, A. The Effect of Polycarboxylate Addition Towards the Workability and High Strength Concrete. *J. Phys. Conf. Ser.* 2020, *1500*, 012068. DOI: 10.1088/1742-6596/1500/1/012068.
22. Ezzat, M.; Xu, X.; El Cheikh, K.; Lesage, K.; Hoogenboom, R.; De Schutter, G. Structure-Property Relationships for Polycarboxylate Ether Superplasticizers by Means of RAFT Polymerization. *J Colloid Interface Sci.* **2019,** *553*, 788–797. DOI: 10.1016/j.jcis.2019.06.088. Epub 2019 Jun 26. PMID: 31255940.
23. Chuang, P-H.; Tseng, Y-H.; Fang, Y.; Gui, M.; Ma, X.; Luo, J. Effect of Side Chain Length on Polycarboxylate Superplasticizer in Aqueous Solution: A Computational Study. *Polymers* **2019,** *11*, (2), 346. https://doi.org/10.3390/polym11020346.
24. González-Sánchez et al. Combination of Polymeric Superplasticizers, Water Repellents and Pozzolanic Agents to Improve, Air Lime-Based Grouts for Historic Masonry Repair. *Polymers* **2020,** *12*, 887. DOI: 10.3390/polym12040887
25. Feng, H. et al. Effects of Molecular Structure of Polycarboxylate Superplasticizers on Their Dispersion and Adsorption Behaviour in Cement Paste with Two Kinds of Stone Powder. *Constr. Build. Mater.* **2018,** *170. DOI:* 10.1016/j.conbuildmat.2018.02.195
26. Mani, S. et al. Study on Behavior of Ground Granulated Blast Furnace Slag (GGBS) as Partial Replacement of Cement in Concrete with Addition of PCE. *AIP Conf. Proc.* **2020,** *2240*, 060002. https://doi.org/10.1063/5.0011112.
27. Sathyan, D.; Anand, K. B. Influence of Superplasticizer Family on the Durability Characteristics of Fly Ash Incorporated Cement Concrete. *Constr. Build. Mater*. **2019,** *204*, 864–874. DOI: 10.1016/j.conbuildmat.2019.01.171.
28. Jonbi, J. et al. Effect of Added the Polycarboxylate Ether on Slump Retention and Compressive Strength of the High-Performance Concrete. *MATEC Web Conf.* **2018,** *195*, 01020. https://doi.org/10.1051/matecconf/201819501020
29. John, E.; Gettu, R. Effect of Temperature on Flow Properties of Superplasticized Cement Paste. *ACI Mater. J.* **2014,** *111*, 67–76.
30. Choudhary, V.; Gupta, A. Polymer/Carbon Nanotube Nanocomposites, Carbon Nanotubes—Polymer Nanocomposites. IntechOpen: Siva Yellampalli, Aug 17th 2011. DOI: 10.5772/18423.
31. Guo, Z. et al. Multifunctional Desulfurization Elastomeric Polymer Nanocomposites. 61st Annual Report, American Chemical Society, 2016.
32. Wohlleben, W. et al. NanoRelease: Pilot Interlaboratory Comparison of a Weathering Protocol Applied to Resilient and Labile Polymers with and Without Embedded Carbon Nanotubes. *Carbon* **2016.** http://dx.doi.org/10.1016/j.carbon.2016.11.011.
33. Ghosal, M.; Chakraborty, A. K. A Study on Performance of Carbon-Based Nano-Enabled Cement Composites and Concrete. RILEM Book Series, Vol. 29; Springer: Cham, 2021. https://doi.org/10.1007/978-3-030-51485-3_29.
34. https://en.wikipedia.org/wiki/Polymer.
35. Rudin, A.; Choi, P. Chapter 1—Introductory Concepts and Definitions. . In *The Elements of Polymer Science & Engineering*; Rudin, A., Choi, P., Eds., 3rd ed.; Academic Press, 2013; pp 1–62. ISBN 9780123821782. https://doi.org/10.1016/B978-0-12-382178-2.00001-8.

36. *Bio Surfaces: A Materials Science and Engineering Perspective*. Edited by Balani, K., Verma, V., Agarwal, A., Narayan, R. © 2015 The American Ceramic Society, Published 2015 by John Wiley & Sons, Inc, 2015.
37. Mustafa N.S.et al., Reviewing of General Polymer Types, Properties and Application in Medical Field. *Int. J. Sci. Res. (IJSR)* **2016,** *5* (8), 212–222. DOI: 10.21275/ART2016772.
38. Polymer Stress-Strain Curve. 2020, Nov 27. Retrieved July 9, 2021, from https://eng.libretexts.org/@go/page/7867.
39. Kamat, V. P. Making Mission India Single-Use Plastics Free Possible: Perspectives, Prospects and Actions, *Rambhau Mhalgi Prabodhini Indian J. Democratic Governance* **2021,** *II* (Special Issue), | Thane, India.
40. Burford, R. Polymers: A Historical Perspective. *J. Proc. R. Soc. New South Wales* **2019,** *152*, (Part 2), 242–250. ISSN 0035-9173/19/020242-09

CHAPTER 8

Protein–Polyelectrolyte Complexes: Structure and Properties

HRISHIKESH TALUKDAR[1] and SARATHI KUNDU[2]

[1]*Department of Physics, Anandaram Dhekial Phookan College, Haibargaon, Nagaon, Assam, India*

[2]*Soft Nano Laboratory, Physical Sciences Division, Institute of Advanced Study in Science and Technology, Paschim Boragaon, Garchuk, Guwahati, Assam, India*

ABSTRACT

Polyelectrolytes (PEs) are polymers composed of ionizable functional groups, and provide a number of useful properties for both basic and applied research. Proteins are biopolyelectrolytes (BPEs) and can form complexes with oppositely charged polyelectrolytes via combined influences of electrostatic interaction and entropy gain from counterion release in solution. This complex is known as protein–polyelectrolyte complex (PPC). During complex formation, protein exhibits variation of functional properties with their structural modifications depending on different physiochemical parameters. Owing to its specific structures and properties, PPC is found to be a potential material in the fields of biomaterial, biosensing, food chemistry, etc. However, basic understanding on structure–property correlation of PPC is still needed to be explored for investigating over structures, interactions, and productions of efficient materials for different applications. Therefore, the primary objective of this chapter is to ascertain the understanding of structure–property correlation of PPC material.

Foundation and Growth of Macromolecular Science: Advances in Research for Sustainable Development. Meegle S. Mathew, PhD, Józef T. Haponiuk, PhD & Sabu Thomas, PhD, DSc (Eds.)

8.1 INTRODUCTION

PEs are long-chained molecules bearing charges on their backbone. They dissociate in solution as polyanions or polycations and gets associated with counterions.[1,2] They are also known as polysalts because they exhibit nearly similar behaviors of polymers (such as high viscosity, huge response under the small perturbation, etc.) and salts. Thus, PEs behave as charged macromolecules for having the properties of both polymer and salt. Like electrolytes, different physicochemical parameters can also easily control the interactions among the constituent monomers and accordingly the properties can be tuned. These macromolecules exhibit attractive properties owing to their dual character of long chain and high charge. The intricate balance between the inter- and intraparticle interactions of different PEs causes different organizations and structures in nanometer length scale and modifies their properties such as mechanical, optical and electrical properties, thermal stability, T_G, etc.[3–5] Depending on the charge, charge density, position of ions sites, etc., PEs are classified into different categories which are shown in Figure 8.1. The behavior of PE inside solvent is different primarily due to the counterion condensation and Debye–Huckle charge screening. The electrostatic interaction between the counterion and polymer chain and loss of translational entropy by the counterions are the primary reasons for counterion condensation.[6–8] The counterion condensation can be expressed by eq 8.1 where l_B and l_i are the Bjerrum length and the separation between two neighboring charged ions in the chain respectively. ε and ε_0 are the permittivity of the solvent medium and absolute permittivity, respectively; e, k_B, and T are the electronic charge, Boltzmann constant, and absolute temperature, respectively. The Bjerrum length is the separation between two elementary charges in the polymer chain if the electrostatic energy becomes comparable with thermal energy in the chain and is expressed as:

$$l_B = \frac{e^2}{4\pi\varepsilon\varepsilon_0 k_B T} \tag{8.1}$$

If the actual l_i is greater than the l_B, then all the counterions diffuse away from the chain. If it is smaller, then a certain fraction of counterions stays in the immediate neighborhood of the PE chain and tries to increase the value of l_i. The limitation of the counterion condensation is the equality of l_i and l_B. Therefore, it occurs until the l_i becomes equal to the l_B. The counterion condensation reveals the fraction of charged monomers which depends on the ratio of l_i and l_B in the chain. Thus, instead of actual charge density, this effective charge density of the chain incorporating the effect of counterion condensation becomes an important parameter for all properties of the PEs.

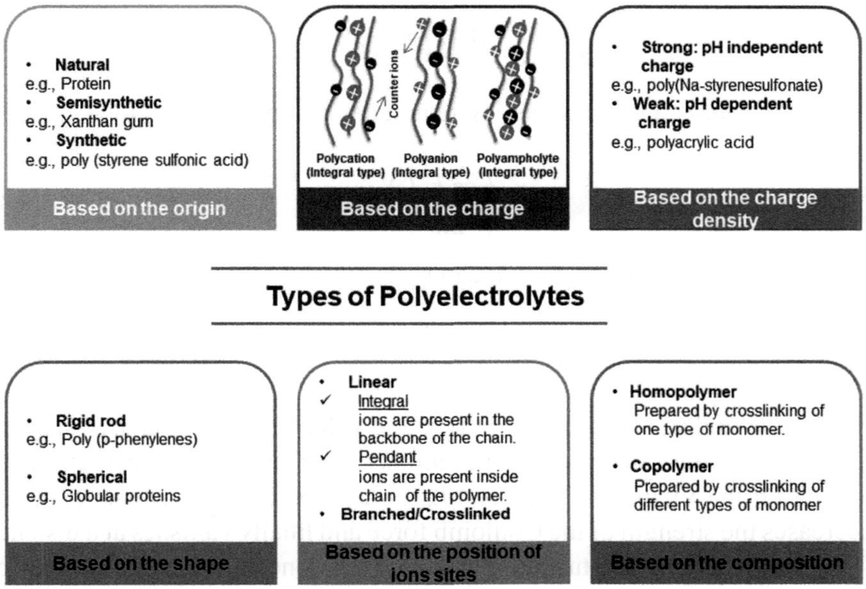

FIGURE 8.1 Types of PEs based on their origin, charge type, charge density, shape, position of ion sites, and compositions.

Source: Adapted from Ref. [101]

The second phenomenon, that is, Debye–Hückle charge screening reduces the coulombic interaction associated with the PE chains. The charges on the chain are often encircled by a shell of counterions and thus cannot create the associated Coulomb force. This type of cloud formation of opposite charges can be found in all electrolytes, polyions, and low molecular mass ions up to a certain length scale which is called Debye screening length ξ_D.[8,9] The potential energy generated by an isolated charge particle in a medium of dielectric constant ε can be expressed as in eq 8.2,

$$V(r) = \frac{e}{4\pi\varepsilon\varepsilon_0 r} \tag{8.2}$$

whereas, this potential is screened by the presence of small ion in an electrolyte solution and the spatial dependence turns into[10]

$$V(r) = \frac{e}{4\pi\varepsilon\varepsilon_0 r}\exp\left(-\frac{r}{\xi D}\right) \tag{8.3}$$

Equation 8.3 can be deduced from the solution of Poisson–Boltzmann equation for an electrolyte solution and can be expressed as in eq 8.4,

$$\frac{1}{r^2}\frac{d}{dr}\left(r^2\frac{d}{dr}\right)V(r)=\frac{1}{\xi_D^{\ 2}}V(r) \tag{8.4}$$

where, ξ_D, that is, Debye screening length can be obtained from the eq 8.5[11]

$$\xi_D=\left(\frac{\varepsilon\varepsilon_0 k_B T}{2Ie^2}\right)^2 \tag{8.5}$$

where, I is the ionic strength that can be expressed by eq 8.6,

$$I=\frac{1}{2}\sum c_i z_i^2 \,. \tag{8.6}$$

Here, c_i and z_i are the concentration and valency of the ith charge species. Equation 8.3 indicates that the accumulation of oppositely charged counterions on the vicinity of ions results to a screening effect that gradually decreases the strength of the Coulomb force and finally vanishes at a distance ξ_D. Therefore, ξ_D can be tuned by changing the concentration and valency of the low molecular weight ion species. If the separation between central ion and oppositely charged counterion is small, that is, $r \ll \xi_D$ then there will be no screening and Coulomb potential will be fully active which can be seen from eq 8.3. The screening effect becomes higher in the presence of low molecular weight salt ions in the solution. The screening strength increases with the concentration of ions and is strong enough to make the PE chain completely free and behave like a neutral polymer at higher salt concentration.[8–11] PEs exhibit different phenomena like aggregation, gel formation, phase separation, condensation, reentrant condensation, crystallization, etc. in the presence of different external agents (ions, oppositely charged PE, charged particle, surfactant, etc.) which are directly related to the interaction behaviors among different constituent materials.

8.1.1 PROTEIN–POLYELECTROLYTE COMPLEXES AND THEIR THIN FILMS

BPEs like globular proteins, DNA, RNA, nucleic acids, chitosan, dextran, starch, etc. can form new BPE complex materials via electrostatic interaction. These materials exhibit interesting properties from both biological and physical point of view depending on the structure of the BPEs as well as the interactions, and hence found to have potential application in the field of food chemistry, biotechnology, pharmacology, etc.[12–15] Protein is one of the fascinating BPEs for showing intrinsic fluorescence and can form complexes with PEs depending on the pH of the solution as well as on the isoelectric point (zero net charge at a specific pH) of protein, such complex is called PPC.[15,16] Proteins, in contrast, typically have ionizable surface moieties with both positive and negative charges. The net charge of the protein molecules depends on pH of the aqueous solution. Thus, pH of aqueous solution is one of the primary factors for enabling complexation with other PEs (such as homo and block PEs) to form bulk and micellar complexes. At pH below and above the pI, a protein molecule carries net negative and positive charge, respectively. But the charge distribution may not be uniform and thereby such approximation is crude at best.[17–19] Therefore, pH can be considered as a vital tuning parameter to modify the protein–polyelectrolyte complexation by modifying the interaction strength between them. PPC has widely varied stoichiometric architecture and phase states. The protein molecules can be treated as interesting biomaterial in laboratory and in vivo conditions due to the unperturbed biofunctionality nature under most of the conditions.[15,16] The remarkable achievement in the field of biophysics and the active and growing area in different industrial fields has promoted a new look toward modern research. One of the most challenging problems in protein formulation till date is the protein aggregation. The pristine structure of a protein is highly sensitive to the physicochemical parameters such as heat, pressure, organic solvents, agitation, and protease resulting in denaturation of protein's three-dimensional structures.[20–22] The denatured proteins not only lose their functional properties but also modify their optical properties because of the exposure of amino acids from hydrophobic cleft into the aqueous solvent.[21–23] The development of therapeutic antibodies demands high concentration of protein solutions.[24–26] However, the high concentration of protein solution has a tendency to form unwanted aggregates due to the separation between the protein molecules, as the separation is of the same order of individual protein molecular size.[27] One promising advantage of PPC is that it can remove the drawbacks of aggregation states of proteins, as PEs can be used as potential agents for protein stabilizer and functionalizing agents.[28–30] During the formation of PPC, different physical properties such as conformational, optical, structural, etc. vary depending on the different physiochemical parameters like concentration of PEs, polarity of the solvents, ionic environment, etc.[26,28–30] Globular proteins like BSA, HSA, lysozyme, OV, etc. can show intrinsic optical absorption and emission behaviors. Such absorption and emission behaviors of protein are due to the presence of optically active aromatic amino acids, namely tryptophan, phenylalanine, and tyrosine. Among them the tryptophan emission intensity is most dominating in nature, and is very sensitive to its microenvironment such as conformational change, substrate binding, solvent polarity, pH of the solution, ionic environment, etc.[31–33] Fluorescence is also one of the methods to study the PPC.[32] It is possible to obtain the interaction behavior of protein with PE from the study of tryptophan fluorescence. PEs can be used as coating material for proteins so that the surface properties of PPC allow interaction with particular biological systems. Therefore, surface properties of PPC are very important for making them biocompatible. Protein molecules exhibit variation in their functional properties with their structural modifications during conjugation with different PEs. Thus, understanding on modification of different physicochemical properties (structural, conformational, and functional properties) is necessary to investigate protein aggregation, adsorption, and crystallization, as also some protein-related neurodegenerative diseases such as Alzheimer's, Parkinson's, etc.[34,35] Protein behaves as a complex 3D structure with various folding structural conformation levels.[36,37] The conformation of protein is estimated mostly by the different primary and secondary structures of proteins. The primary structure is constructed by the long polypeptide chain, which is composed of different amino acids sequence. There are several secondary structures of protein such as alpha-helix, beta-sheets, beta-turn, random coils, etc., which is due to H-bonding between peptide groups of primary chains. Tertiary structures refer to the 3D structure formed by folding the secondary structures into compact globular form through nonspecific hydrophobic interactions, salt bridge, disulfide bonds among the side chains, etc. The quaternary structures are the three-dimensional structure formed by the different tertiary subunits of multiple peptide chains together through noncovalent interactions, disulfide bonds, etc.[36] There are different instrumental techniques to obtain the protein conformations, such as FTIR,[37,38] CD,[39,40] and fluorescence spectroscopy[32,41] by observing the primary and secondary structures of proteins. Different amide bonds related to the primary structure of proteins are obtained from the FTIR studies, which are mentioned in Table 8.1. The protein adsorption is determined by its conformational modifications, which in turn is affected by variation in physiochemical parameters of protein.[42] The 2D structures of PPC are gaining remarkable attention for the different biological applications such as biosensors for the detection of inter alia glucose, cholesterol, antibodies, etc.[43,44] PPC thin film is also beneficial for providing protection of proteins with enhanced stability and also for preserving their activity when exposed to harsh environment.[43,44] These films can be obtained by alternate deposition of PE and protein on charged substrates using layer-by-layer (LBL) technique, and also by spin coating of PPC solution on solid substrates. Trapping of proteins inside highly hydrated PE multilayer formed by LBL deposition method can keep the protein conformation nearly unchanged and protect it from denaturation.[45–48] The immobilization of protein at interfaces has great potential for several applications in biomedical and chemical engineering fields.[44,45,47,48] van der Straeten and his coworkers prepared PPC of lysozyme with polystyrene sulfonate (PSS) at different ionic strengths and pH values through LBL assembly.[45] They found that the layer made with PPC shows higher protein adsorption and hydration level. A high amount of protein was immobilized in a condition where adequately favorable NaPSS conformation coexists with adequate NaPSS–lysozyme electrostatic interactions.[45] For successful assembly of globular proteins, PEs layer seems to be important where PE efficaciously acts as a bridge among protein molecules. The bridging capacity of PEs is influenced by its chain length.[45–47] For instance, when chain length of PE is shorter, it reduces the bridging capacity and steric stabilization of protein solutions. Houska et al.[49] found that the molecular weight of the PE did not have any impact on adsorption of protein layer in the multilayer assembly of albumin–NaPSS and albumin–heparin single PE multilayers, but the effect of chain length was clearly observed in those layers. The bioderived PE, polydopamine can form LBL assembly with three model proteins lysozyme, myoglobin, and α-lactalbumin because of the combined effect of covalent bonding and electrostatic interaction.[50] There are some experimental parameters such as pH, ionic strength, temperature, etc. which can directly control the interaction between protein and PE layer; as a result, morphology, structures, and thickness of the films are modified.[51] The titratable functional group of PE is sensitive to pH; thus, any alteration in the pH can alter the effective charge density of PE, and thus the PE–PE interactions.[45–47,51]

TABLE 8.1 Characteristics of Peptide Bands of Protein from the FTIR Spectroscopy.[37]

Bands	Wavenumber range (cm^{-1})	Bond	Vibration
Amide A	3300	N–H	Stretching
Amide B	3100	N–H	Stretching
Amide I	1600–1700	C=O	Stretching

TABLE 8.1 *(Continued)*

Bands	Wavenumber range (cm^{-1})	Bond	Vibration
Amide II	1480–1580	C–N, N–H	Stretching, bending
Amide III	1230–1300	C–N, N–H	Stretching, bending
Amide IV	625–767	OCN	Bending
Amide V	640–800	N–H	Out of plane bending
Amide VI	537–606	C=O	Out of plane bending
Amide VII	200		Skeletal torsion

Source: Reprinted from Ref. [37]. https://creativecommons.org/licenses/by/4.0/

Dai et al.[52] reported that the PAA layer binds positively-charged lysozyme more strongly at pH ≈ 7.4 (pH < pI of lysozyme) than the negatively charged BSA (pH > pI of BSA). If both the components of PEs are weak, then protein adsorption occurs at narrow pH range which was observed in chitosan/alginate films by Yuan et al.[53] Hbg and NaPSS were reported to form thicker multilayer at higher ionic strength. However, the significant loss of attractive forces results in the release of proteins within the multilayer PMMA in 0.6 M NaCl.[54] Therefore, ionic strength has a great role in selectivity of protein in multilayers and formation of protein charge patches in PEMs.[55] Thus, the correlation between structure and properties of PPCs can be modified by tuning the solution's ionic strength/pH. Spin assembly is one of the simple and efficient mechanisms for PPC thin film formation. P. A. Chiarelli and his coworkers established that the spin assembly of PE multilayers up to 10 bilayers of the PEs poly(ethylenimine) and poly[1-[4-(3-carboxy-4-hydroxyphenylazo) benzene sulfonamido]-1,2-ethanediyl, sodium salt].[56] The dilute solution of PE on a spinning solid substrate is a unique method for fabricating thin films with single layer thickness. Thin films created in this method have been found to be consistent and reproducible in having the same thickness. Moreover, there are additional advantages in spin-mediated thin film formation, such as preparation time efficiency and control over deposition rate or thickness. Thin film characterizations like film thickness measurement, density variation in out of plan direction, and the roughness of the PPC thin films have been characterized using XRR.[57,58] The morphological and conformational information of PPC thin films is well observed using AFM and FTIR.[59–61] As globular proteins exhibit intrinsic optical property, therefore the optical property of PPC in 2D conformation is studied using UV-vis and fluorescence spectroscopy.[31,62]

8.2 PROPERTIES OF PROTEIN–POLYELECTROLYTE COMPLEXES

Understanding the properties of PPC is very important to involve them in technological applications. The development and advancement of biomaterials using PPC is an extensive need in modern systems that extract and distribute in different biomedical applications. The performance of these systems depends on the materials' structures and properties. Therefore, there is a continuous search for engineered biomaterials with improved properties for high tech applications. Some of the properties of PPC are explained herewith.

8.2.1 *STRUCTURE AND CONFORMATION*

Although there are plenty of studies on structure and conformation of PPC, but proper understanding on structure property correlation of PPC still remains the controlling limit for various applications as globular protein is ultrasensitive to its microenvironment. Thus, the researchers need to study on understanding of structure–property correlation of globular protein in different experimental conditions for fabrication of efficient biomaterials. There are lot of evidences existing for interactions of protein below and above isoelectric points with both synthetic and natural PEs. This results in the formation of soluble complexes, complex coacervation, or the formation of amorphous precipitation depending on pH, ionic environment, mixing ratio, etc. Microscopically intermolecular interactions such as H-bonding, van der Waal interaction, hydrophobic interaction, and dipolar interactions are also involved in the formation of PPC. Analogous to PE complexation, the electrostatic interaction and translational entropy gain from the counterions release are responsible for PPC formation.[63] In addition to ionized charged groups, characteristics such as size and persistence lengths of PEs and proteins can also affect the complex structure–properties correlation. Longer PE and floppier chains provide interaction among protein globules and conform to increase adsorption on the PE surface, respectively. Both PEs chains can lead to denser complexes.[64–66] The composition, structure, and morphology of the PPC are associated with the nature and strength of interaction between PE and globule proteins.[64] Moreover, the transition of morphology from globular to mesh-like complexes is induced by the higher composition inside the PPC.[64] Additionally, hydrophobic interactions between PE backbone and protein surfaces cause reinforcement and inhibit complexation in some cases.[67,68] The role of PPC interactions need to be considered carefully for understanding structure property relation[69–71] and for the development of specific biomedical applications such as therapeutic protein purification, protein stability and delivery, etc.[72,73]

Small-angle neutron scattering (SANS),[74,75] dynamic light scattering (DLS),[76] and rheology[76] have been employed for obtaining structures of PPC. The complex structure of lysozyme–NaPSS has been reported by employing deuterium-labeled PE chain where the SLD of unlabeled NaPSS and lysozyme are nearly the same.[74,76] Thus, the possibility of studying lysozyme–PSS system could be enhanced, as scattering from only d-NaPSS was obtained in a solvent. The conformational modification of lysozyme in the PPC was quantified by observing the secondary band structure of lysozyme using FTIR in presence of d-NaPSS chains with varying concentrations.[77] The lysozyme–NaPSS complexes have been found with two regimes such as gel-like and liquid-like. The former constitutes of a mesh structure in which the lysozyme globules act as nodes interconnecting different NaPSS chains, while the latter regime comprises of a densely packed spherical aggregates of lysozyme and NaPSS chains compared to unfolded lysozyme molecule which are shown in Figure 8.2i(a–c).[77] The shift from globular to the gel-like structure of PPC was obtained depending on the length of NaPSS chain.[77] Similar type of structural shift was also observed by elevating NaPSS concentration (degree of polymerization N = 800) at a constant protein and ion concentrations, or reducing the ion concentration at a fixed protein and NaPSS concentration.[78] The structural shift from a globular to a gel-like form and the conversion of a pure PE (in solution) from the dilute to the semidilute form were found to be alike at the critical concentration for chain overlap. This transition is also size dependent of PE chains. The electrostatic persistence length of PEs is affected by the protein globule inside PPC, which increases the critical overlap concentration of PE. However, insignificant protein–PE interactions do not affect PE chain sizes.[63] Similarly, fractal and gel-like (mesh-like) structures were observed for BSA–NaPSS complexes at relatively high and low pH, respectively.[79] These are explained by SANS measurements and shown in Figure 8.2ii(d–f). However, the protein–PE concentration ratios did not have any impact on the structure of the complexes and coacervates. This observation was attributed to the additional amount of protein or PEs that remained in the solution instead of being incorporated in the complexes.

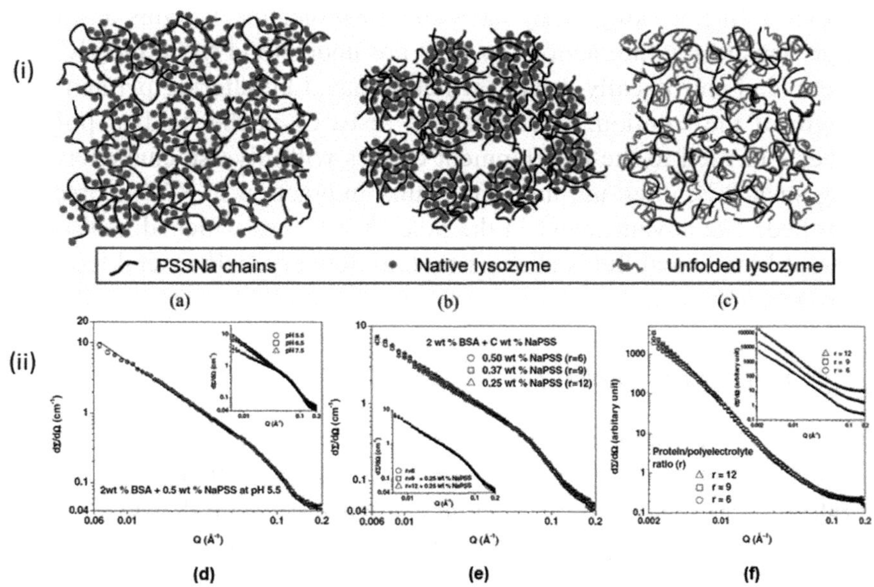

FIGURE 8.2 (i) Schematic representation of lysozyme–NaPSS complexes at a charge ratio of 3.33 between PE and protein (**a**) long PE chains, (**b**) short PE chains, and (c) unfolded lysozyme structure. (ii) SANS experiment from BSA–NaPSS complexes (d) at pH 5, (e) Various BSA/NaPSS ratios obtained from 2 wt.% BSA at different NaPSS concentration (0.25–0.5 wt.%), and (f) Coacervate prepared at various BSA/NaPSS ratios.[79]

Source: Reprinted with permission from Ref. [79] Copyright © 2008 American Physical Society, and Ref [65]. Copyright © 2005 American Chemical Society

These have highlighted the fractal structure, independent complex structure on protein and PE concentration and independent coacervate structure on BSA/NaPSS ratio of PPC. The PPC structure is primarily controlled by the persistence length of the PE.[80] A comparison of SANS study obtained from lysozyme–pectin complexes and from work lysozyme–NaPSS complexes[78] described that an increase in chain flexibility results in higher H_2O content in the PPC which in turn reduces complexes' compactness.[79] In comparison to stiffer chains having larger persistence length, the flexible chains with small persistence length can provide various configurations which allow the complex structures with lower free energies. The rheological property and diffusion mode in PPC is also altered with the structural modification caused by variation in PE persistence length.[80] Poly(diallyldimethylammonium chloride) (PDADMAC) and chitosan show significantly different properties after forming complexation with BSA due to having different persistence lengths (PDADMAC-2.5 nm;

chitosan-6 nm); although they have similar molecular weights and linear charge densities.[75] Interactions in PPC are understood to be mostly electrostatic and entropically driven, which is tuned by altering pH and ionic strength of the solution. The advanced research in the field of polymer synthesis has led to the development of PEs with diverse characteristics. Using such PEs, inter and intra molecular interaction in PPC is precisely controlled. The advancement in the field of PPC has widened its scope in different biomedical fields such as protein delivery, self-assembled artificial bioreactors, etc.

8.2.2 BIOLOGICAL ACTIVITY

An overwhelming number of biointerfaces are being developed with the goal of maintaining protein biologically active for its application in biocatalysis, biosensing, biobanking, cell and virus manipulation, drug delivery,[81] etc. In a versatile spontaneous protein adsorption and covalent linking methods, a monolayer is restricted which can lead to denaturation of protein. Therefore, an alternate solution is used for LBL deposition for trapping of protein inside highly hydrated PE which stabilizes the protein conformation and hence prevents the denaturation.[44,45,82] The method of PPC formation can be applied as one of the promising techniques for LBL construction owing to their standardized charge and PE corona. It was found that LBL assembly between lysozyme and PMMA at pH 8.4 was ineffective,[83] but at pH 4, lysozyme was effectively immobilized by adsorption with NaPSS. The stability of protein as well as its protection against harsh environmental conditions (high temperature and proteolysis) can be obtained using PE complexes including LBL, multilayer capsules, and bulk PE complexes. For instance, PAH–PAA complex is immobilized with NaPSS by the LBL method which provides new film morphologies.[84] In PPC, PSS–lysozyme complexation could effectively inhibit the heat-induced aggregation of lysozyme,[85] denaturing of protein via immobilization,[44] etc. The net charge and morphology of PPC are highly dependent on parameters like ion concentration, pH, and polymer molar mass.[51,86] Encapsulation of enzyme in PEM capsules is an effective method for preventing protease degradation as compared to the enzyme solubilized in aqueous solution. This protective barrier also enhances stability and activity of the enzyme.[87] PPCs are gaining immense attention and popularity among researchers due to their ability to reduce protein aggregation, preserve enzymatic activity through structure stabilization, and protein separation and purification. PPCs also allow PEs responsive properties to be utilized for salting-out or salting-in strategies.[86,88]

In LBL assemblies, the stabilizing effect of PPCs and hydrated nature of the PEs are exploited for keeping the biological activity of the immobilized protein.[45,86] A number of PEs like NaPSS, PDADMAC, CMC, PAH, PAA, etc. are the most usable PEs in current research for protein immobilization. Straeten et al. developed strategies for building enzyme-based bioreactors by using lysozyme in LBL assemblies.[89] They demonstrated immobilization of lysozyme by forming complexation with NaPSS and then integrating into LBL assemblies using PAH. The resulting PPCs and LBL assemblies were then characterized along with the assessment of enzymatic activity of the prepared multilayers. Turbidimetric assay provided the information about the formation of complexation as a function of net negative and positive charge ratio which is defined as the structural charge ratio, where lysozyme charge is +8 and NaPSS charge is −36.[89,90] The charge ratio becomes positive or negative depending on the amount of lysozyme or PSS in the complexation. Lysozyme charge compensation by PSS occurs close to (−net)/(+net) = 1.5 which is in concurrence with available literature,[91,92] but any further addition of PSS causes PPCs to exhibit negative charge. Thus, beyond specific charge compensation, the complexes take a core-corona structure where internal domain is neutral with negatively charged PSS chains dangling outside.[90] The immobilization of PPCs after assembling with PAH was monitored by QCM with dissipation monitoring. The normalized frequency shift decreases for both PAH–PPCs and lysozyme–PSS assembly, which confirms that the electrostatic interaction is responsible for the assembly. The deposition of a single protein layer causes the activities of both LBL assemblies to be equivalent to one of the monolayers of adsorbed lysozyme molecules. Here, a decrease in the lysozyme amount causes an increase in the enzymatic activity. In fact, the specific activity is found to be $8314 \pm 1144\ \mu g^{-1}$ and $4814 \pm 755\ \mu g^{-1}$ for $[PAH–PPCs\text{-}2]_5$ and $[Lys–PSS]_{4.5}$, respectively. The difference in the specific activity can be attributed to the presence of larger amount of PE in the multilayer and the higher hydration level present in $[PAH–PPCs\text{-}2]_5$. This promotes steady increase in the enzymatic activity at each lysozyme deposition step, while when the last layer is a PE there is a drop in the enzymatic activity (≈ zero). Thus, such new findings of PPCs widen the scope of applications along with adding versatility to the LBL systems.

8.2.3 OPTICAL PROPERTIES

BPEs like proteins exhibit intrinsic fluorescence properties. Globular proteins like BSA and lysozyme have unique physicochemical properties, which make them interesting material for both basic and applied research fields. Such properties can be improved by altering the structures of protein.[33,35–37] In BSA and lysozyme, the numbers of optically dominating tryptophan residues are two and six, respectively. The hydrophobic cleft and the surface of BSA have one tryptophan residue each. But in lysozyme, hydrophobic cleft has two tryptophan residues and four are present on the surface.[38,39] The optical emission behavior of a protein molecule is found to be altered when protein conformations are modified by its interactions with charged polymers, ions and nanoparticles, etc.[93,94]

By using fluorescence emission technique, BSA and the BSA–PE complexes were studied mixing with divalent ions and lower polymer concentration.[95] The fluorescence behavior revealed that the quenching as well as blue shift in BSA emission takes place due to the formation of intra-polymer complexes between BSA and PE.[31,96] Thus, PPC can provide modified fluorescence behavior depending on the interactions with ions, PE, etc.[97] Further, PPC thin films have been reported for showing unique features upon modification. Therefore, PPC thin films need more extensive studies regarding its optical responses. In this section, we have shown the modification of fluorescence emission behavior of BSA and lysozyme as mixed with negatively charged PEs in thin film conformation.[98,99] The PEs, PAA and PSS were chosen as their molecular weight and dissociation nature inside the solvent are different. Moreover, PAA and PSS are weak and strong PEs, respectively. These PEs can also be used as model PEs. PPC was prepared by mixing PAA and PSS with BSA and lysozyme proteins separately using appropriate concentrations. The complexation of BSA and lysozyme with PAA is designated as PAAB and PAAL, respectively. The spectroscopic studies of the complexes were carried out by using UV-vis absorption and fluorescence emission spectrophotometer. The intrinsic absorption peak of BSA and lysozyme exhibits at $\approx$ 273 and $\approx$ 271 nm, respectively; however, after complex formation, that is, for PAAB and PAAL, the absorption peak is found at $\approx$ 272 nm.[98] Thus, after complex formation the absorption wavelength of BSA and lysozyme remains almost unchanged which is shown in Figure 8.3i(a). The emission peak for PAAB and PAAL was obtained at 342 nm, whereas there was no peak for PAA solution. The emission spectra of pure BSA and lysozyme revealed that the emission peak occurred at $\approx$ 345 nm as shown in the inset of Figure 8.3i(b).[98] However, the emission behavior in case of 2D (thin film) structure of PAAB and PAAL films is different although absorption peak position is unchanged which is explained in Figure 8.3ii(a–d). PPC thin films were obtained using spin coating method on solid substrates (quartz and silicon). The thickness variations from 30 to 60 nm and their out-of-plane structure variations are obtained using XRR technique.[98] The out-of-plane structure is obtained through the variation of electron density from substrate to air side. Figure 8.3ii (a and c) shows the absorption spectra as obtained from PAAB and PAAL films, respectively of different thicknesses.[98]

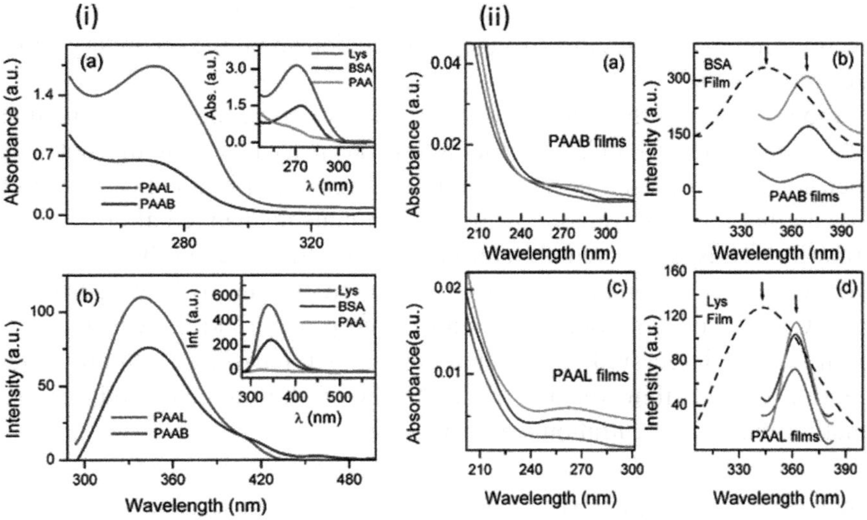

FIGURE 8.3 UV-vis absorption and fluorescence emission spectra obtained from the solution as well as thin films of PPC. (i): (a) absorbance spectra PAAB (BSA + PAA) and PAAL (lysozyme + PAA) solution. Inset: absorbance spectra of pure PAA, BSA, and lysozyme solution (b) fluorescence spectra in solution of PAAB and PAAL. Inset: fluorescence spectra in solution of pure PAA, BSA, and lysozyme. (ii): (a) absorbance and (b) emission spectra of the PAAB films of varying thicknesses. (c) Absorbance and (d) emission spectra of the PAAL films of varying thicknesses. Emission from pure BSA and lysozyme is depicted by the dashed lines for showing the larger red-shift. Red, blue and, green lines represent thinner film, intermediate thickness, and thicker film of PPC, respectively.[98]

Source: Reprinted with permission from Ref. [98]. Copyright © 2016 Elsevier B.V.

Mostly the absorption peak of BSA and lysozyme was obtained from the PPC films and the peak value gradually surges with the film thicknesses. The emission peak from all the PAAB films is observed at 368 nm, whereas from all PAAL films it is observed at 361 nm.[98] The peak intensity was maximum for the thicker films in both the cases, which is shown in Figure 8.3ii (b and d). However, emission peaks were found at ≈ 344 nm for pure BSA and lysozyme films, which are shown by the dotted line in the Figure 8.3ii(b and d) respectively. The emission spectroscopy revealed the peak-shift of ≈ 26 nm or 23 nm for PAAB film than that of PAAB bulk or BSA solution, and about ≈ 19 nm or 16 nm peak-shift for PAAL film than that of PAAL bulk or lysozyme solution. Thus, a large red-shift is observed in emission peak of all the PAAB and PAAL films as compared to PAAB and PAAL solutions or from the pure proteins.[98] However, such peak-shift is independent of film thickness.[98] The protein conformation in 2D structure after complex formation was recognized via the amide-I band (1700–1600 cm^{-1}) in a FTIR spectra. This spectral region provides information about the secondary structures of proteins.[100]

Table 8.2 shows the peaks associated with different secondary structures within the Amide-I band. It is found that the peak positions did not vary much for certain peaks; in fact, it remained unaltered in most cases which indicate that the conformation of protein remains unchanged after complex formation. Thus, larger red-shift from PPC thin films[98] could be because of other reason. Emission behavior from denatured protein (by heating from 30 to 90°C) and concentration variation of both protein and PAA inside PPC doesn't show larger red-shift. Therefore, larger red-shift in fluorescence emission of PPC thin film is not due to the conformational modification, denaturation or concentration variation of protein.[98]

TABLE 8.2 Assignments of ATR–FTIR Peaks of Secondary Structure of Amide-I Band Obtained from the Thin Films of Pure Globular Proteins (BSA and Lysozyme) and PPC (PAAB and PAAL).[98]

Band	BSA	Lysozyme	PAAB	PAAL
Side chain vibration	1604	1602	1604	1604
Intermolecular β strand	1612	1608	1610	1611
β-Sheet	1618	1616	1617	1617
Intramolecular β strand	1625	1622	1624	1625
β-Sheet	1630	1632	1629	1629
β-Sheet	1637	1636	1638	1637
Random coil	1648	1647	1648	1648
α Helix	1653	1650	1654	1654
3_{10} Helix	1663	1656	1664	1664
β-Turn	1675	1673	1676	1676
β-Turn	1684	1679	1684	1684
β-Turn	1690	1686	1691	1690
Antiparallel β-Sheet	1694	1693	1694	1695

Source: Reprinted/Adapted w ith permission from Ref. [98]. Copyright © 2016 Elsevier B.V.

Similarly, PPC thin film formed by BSA and lysozyme in the existence of optically active and relatively higher molecular mass strong anionic PEs PSS, designated as PSSB and PSSL shows larger red-shift in fluorescence emission. Conformation of BSA in presence of PSS is unchanged which is analyzed through amide-I and amide-II bands[99]; whereas, in case of lysozyme, the amide-I band is shifted toward lower vibrational energy, that is, conformation of lysozyme changes as agglomeration takes place between the two constituents (PSS and lysozyme) of PPC.[99] This signifies that the protein conformation after complex formation with PSS is modified and the secondary structure peak position is shown in Table 8.3.

TABLE 8.3 Secondary Peaks of Amide-I Band Obtained from the Thin Films of Pure Globular Proteins (BSA and Lysozyme) and PPC (PSSB and PSSL) Using FTIR in ATR Mode.[99]

Band	BSA	Lysozyme	PSSB	PSSL
Side chain vibration	1606	1605	1607	
Intermolecular β-strand	1609	1611	1615	1634
Intermolecular β-strand	1619	1618	1618	1639
Intermolecular β-strand	1622	1621	1622	1645
Intermolecular β-strand	1627	1625	1626	1652
β-sheet	1633	1630	1631	1656
β-Sheet	1635	1634	1635	1660
β-Sheet	1638	1638	1638	1668
β-Sheet	1642	1641	1642	1674
β-Sheet	1646	1645	1645	1678
Random coil	1651	1648	1649	1687
α-Helix	1656	1654	1653	1699
3_{10} Helix	1665	1664	1663	1704
β-Turn	1673	1671	1672	1706
β-Turn	1677	1676	1676	1712
β-Turn	1681	1680	1680	1721
β-Turn	1687	1685	1685	
Antiparallel β-Sheet	1693	1690	1691	

Source: Reproduced/Adapted with permission from Ref. [99]. Copyright © 2017 Elsevier B.V.

Our study reveals that after the complexation of BSA and lysozyme with PAA the secondary structure, that is, conformation of BSA and lysozyme remains unchanged. In case of complexation prepared with PSS, although the conformation of BSA is almost unaltered but lysozyme conformation is highly modified. However, in both the cases, a larger red-shift occurs from the thin films of PPCs. Modified optical emission behaviors of proteins after complex formation with PEs in thin film conformation are unusual and are irrespective of protein conformation. Therefore, the most probable reason behind the unusual emission response of PPC thin film is nonradiative decay of protein fluorophore after interaction with anionic PE in dry state resulting in larger Stokes shift compared to the pristine protein thin film.[98,99] Such modified optical behaviors, that is, larger red-shift in optical emissions of PPCs in thin film confirmation are explored in this study which may have potential applications in the field of bioimaging, biosensors, etc.[98,99] The schematic representation of larger red-shift in PPC thin film is shown in Figure 8.4.

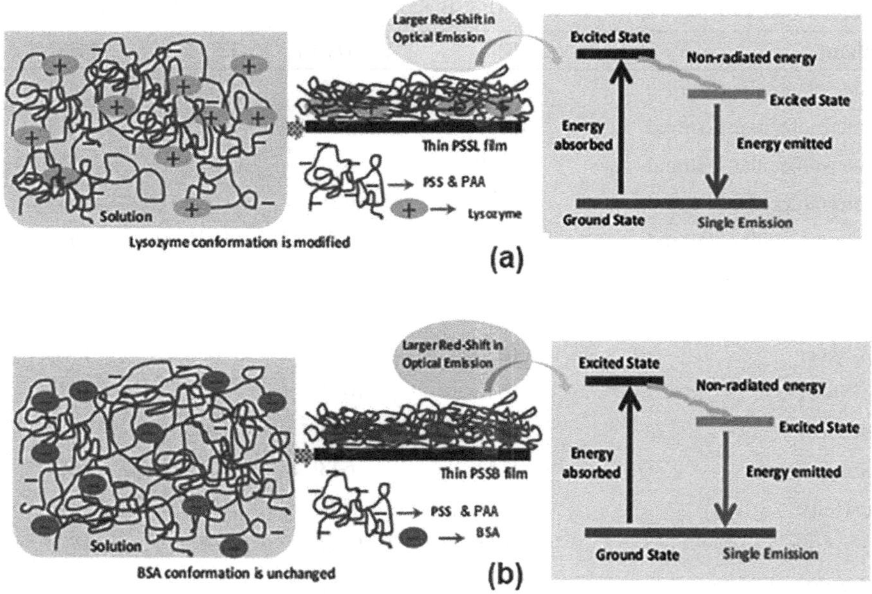

FIGURE 8.4 Diagrammatic representations for mechanism of larger red-shifts in PPC (PSSL and PSSB) thin films. Nonradiative decay of protein fluorophore after interaction with anionic PE in thin film resulting in larger Stokes shift compared to the pristine protein thin film which turn into larger red-shift in emission spectra.

Source: Reprinted with permission from Ref. [98]. Copyright © 2016 Elsevier B.V.

8.3 CONCLUSION AND FUTURE PERSPECTIVE

The present chapter provides a wide study on PPCs which are useful for both basic understanding and technological enhancement. This study provides the information about the techniques for studying PPCs and also describes their several structural, conformational, and optical properties. SANS study showed the variation of complex formation between lysozyme and PSS, and the structure of PPC. This depends mostly on the electrostatic and hydrophobic interactions which can be tuned by varying the PSS chain length and charge ratio of the components inside the systems. Such typical structure of PPC can be mimicked into biological systems. PPC thin films produce larger red-shift (around 25 nm) in protein fluorescence, which is irrespective of protein conformation confirmed by the FTIR study. Such materials can be used as fluorescence-based sensing material for different hazards molecules which are very harmful for health and environment. The protein–polyelectrolyte interaction can also be useful in different areas, for example, protein can be binded with different biocompatible PEs (carboxymethyl cellulose, poly (L-lysin), chitosan, starch, etc.) in the form of protein–polyelectrolyte complex thin film to prevent the denaturation of protein via immobilization. In this process, not only biological activity is maintained but also the specific activity like optical behavior is significantly improved. However, there is still the need for better understanding of the structure–property relationship of these complexes at microscopic level as such acknowledgment will enhance administration over the prevailing interactions, self-assemblies, and production of complexes, and also for their production at commercial scale. Thus, this chapter provides a glance and future prospects of PE materials.

ACKNOWLEDGMENT

Authors acknowledge Department of Science and Technology, Govt. of India for the financial support.

KEYWORDS

- **polyelectrolyte**
- **protein**
- **protein–polyelectrolyte complexes**
- **structures**
- **optical property**

REFERENCES

1. Oosawa, F. *Polyelectrolytes*; Mercel Dekker, 1971.
2. Strobl, G. R. *The Physics of Polymers: Concepts for Understanding Their Structures and Behaviour*; Springer, 2007.
3. Gilles, F. M.; Tagliazucchi, M.; Azzaroni, O.; Szleifer, I. Ionic Conductance of Polyelectrolyte-Modified Nanochannels: Nanoconfinement Effects on the Coupled Protonation Equilibria of Polyprotic Brushes. *J Phys Chem C* **2016,** *120*, 4789–4798.
4. Khan, N.; Brettmann, B. Intermolecular Interactions in Polyelectrolyte and Surfactant Complexes in Solution, Polymers, **2019,** *11*, 51–79.
5. Urbánek, P.; di Martino, A.; Gladyš, S.; Kuritka, I.; Minarík, A.; Pavlova, E.; Bondarev, D. Polythiophene-Based Conjugated Polyelectrolyte: Optical Properties and Association Behavior in Solution. *Synth. Metals* **2015,** *202*, 16–24.
6. Hsu, H. P.; Lee, E. Counterion Condensation of a Polyelectrolyte. *Electrochem. Commun.* **2012,** *15*, 59–62.
7. Muthukumar, M. Theory of Counter-Ion Condensation on Flexible Polyelectrolytes: Adsorption Mechanism. *J. Chem. Phys.* **2004,** *120*, 9343–9350.
8. Lei, Q.; Li, K.; Bhattacharya, D.; Xiao, J.; Kole, S.; Zhang, Q.; Strzalka, J.; Lawrence, J. Revati Kumar, J.; Arges, C. G. Counterion Condensation or Lack of Solvation? Understanding the Activity of Ions in Thin Film Block Copolymer Electrolytes. *J. Mater. Chem. A.* **2020,** *8*, 15962–15975.
9. Rathee, V. S.; Sikora, B. J.; Sidky, H.; Whitmer, J. K. Simulating the Thermodynamics of Charging in Weak Polyelectrolytes: The Debye–Hückel Limit. *Mater. Res. Express* **2018,** *5*, 014010–014022.
10. Bordi, F.; Colby, R. H.; Cametti, C.; De Lorenzo, L.; Gili, T. Electrical Conductivity of Polyelectrolyte Solutions in the Semidilute and Concentrated Regime: The Role of Counterion Condensation. *J. Phys. Chem. B* **2002,** *106*, 6887–6893.
11. Tadmor, R.; Zapata, E. H.; Chen, N.; Pincus, P.; Israelachvili, J. N. Debye Length and Double-Layer Forces in Polyelectrolyte Solutions. *Macromolecules* **2002,** *35*, 2380–2388.
12. Derkach, S. R.; Voronko, N. G.; Sokolan, N. I. et al. The Rheology of Hydrogels Based on Chitosan–Gelatin (Bio)Polyelectrolyte Complexes. *J. Disper. Sci. Tech.* **2017,** *38*, 1427–1434.
13. Chen, T.; Li, S.; Zhu, W.; Liang, Z.; Zeng, Q. Self-Assembly pH-Sensitive Chitosan/Alginate Coated Polyelectrolyte Complexes for Oral Delivery of Insulin. *J. Microencapsul.* **2019,** *36*, 96–107.
14. Horn, J. M.; Kapelner, R. A.; Obermeyer, A.C. Macro- and Microphase Separated Protein-Polyelectrolyte Complexes: Design Parameters and Current Progress. *Polymers* **2019,** *11*, 578–604.
15. Kayitmazer, A. B.; Seeman, D.; Minsky, B. B.; Dubina, P. L.; Xu, Y. Protein–Polyelectrolyte Interactions. *Soft Matter*. **2013,** *9*, 2553–2583.
16. Derkach, S. R.; Voron'ko, N. G.; Sokolan, N. I.; Kolotova, D. S.; Kuchina, Y. A. Interactions Between Gelatin and Sodium Alginate: UV and FTIR Studies. *J. Disper. Sci. Tech.* **2020,** *41*, 690–698.

17. Voronko, N. G.; Derkach, S. R.; Kuchina, Y. A.; Sokolian, N. I. The Chitosan–Gelatin (Bio)Polyelectrolyte Complexes Formationin an Acidic Medium. *Carbohyd. Poly.* **2016,** *138*, 265–272.
18. Stajner, L.; Pozar, J.; Civic, D. K. Complexation Between Lysozyme and Sodium Poly(Styrenesulfonate): The Effect of pH, Reactant Concentration and Titration Direction. *Colloid Surf. A: Physicochem. Eng. Asp.* **2015,** *483*, 171–180.
19. Shang, G.; Holkar, A.; Srivastava, S. Protein–Polyelectrolyte Complexes and Micellar Assemblies. *Polymers* **2019,** *11*, 1097–1131.
20. Tsai, A. M.; Zanten, J. H. V.; Betenbaugh, M. J. Electrostaic Effect in the Aggregation of Heat-Denatured RNase A and Implication for Protein Additive Design. *Biotechnol. Bioeng.* **1998,** *59*, 273–277.
21. Kurinomaru, T.; Shiraki, K. Aggregative Protein–Polyelectrolyte Complex for High-Concentration Formulation of Protein Drugs. *Internet J. Biolog. Macrom.* **2017,** *100*, 11–17.
22. Huang, A.; Yao, H.; Olsen, B. D. SANS Partial Structure Factor Analysis for Determining Protein–Polymer Interactions in Semidilute Solution. *Soft Matter*. **2019,** *15*, 7350–7359.
23. Joyce, A. M.; Brodkorb, A.; Kelly, A. L.; Mahony, J. A. O. Separation of the Effects of Denaturation and Aggregation on Whey-Casein Protein Interactions During the Manufacture of a Model Infant Formula. *Dairy Sci. Technol.* **2017,** *96*, 787–806.
24. Leveque, D.; Delpeuch, A.; Gouriex, B. New Anticancer Agents: Role of Clinical Pharmacy Services. *Anticancer Res.* **2014,** *34*, 1573–1578.
25. Gardulf, A. Immunoglobulin Treatment for Primary Antibody Deficiencies. *Bio Drugs* **2007,** *21*, 105–116.
26. Hart, J.; Peschel, A.; Johannsmann, D.; Garidel, P. Characterizing Protein–Protein-Interaction in High-Concentration Monoclonal Antibody Systems with the Quartz Crystal Microbalance. *Phys. Chem. Chem. Phys.* **2017,** *19*, 32698–32707.
27. Laue, T. Proximity Energies: A Framework for Understanding Concentrated Solutions. *J. Mol. Recognit.* **2012,** *25*, 165–173.
28. Kayitmazer, A. B.; Seeman, D.; Minsky, B. B.; Dubin, P. L.; Xu, Y. Protein–Polyelectrolyte Interactions. *Soft Matter*. **2013,** *9*, 2553–2558.
29. Xu, Y.; Mazzawi, M.; Chen, K.; Sun, L.; Dubin, P. L. Protein Purification by Polyelectrolyte Coacervation: Influence of Protein Charge Anisotropy on Selectivity. *Biomacromolecules* **2011,** *12*, 1512–1522.
30. Cao, Y.; Fang, Y.; Nishinari, K.; Phillips, G. O. Effects of Conformational Ordering on Protein/Polyelectrolyte Electrostatic Complexation: Ionic Binding and Chain Stiffening. *Sci. Rep.* **2016,** *6*, 23739–23749.
31. Lakowicz, J. R. *Principle of Fluorescence Spectroscopy*; Springer, 2006.
32. Ghisaidoobe, A. B. T.; Chung, S. J. Intrinsic Tryptophan Fluorescence in the Detection and Analysis of Proteins: A Focus on Förster Resonance Energy Transfer Techniques. *Int. J. Mol. Sci.* **2014,** *15*, 22518–22538.
33. Au, K. M.; Armes, S. P. Heterocoagulation as a Facile Route to Prepare Stable Serum Albumin-Nanoparticle Conjugates for Biomedical Applications: Synthetic Protocols and Mechanistic Insights. *ACS Nano* **2012,** *6*, 8261–8279.
34. Piazza, R. Protein Interactions and Association: An Open Challenge for Colloid Science. *Curr. Opin. Coll. Interface Sci.* **2004,** *8*, 515–522.
35. Oakley, A. E.; Collingwood, J. F.; Dobson, J.; Love, G.; Perrott, H. R.; Edwardson, J. R.; Elstner, M.; Morris, C. M. Individual Dopaminergic Neurons Show Raised Iron Levels in Parkinson Disease. *Neurology* **2007,** *68*, 1820–1825.

36. Dorna, M.; Silva, M. B. E.; Buriol, L. S.; Lamb, L. C. Three-Dimensional Protein Structure Prediction: Methods and Computational Strategies. *Comput. Biol. Chem.* **2014,** *53*, 251–276.
37. Bouhekka, A.; Bürgi, T. In situ ATR-IR Spectroscopy Study of Adsorbed Protein: Visible Light Denaturation of Bovine Serum Albumin on TiO_2. *Appl. Surf. Sci.* **2012,** *261*, 369–374.
38. Arunkumar, R.; Drummond, C. J.; Greaves, T. L. FTIR Spectroscopic Study of the Secondary Structure of Globular Proteins in Aqueous Protic Ionic Liquids. *Front Chem.* **2019,** *7*, 2296–2317.
39. Güler, G.; Vorob'ev, M. M.; Vogel, V. Werner Mäntele, Proteolytically-Induced Changes of Secondary Structural Protein Conformation of Bovine Serum Albumin Monitored by Fourier Transform Infrared (FT-IR) and UV-Circular Dichroism Spectroscopy. *Spectrochim. Acta A Mol. Biomol. Spectrosc.* **2016,** *161*, 8–18.
40. Mátyus, L.; Szöllősi, J.; Jenei, A. Steady-State Fluorescence Quenching Applications for Studying Protein Structure and Dynamics. *J. Photochem. Photobiol. B: Biol.* **2006,** *83*, 223–236.
41. Brahma, A.; Mandal, C.; Bhattacharyya, D. Characterization of a Dimeric Unfolding Intermediate of Bovine Serum Albumin Under Mildly Acidic Condition. *Biochim. Biophys. Acta* **2005,** *1751*, 159–169.
42. Yabuki, Y. Polyelectrolyte Complex Membranes for Immobilizing Biomolecules, and Their Applications to Bio-Analysis. *Analyt. Sci.* **2011,** *27*, 695–702.
43. Malay , Ö.; Yalçın , D.; Batıgün , A.; Bayraktar, O. Characterization of silk fibroin/ hyaluronic acid polyelectrolyte complex (PEC) films, J Therm Analy Calor, **2008,** *94*, 749–755.
44. Zhao, M.; Zacharia, N. S. Protein Encapsulation via Polyelectrolyte Complex Coacervation: Protection Against Protein Denaturation. *J. Chem. Phys.* **2018,** *149*, 163326–163335.
45. Straeten, A. L. V.; Skicki, A. B.; Jonas, A. M.; Fustin, C. A.; Gillain, C. D. Integrating Proteins in Layer-by-Layer Assemblies Independently of Their Electrical Charge. *ACS Nano* **2018,** *128*, 8372–8381.
46. Rodrigues, R. C.; Ortiz, C.; Murcia, A. B.; Torresd, R.; Ferna´ndez-Lafuente, R. Modifying Enzyme Activity and Selectivity by Immobilization. *Chem. Soc. Rev.* **2013,** *42*, 6290–6307.
47. Fan, Y.; Su, F.; Li, K.; Ke, C.; Yan, Y. Carbon Nanotube Filled with Magnetic Iron Oxide and Modified with Polyamidoamine Dendrimers for Immobilizing Lipase Toward Application in Biodiesel Production. *Sci. Rep.* **2017,** *7*, 45643–45655.
48. Prieto-Simo, B.; Saint, C.; Voelcker, N. H. Electrochemical Biosensors Featuring Oriented Antibody Immobilization via Electrografted and Self-Assembled Hydrazide Chemistry. *Anal Chem*, **2014,** *86*, 1422–1429.
49. Houska, M.; Brynda, E. Interactions of Proteins with Polyelectrolytes at Solid/Liquid Interfaces: Sequential Adsorption of Albumin and Heparin. *J. Coll. Interface Sci.* **1997,** *188*, 243–250.
50. Bernsmann, F.; Frisch, B.; Ringwald, C.; Ball, V. Protein Adsorption on Dopamine–Melanin Films: Role of Electrostatic Interactions Inferred from ζ-Potential Measurements Versus Chemisorption. *J. Coll. Interface Sci.* **2010,** *344*, 54–60.
51. Cooper, C. L.; Dubin, P. L.; Kayitmazar, A. B.; Tarksen, S. Polyelectrolyte-Protein Complexes. *Curr. Opin. Collo. Interface Sci.* **2005,** *10*, 52–78.

52. Dai, J. H.; Bao, Z.; Sun, L.; Hong, S. U.; Baker, G. L.; Bruening, M. L. High-Capacity Binding of Proteins by Poly(Acrylic Acid) Brushes and Their Derivatives. *Langmuir* **2006,** *22*, 4274–4281.
53. Yuan, W. Y.; Dong, H.; Li, C. M.; Cui, X.; Yu, L.; Lu, Z.; Zhou, Q. pH-Controlled Construction of Chitosan/Alginate Multilayer Film: Characterization and Application for Antibody Immobilization. *Langmuir* **2007,** *23*, 13046–13052.
54. Kozlovskaya, V.; Kharlampieva, E.; Mansfield, M. L.; Sukhishvili, S. A. Poly(methacrylic acid) Hydrogel Films and Capsules: Response to pH and Ionic Strength, and Encapsulation of Macromolecules. *Chem. Mater.* **2006,** *18*, 328–336.
55. Muller, M.; Kessler, B.; Houbenov, N. pH Dependence and Protein Selectivity of Poly(Ethyleneimine)/Poly(Acrylic Acid) Multilayers Studied by In Situ ATR-FTIR Spectroscopy. *Biomacromolecules* **2006,** *7*, 1285–1294.
56. Chiarelli, P. A. Polyelectrolyte Spin-Assembly. *Langmuir* **2002,** *18*, 168.
57. Als-Nielsen, J.; Möhwald, H. *Handbook of Synchrotron Radiation*; Elsevier, 1991.
58. Skoda, M. W. A. Recent Developments in the Application of X-ray And neutron Reflectivity to Soft-Matter Systems. *Curr. Opin. Collo. Interface Sci.* **2019,** *42*, 41–54.
59. vander Straetena, A.; Bratek-Skickia, A.; Germaina, L.; d'Haesea, C.; Eloy, P.; Fustina, C. A.; Gillain, C. D. Protein-Polyelectrolyte Complexes to Improve the Biological Activity of Proteins in Layer-By-Layer Assemblies. *Nanoscale* **2017,** *9*, 17186–17192.
60. She, Z.; Antipina, M. N.; Li, J.; Sukhorukov, G.B. Mechanism of Protein Release from Polyelectrolyte Multilayer Microcapsules. *Biomacromolecules* **2010,** *11*, 1241–1247.
61. Johansson, C.; Hansson, P.; Malmsten, M. Mechanism of Lysozyme Uptake in Poly(Acrylic Acid) Microgels. *J. Phys. Chem. B* **2009,** *113*, 6183–6193.
62. Antosiewicz, J. M.; Shugar, D. UV–Vis Spectroscopy of Tyrosine Side-Groups in Studies of Protein Structure. Part 2: Selected Applications. *Biophys. Rev.* **2016,** *8*, 163–177.
63. Gao, S.; Holkar, A.; Srivastava, S. Protein–Polyelectrolyte Complexes and Micellar Assemblies. *Polymers* **2019,** *11*, 1097–1131.
64. Cousin, F.; Gummel, J.; Combet, S.; Boué, F. The Model Lysozyme-PSSNa System for Electrostatic Complexation: Similarities and Differences with Complex Coacervation. *Adv. Coll. Interface Sci.* **2011,** *167*, 71–84.
65. Cousin, F.; Gummel, J.; Ung, D.; Boué, F. Polyelectrolyte-Protein Complexes: Structure and Conformation of Each Specie Revealed by SANS. *Langmuir* **2005,** *21*, 9675–9688.
66. Gummel, J.; Cousin, F.; Boue, F. Structure Transition in PSS/Lysozyme Complexes: Achain-Conformation-Driven Process, as Directly Seen by Small Angle Neutron Scattering. *Macromolecules* **2008,** *41*, 2898–2907.
67. Sedlák, E.; Fedunová, D.; Veselá, V.; Sedláková, D.; Antalík, M. Polyanion Hydrophobicity and Protein Basicity Affect Protein Stability in Protein-Polyanion Complexes. *Biomacromolecules* **2009,** *10*, 2533–2538.
68. Kamiya, N.; Klibanov, A. M. Controling the Rate of Protein Release from Polyelectrolyte Complexes. *Biotechnol. Bioeng.* **2003,** *82*, 590–594.
69. Brangwynne, C. P.; Eckmann, C. R.; Courson, D. S.; Rybarska, A.; Hoege, C.; Gharakhani, J.; Juelicher, F.; Hyman, A. A. Germline P Granules Are Liquid Droplets That Localize by Controlled Dissolution/Condensation. *Science* **2009,** *324*, 1729–1732.
70. Garfinkle, S. E.; Kim, Y.; Szczepaniak, K.; Chen, C. C. H.; Eckmann, C. R.; Myong, S.; Brangwynne, C. P. The Disordered P Granule Protein LAF-1 Drives Phase Separation Into Droplets with Tunable Viscosity and Dynamics. *Proc. Natl. Acad. Sci. USA* **2015,** *112*, 7189–7194.

71. Brangwynne, C. P.; Tompa, P.; Pappu, R. V. Polymer Physics of Intracellular Phase Transitions. *Nat. Phys.* **2015,** *11*, 899–904.
72. Woitovich, N. V.; Brassesco, M. E.; Picó, G. A. Polyelectrolytes–Protein Complexes: A Viable Platform in the Downstream Processes of Industrial Enzymes at Scaling Up Level. *J. Chem. Technol. Biotechnol.* **2016,** *91*, 2921–2928.
73. Biesheuvel, P. M.; Wittemann, A. A. Modified Box Model Including Charge Regulation for Protein Adsorption in a Spherical Polyelectrolyte Brush. *J. Phys. Chem. B* **2005,** *109*, 4209–4214.
74. Cousin, F.; Gummel, J.; Clemens, D.; Grillo, I.; Boué, F. Multiple Scale Reorganization of Electrostatic Complexes of Poly(Styrenesulfonate) and Lysozyme. *Langmuir* **2010,** *26*, 7078–7085.
75. Kayitmazer, A. B.; Strand, S. P.; Tribet, C.; Jaeger, W.; Dubin, P. L. Effect of Polyelectrolyte Structure on Protein-Polyelectrolyte Coacervates: Coacervates of Bovine Serum Albumin with Poly(Diallyldimethylammonium Chloride) Versus Chitosan. *Biomacromolecules* **2007,** *8*, 3568–3577.
76. Boué, F.; Cousin, F.; Gummel, J.; Oberdisse, J.; Carrot, G. Harrak, E. Small Angle Scattering from Soft Matter-Application to Complex Mixed Systems. *Comptes. Rendus. Phys.* **2007,** 8, 821–844.
77. Cousin, F.; Gummel, J.; Ung, D.; Boué, F. Polyelectrolyte-Protein Complexes: Structure and Conformation of Each Specie Revealed by SANS. *Langmuir* **2005,** *21*, 9675–9688.
78. Gummel, J.; Cousin, F.; Boue, F. Structure Transition in PSS/Lysozyme Complexes: Achain-Conformation-Driven Process, as Directly Seen by Small Angle Neutron Scattering. Macromolecules **2008,** *41*, 2898–2907.
79. Chodankar, S.; Aswal, V. K.; Kohlbrecher, J.; Vavrin, R.; Wagh, A. G. Structural Study of Coacervation in Protein–Polyelectrolyte Complexes. *Phys Rev E.* **2008,** *78*, 031913–031921.
80. Schmidt, I.; Cousin, F.; Huchon, C.; Boué, F.; Axelos, M. A. V. Spatial Structure and Composition of Polysaccharide-Protein Complexes from Small Angle Neutron Scattering. *Biomacromolecules* **2009,** *10*, 1346–1357.
81. Doonan, C.; Ricco, R.; Liang, K.; Bradshaw, D.; Falcaro, P. Metal–Organic Frameworks at the Biointerface: Synthetic Strategies and Applications. *Acc. Chem. Res.* **2017,** *50*, 1423–1432.
82. Rodrigues, R. C.; Ortiz, C.; Berenguer-Murcia, A.; Torresd, R.; Ferna´ndez-Lafuente, R. Modifying Enzyme Activity and Selectivity by Immobilization. *Chem. Soc. Rev.* **2013,** *42*, 6290–6307.
83. He, Y.; Hong, C.; Li, J.; Howard, M. L. T.; Li, Y.; Turvey, M. E.; Uppu, D. S. S. M.; Martin, J. R.; Zhang, K.; Irvine, D. I.; Hammond, P. T. Synthetic Charge-Invertible Polymer for Rapid and Complete Implantation of Layer-by-Layer Microneedle Drug Films for Enhanced Transdermal Vaccination. *ACS Nano.* **2018,** *12*, 10272–10280.
84. Guo, Y.; Geng, W.; Sun, J. Layer-by-Layer Deposition of Polyelectrolyte-Polyelectrolyte Complexes for Multilayer Film Fabrication. *Langmuir* **2009,** *25*, 1004–1010.
85. Wu, F. G.; Jiang, Y. M.; Sun, H. Y.; Luo, J. J.; Yu, Z. W. Complexation of Lysozyme with Sodium Poly(Styrenesulfonate) via the Two-State and Non-Two-State Unfoldings of Lysozyme. *J. Phys. Chem. B.* **2015,** *119*, 14382–14392.
86. Kayitmazer, A. B.; Seeman, D.; Minsky, B. B.; Dubina, P L.; Xu, Y. Protein–Polyelectrolyte Interactions. *Soft Matter.* **2013,** *9*, 2553–2583.

87. Ghiorghita, C. A.; Bucatariu, F.; Dragan, E. S. Influence of Cross-Linking in Loading/Release Applications of Polyelectrolytemultilayer Assemblies: A Review. *Mater. Sci. Eng. C* **2019,** *105*, 110050–110069.
88. Schmitt, C.; Turgeon, S. L. Protein/Polysaccharide Complexes and Coacervates in Food Systems. *Adv. Colloid Interface Sci.* **2011,** *14*, 63–70.
89. Chollakup, R.; Smitthipong, W.; Eisenbach, C. D.; Tirrell, M. Phase Behavior and Coacervation of Aqueous Poly(Acrylic Acid)-Poly(Allylamine) Solutions. *Macromolecules* **2010,** *43*, 2518–2528.
90. Schmidt, I.; Cousin, F.; Huchon, C.; Boue, F.; Axelos, M. A. V. Spatial Structure and Composition of Polysaccharide-Protein Complexes from Small Angle Neutron Scattering. *Biomacromolecules* **2009,** *10*, 1346–1357.
91. Cousin, F.; Gummel, J.; Combet, S.; Boue, F. The Model Lysozyme-PSSNa System for Electrostatic Complexation: Similarities and Differences with Complex Coacervation. *Adv. Colloid Interface Sci.* **2011,** *167*, 71–84.
92. Yamaguchi, A.; Kobayashi, M. Quantitative Evaluation of Shift of Slipping Plane and Counterion Binding to Lysozyme by Electrophoresis Method. *Colloid Polym. Sci.* **2019,** *294*, 1019–1026.
93. Saptarshi, S. R.; Duschl, A.; Lopata, A. L. Interaction of Nanoparticles with Proteins: Relation to Bio-Reactivity of the Nanoparticle. *J. Nanobiotech.* **2013,** *11*, 26–37.
94. Talukdar, H.; Kundu, S. Förster Resonance Energy Transfer-Mediated Globular Protein Sensing Using Polyelectrolyte Complex Nanoparticles. *ACS Omega* **2019,** *4*, 20212–20222.
95. Jose, M.; Tome, M.; Esquembre, R.; Mallavia, R.; Mateo, C. R. Formation of Complexes Between the Conjugated Polyelectrolyte Poly{[9,9-bis(6′-N,N,Ntrimethylammonium) hexyl]fluorene-phenylene} Bromide (HTMA-PFP) and Human Serum Albumin. *Biomacromolecules* **2010,** *11*, 1494–1501.
96. Wu, D.; Schanze, K. S. Protein Induced Aggregation of Conjugated Polyelectrolytes Probed with Fluorescence Correlation Spectroscopy: Application to Protein Identification. *ACS Appl. Mater. Interfaces* **2014,** *6*, 7643–7651.
97. Pu, K. Y.; Liu, B. Fluorescence Turn-on Responses of Anionic and Cationic Conjugated Polymers Toward Proteins: Effect of Electrostatic and Hydrophobic Interactions. *J. Phys. Chem. B* **2010,** *114*, 3077–3084.
98. Talukdar, H.; Kundu, S.; Basu, S. Larger Red-Shift in Optical Emissions Obtained from the Thin Films of Globular Proteins (BSA, Lysozyme)—Polyelectrolyte (PAA) Complexes. *Appl. Surf. Sci.* **2016,** *382*, 121–127.
99. Talukdar, H.; Kundu, S. Thin Films of Protein (BSA, Lysozyme)—Polyelectrolyte (PSS) complexes Show Larger Red-Shift in Optical Emissions Irrespective of Protein Conformation. *J. Mol. Struc.* **2017,** *1143*, 84–90.
100. Kong, J.; Yu, S. Fourier Transform Infrared Spectroscopic Analysis of Protein Secondary Structures. *Acta Biochim. Biophys. Sin.* **2007,** *39*, 549–559.
101. V. S. Meka, et al., A Comprehensive Review on Polyelectrolyte complexes, Drug Discovery Today, 22, 2017, 1697–1706.

CHAPTER 9

Introduction to DNA Nanomechanics: Theory and Simulations

ASHOK GARAI

Department of Physics, The LNM Institute of Information Technology, Jaipur, Rajasthan, India

ABSTRACT

DNA double helix, commonly known as a macromolecule, is a biopolymer. The present chapter focuses on its mechanical properties at the microscopic level. Further, different theoretical and computational methods along with experimental techniques are discussed to determine these properties. The structural transitions of DNA under force, ionic effects, and protein binding effects on elastic properties of DNA are also reviewed. An alluring future perspective of this exciting topic on nanotechnology, nanoelectronics, and clinical research are also be discussed.

9.1 INTRODUCTION

DNA (deoxyribonucleic acid) nanomechanics is an emerging and exciting area of research. To get in-depth knowledge about it, we must know what nanomechanics is and what DNA is? Notably, the fundamental length scale relevant to molecular biology and related biochemical structures is expressed in nanometer-scale (1 nm = 10^{-9} m).[1]. Hence, studying a system's mechanical properties at such a length scale can be called as nanomechanics.[2–4] DNA carries the genetic information for the development, growth, and reproduction of all organisms and is the most sophisticated engineering material

Foundation and Growth of Macromolecular Science: Advances in Research for Sustainable Development. Meegle S. Mathew, PhD, Józef T. Haponiuk, PhD & Sabu Thomas, PhD, DSc (Eds.)

accessible in nanotechnology.[5–7] Inside a living eukaryotic cell, DNA is organized into a chromosome. A chromosome is consisted up of chromatin. The fundamental subunit of chromatin is the nucleosome, which consists of a DNA segment wrapping a little less than two turns around a histone protein. Thus, long DNA is compacted and packaged into nucleosomes and forms beads on a string structure.[8] It is inevitable to know the mechanical properties of DNA to perspicuously understand how DNA is organized and processed during various active biological processes, for example, DNA transcription, DNA replication, recombination, and repair inside the cell.[1] Over the last two and half decades, the technological advancement in single-molecule force spectroscopic experiments has enlightened us about DNA molecules' structural and mechanical properties.[9–18] Various theoretical and numerical studies have done the mammoth task to enrich the concepts of DNA nanomechanics.[4,19–26] Thus, research in this field provides new insights into emerging DNA nanotechnology and is essential for understanding life phenomena.

9.1.1 *STRUCTURAL PROPERTIES OF THE DNA*

Understanding the biological macromolecules' structure and functional properties is one of the critical aspects of macromolecular science. The four primary biological macromolecules are carbohydrates, proteins, lipids, and nucleic acids.[8] Among these, nucleic acids are the genetic factor that is an essential part of all organisms, and loss in their sequence may affect an organism's usual functioning. DNA and ribonucleic acid (RNA) are the two types of nucleic acids. Both can act as hereditary material according to the organism. The fascinating macromolecule in the living cell is DNA, which is a heteropolymer. It consists of two antiparallel strands and forms a double helix.[27,28] It is made up of nucleotides. Each nucleotide consists of a phosphate backbone, pentose sugar, and nitrogenous bases. Nitrogenous bases are of two types: (a) pyrimidines and (b) purines. Pyrimidines (cytosine and thymine) have the small 5-atom aromatic single ring and purines (guanine and adenine) having two aromatic rings (see Fig. 9.1). The short form of the above four bases is A (adenine), T (thymine), C (cytosine), and G (guanine). Thus, the nucleic acids are built up from the four letters of alphabet (A, T, C, and G), and DNA contains a message written in this alphabet of four letters.

Interestingly, A binds with T by two hydrogen bonds ($A = T$), and G binds with C by three hydrogen bonds (G ≡ C), identified by J. Watson and F. Crick, also known as Watson–Crick base pairing.[29] In all the purine and pyrimidine base pairs, G–C pair has the higher binding energy. One can easily construct the complementary strand of DNA if the base sequences of one strand are known.

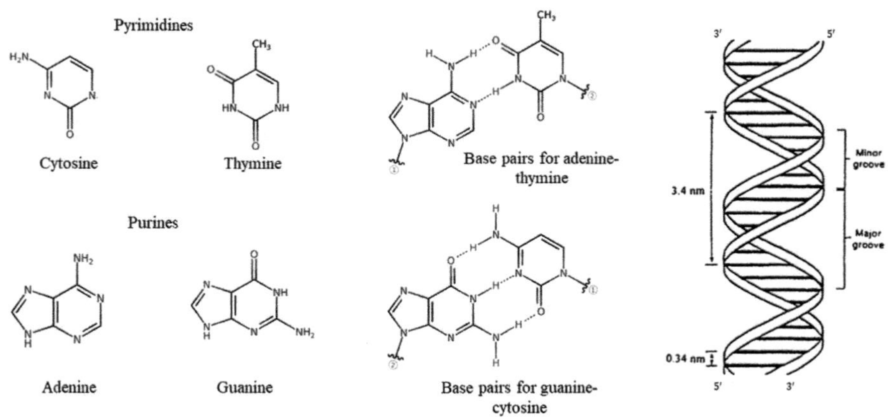

FIGURE 9.1 Chemical structures of different types of nitrogen bases of DNA. The rightmost figure describes the schematic of the helical structure of DNA. Horizontal bars represent the base pairs of the DNA. It also depicts the two antiparallel strands of DNA. It shows major and minor grooves.
Source: Reprinted with permission from Ref. [123]. Copyright © 1999 Elsevier

The bases are connected to the 5-atom sugar rings. These sugar molecules are 2-deoxyribose, that is, a loss of an oxygen atom at the second carbon of the sugar molecule and are familiar with deoxyribose. Tetrahedral phosphate groups (PO_4^-) are connected to the deoxyribose of two adjacent bases and form together with the sugar-phosphate backbone. The DNA double helix's backbone strands have directionality: deoxyribose and phosphate can be found at the 3′ end and 5′ end, respectively. Notably, the backbone is hydrophilic and negatively charged, and the bases are hydrophobic. Thus, DNA belongs to a particular group called "amphiphiles." The amphiphilic character of DNA, the backbone's flexibility, the stacking interaction between bases, and hydrogen bonding between complementary bases play an essential role in forming the double-helical DNA structure. The two strands of the DNA fall apart if we heat the DNA in the range of 70–80°C. The structure of the DNA may change in varying environments. The dominant stable structure in typical living cell environments is B-DNA. It can adopt an A-DNA structure while the water activity is lowered (relative humidity is reduced to less than 75%).[30,31] Both the A- and B-DNA are right-handed double helix, while the Z-DNA has the helix's left-handed orientation,[30–32] as shown in Figure 9.2. Different helical parameters (see Table 9.1) and various DNA helix conformations highly influence the other biological functions. Thus, the structural flexibility of DNA is essential as multiple active biological processes require access to the genetic code hidden in duplex DNA.

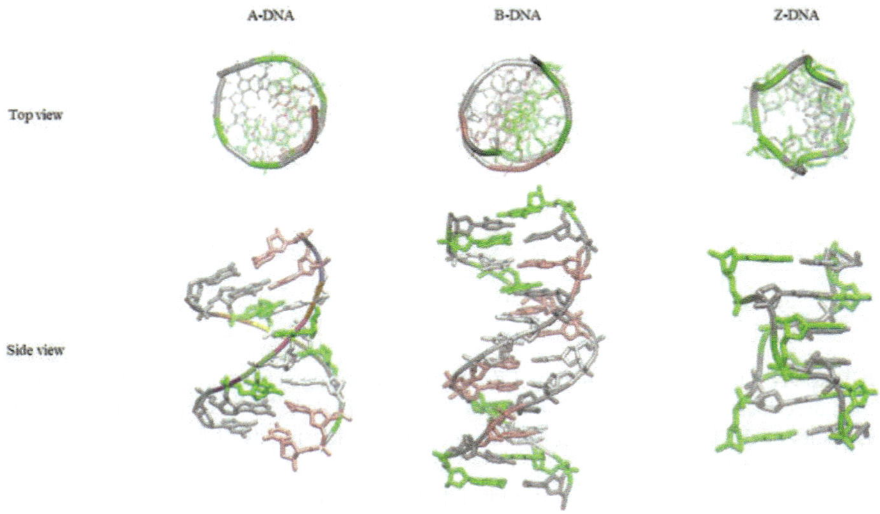

FIGURE 9.2 Different forms of all-atom DNA structure. A- and B-DNA are both right-handed helical structures, while the Z-DNA is a left-handed double helix. The backbone of Z-DNA follows a zigzag path.

TABLE 9.1 Details of Helical Parameters for A-DNA, B-DNA, and Z-DNA.

	A-DNA	B-DNA	Z-DNA
Screw sense	Right	Right	Left
Shape	Wide	Intermediate	Narrow
Major Groove	Narrow and very deep	Wide and less deep	Flat
Minor Groove	Broad and shallow	Narrow and quite deep	Very narrow and deep
Helix diameter (nm)	2.3	2.0	1.8
Rise/bp along axis (nm)	0.23	0.34	0.38

TABLE 9.1 *(Continued)*

	A-DNA	B-DNA	Z-DNA
Base pairs per helix turn	11	10.4	12
Pitch per turn of helix	2.53 nm	3.54 nm	4.56 nm
Tilt of base pairs from normal to the helix axis	19°	1°	9°
Twist angle (°)	33	36	–30

Source: Adapted from Ref. [30] with permission from John Wiley and Sons.

The length of the DNA varies from bacteria to human cells. For example, the size of DNA in an *E. coli* cell is approximately 1.5 mm, and it is about 1.6 μm in λ bacteriophage, and in the human cell, the size of DNA is around 2 m long. DNA is a negatively charged molecule due to the presence of a phosphate group on its backbone. It is observed that each base carries one electron charge, and thus the presence of high electric charge density makes it a strong polyelectrolyte. Notably, the negative charge of DNA plays a crucial role in packing genetic material inside the cell nucleus and electrostatic interactions with various proteins during DNA processing and regulations. Notably, the dominating forces at the nanodomain are thermal and viscous forces. The characteristic force scale associated with molecular biology is $k_B T/\text{nm} \approx 4.1$ pN.[31] A few $k_B T$ is enough to break the noncovalent bonds present in DNA and may give rise to various conformational changes. However, breaking a covalent bond requires more force and approximately 1.6 nN.[33]

9.2 OVERVIEW OF MECHANICAL PROPERTIES OF DNA

DNA is a double helix, and to understand its structural properties, six main helical parameters are to be studied. Three distances (shift, slide, and the rise) and three angles (twist, roll, and tilt) can relate to successive base-pair planes.[24,34] Additionally, some other geometries like shear, stretch, stagger, buckle, propeller, opening, x-displacement, y-displacement, inclination, tip, and pitch are considered to precisely define the location and orientation of the base pair in space relative to its predecessor along the helical axis.[34–36] Figure 9.3 depicts all the above-mentioned helical parameters. Rise (denoted by "*d*") describes the successive base-pair heights, and by summing over consecutive rise values in ds-DNA, one obtains the contour length of the

DNA, that is, $L_t = \sum_j d_j$ (see Fig. 9. 4(a)). End-to-end length of ds-DNA can be defined as $\boldsymbol{R} = \left| \boldsymbol{r}'_{i+n} - \boldsymbol{r}'_i \right|$, where $\boldsymbol{r}'_i$ denotes the position vector of the i-th base pair of ds-DNA (see Fig. 9.4(b)).

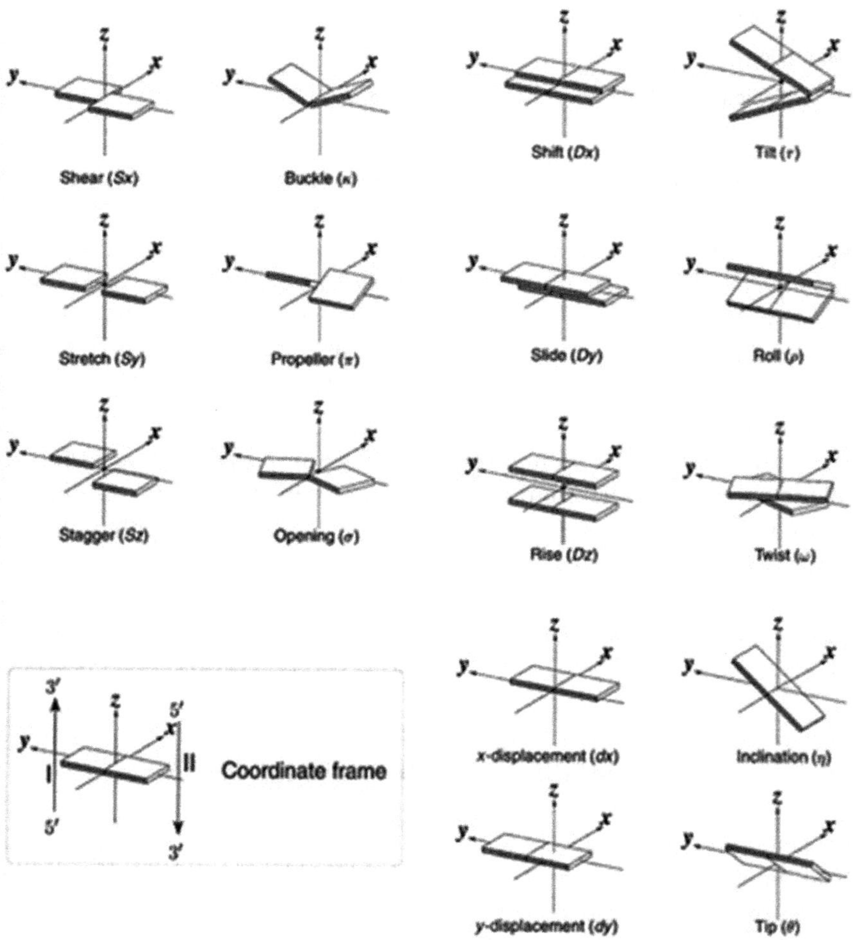

FIGURE 9.3 Schematic diagram showing various helical parameters of DNA. The reference frame chosen to describe these parameters is shown in the lower-left corner. Shear, stretch, stagger, buckle, propeller, and opening are the intrabase pair parameters. Shift, slide, and rise are the translation; tilt, roll, and twist are the rotation parameters required to describe the DNA-step geometry. The other commonly used helical parameters are x-displacement, y-displacement, inclination, and tip.

Source: Reprinted with permission from Ref. [124]. Copyright © 2003 Oxford University Press

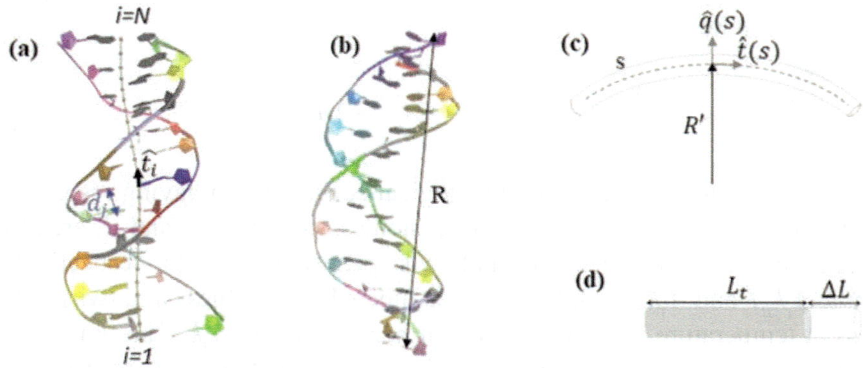

FIGURE 9.4 (a) Shows the helical axis, $\hat{t}_i$ local tangent and d_j represents the rise at the j-th base pair, (b) Depicts the end-to-end vector of DNA, (c) R' represents the radius of curvature of the cylindrical tube's center axis, s refers to the arc length, $\hat{t}(s)$ and $\hat{q}$ be the tangent vector and a unit vector perpendicular to the tube's axis, respectively, (d) L_t represents the contour length of the tube, and ΔL is the extension.

To understand the other physical properties of DNA, consider the DNA a cylindrical molecule whose diameter is much smaller than its total length. It can be assumed to be a uniform tube whose total underformed length is L_t. Considering a little segment, *ds* located at an arc length *s* and $R'(s)$ be the curvature radius of the tube's center axis (see Fig. 9.4(c)). Let $\hat{t}(s)$ is the tangent unit vector and $\hat{q}$ be a unit vector attached everywhere along the underformed tube axis in a direction perpendicular to its axis. If $\hat{t}(s)$ is moved to a nearby point at a nearby distance *ds*, then the difference of the tangent vectors $d\hat{t}(s)$ has magnitude $d\theta$. Thus $d\theta/ds$ is the rate of rotation of $\hat{q}(s)$ along *s* in the deformed state of the tube. Now define some associated phenomenological parameters to understand deformations of the segment of such a thin isotropic elastic tube. Note that the deformation's energy cost must be an integral along *s* of some function of the stretch, bend, and torsion/twist. (a) stretch (*r(s)*): It measures a fractional change in its length of the segment and can be expressed as follows $r = \Delta(ds)/ds$ (see Fig. 9.4(d)). It is a dimensionless scalar quantity. (b) bend (b(s)): It measures the orientation of the unit tangent vector while it moves from the arc length *s* to $s + ds$ and can be expressed as follows $b = d\hat{t}/ds$. It is a vector with dimensions L^{-1}. (c) twist density *c*(*s*): It measures how the relative segments rotate to the adjacent segment about the central axis and can be expressed as $c = d\phi/ds$. It is a scalar quantity and has a dimension of L^{-1}. According to the elastic theory, the elastic energy cost of an infinitesimal deformation of the uniform tube is given by[37]:

$$dH = \frac{1}{2} k_B T \left[A_1 b^2 + A_2 r^2 + A_3 c^2 + 2A_4 rc \right] ds \tag{9.1}$$

where A_1, A_2, and A_3 are phenomenological constants and having respective dimensions L, L^{-1}, and L. A_4 is a dimensionless constant. A_1 and A_3 are also called bend persistence length and twist persistence length, respectively. Thus, $A_1 k_B T$ and $A_3 k_B T$ can be called as tube's bend stiffness and twist stiffness at temperature T, respectively. Further, $A_2 k_B T$ and $A_4 k_B T$ are known as stretch stiffness and stretch-twist stiffness. Note that $A_2 k_B T$ (say it stretch modulus or γ) has the same dimension as a force. For the case of a long polymer (e.g., DNA) which can be assumed to be a rigid and inextensible rod, a one-parameter phenomenological model can be introduced with elastic energy is given by:

$$H = \frac{1}{2} \kappa_B T \int_0^{L_1} ds A_1 b^2 = \frac{1}{2} \int_0^{L_1} ds\, \kappa \left(\frac{d\hat{t}}{ds} \right)^2 \tag{9.2}$$

where k is the bending modulus with dimensions of energy-length, A_3 and A_4 are assumed to be zero, and because of the inextensible nature, r will be negligibly small. Bending modulus embodies both Young modulus (Y) and the geometric shape through the geometric moment I. Thus, $\kappa = YI$. This model is also known as Kratky–Porod or wormlike chain (WLC) model.[20,38] This model is beneficial for the mechanical stretching of DNA.

Now assume a close strand of DNA (without twisting) that lies along a circular loop with a radius of curvature R', and its length is $2\pi R'$. Therefore, the elastic energy cost is given by:

$$H = \frac{\pi \kappa}{R'} \tag{9.3}$$

The interesting point about this expression is that H becomes smaller for increasing R'. Notably, when R' is much larger than A_1, elastic energy cost is negligible relative to the thermal energy $k_B T$. This indicates that such an elastic rod (or biopolymer) can randomly bend (or orient) by thermal fluctuations. Thus, when a biopolymer such as DNA is immersed into a fluid at temperature $T(\neq 0)$, it will be kicked into various orientations because of interactions with the surrounding fluid. Thus, DNA can locally bend and curve due to thermal fluctuations.[39,40] Therefore, elastic forces set the length scale over which such fluctuations are tolerated. Such length would be the result of the competition between the thermal fluctuations and the bending energy cost. This characteristic length is known as the persistence length $(\xi = \kappa / k_B T)$. Notably, if $R' \ll \xi$ then thermal fluctuations will not be able to deform the DNA on length scales less than ξ. Thus, persistence length is the characteristic length over which a polymer remains rigid. In other words, it is the characteristic length over which the tangent–tangent correlations die of exponentially and is given by:

$$g(s) = \langle \hat{t}(0) \cdot \hat{t}(s) \rangle = \exp\left(-\frac{s}{\xi}\right) \tag{9.4}$$

where $g(0) = 1$ and for $s \gg \xi$, two tangent vectors will be independent, $g(s) \to 0$. Thus, the WLC model introduces a single fitting parameter ξ and assumes a quadratic form of bending energies to deformation a semiflexible polymer such as DNA. This model is widely used to calculate various elastic properties of DNA. The persistence length of ds-DNA is approximately 50 nm,[9,10,20] and it is for ss-DNA about 0.75 nm. On the other hand, the persistence length of a microtubule is 6 mm, and for spaghetti, it is 1.5 cm.[8] Thus, a smaller persistence length indicates the polymer is more flexible. However, the elastic properties of DNA also depend on its base sequences.[11,12,41–45]

9.3 METHODS TO UNDERSTAND THE DNA NANOMECHANICS

9.3.1 THEORETICAL MODELS

In the absence of external force and at room temperature, the DNA's length in a fluid medium will fluctuate from its unperturbed contour length. Thus, one expects the equilibrium probability distribution of the contour length in the linear regime as Boltzmann distribution and is given by[24]:

$$P(L) = \sqrt{\frac{\beta\gamma}{2\pi L_t}}\, exp\left[-\frac{\beta\gamma}{2L_t}(\delta L)^2\right] \tag{9.5}$$

where δL is the equilibrium extension in the length of the DNA and β is the inverse of thermal energy. Similarly, equilibrium bend angle (θ) distribution is given by:

$$P(\theta)=\sqrt{\frac{\beta\kappa}{2\pi L_t}}\,exp\left[-\frac{\beta\kappa}{2L_t}\theta^2\right] \tag{9.6}$$

These equations are handy for calculating the stretch modulus (γ), bending modulus (k), and persistence length (ξ) of DNA.[24,26,46,47]

The mean square end-to-end distance of DNA under the WLC model is given by[48]:

$$\langle R^2\rangle = 2\xi L_t - 2\xi^2\left[1-exp\left(-\frac{L_t}{\xi}\right)\right] \tag{9.7}$$

In the limit of $L_t \gg \xi$ the eq 9.7 becomes $\langle R^2\rangle \cong 2\xi L_t$ and when $\xi \gg L_t$, eq 9.7 yields $\langle R^2\rangle \cong L_t^2 - \frac{L_t^2}{3\xi} + ...$ Thus, eq 9.7 is a smooth crossover between these two limits. This equation can also be used to calculate the persistence length of the DNA by measuring the mean square end-to-end distance of the DNA. Notably, in the WLC model, the polymer segments are not entirely rigid but bend, unlike the freely jointed chain (FJC) model.[48] In the FJC model, the rigid elements of the polymer are known as Kuhn segments having lengths ζ and $\zeta = 2\xi$.[48] Recently, Marin-Gonzalez et al., introduced a new parameter "crookedness of DNA (β_c)," quantitatively defined as $cos\beta_c = R / L_t$, useful to study DNA mechanics at short-length scale.[49] Low crookedness indicates high DNA stretching stiffness. Their model is quite helpful in determining the stretch modulus of various DNA sequences.

While the DNA is under the application of force (say, f), then the DNA will be stretched, and the relative extension becomes $\langle L\rangle / L_t$. In the small force regime and within the framework of the WLC model, one obtains

$$\langle L\rangle = \beta L_t \zeta f \tag{9.8}$$

In the low force regime, DNA behaves like a Hookean spring with spring constant $(\beta L_t \zeta)^{-1}$ Interestingly, the spring constant depends on temperature proportionally that reveals its entropic origin. Thus, in the low force regime, DNA gets stretched linearly with force, and here temperature plays an important role. For λ– phage DNA $L_t \approx 16.6\,\mu m$ and at room temperature, the effective spring constant becomes approximately 2.3 fN/μm. FJC model is limited to explaining the entropic elastic behavior of DNA only at low forces and fails at intermediate and high forces (9.10), whereas the WLC model describes the elastic behavior of DNA at low and intermediate forces.[2] For the inextensible WLC model, an interpolated formula to represent the entropic elasticity of the polymer in the limit $\langle L \rangle < L_t$ is given by

$$f = \frac{1}{\xi\beta}\left[\left(1 - \frac{\langle L \rangle}{L_t}\right)^{-2} - \frac{1}{4} + \frac{\langle L \rangle}{L_t}\right] \tag{9.9}$$

Near full extension, that is, when $\langle L \rangle$ approaches L_t, the extensible WLC model that captures additionally the enthalpic stretching of DNA, the force-extension relation is given by

$$\frac{\langle L \rangle}{L_t} = 1 - \frac{1}{2}\left(\frac{1}{\beta\xi f}\right)^{\frac{1}{2}} + \frac{f}{\gamma} \tag{9.10}$$

The first two terms of the equation consider the limit when $\langle L \rangle$ approaches L_t and the third term describes the enthalpic contribution of the elasticity. These equations are handy for determining the stretch modulus and persistence length of DNA while it is under force. However, the WLC model is incapable of explaining the force-extension behavior of DNA at a high force regime. Further, a few experimental studies have challenged the WLC model's applicability at a shorter DNA length scale.[11–13,17,22,50–55]. A few modifications of the WLC model have been proposed in the literature to capture the short-length elastic behavior of DNA[13,56] and high-force behavior of DNA.[57,58]

9.3.2 COMPUTER SIMULATIONS

Computer simulations play a pivotal role in determining the biomolecule's structural properties and function.[59] Two leading families of computer simulations are molecular dynamics (MD) and Monte Carlo (MC) techniques.[60] Additionally, many hybrid methods are adapted using these two families.[60–62] Computer simulations make a bridge between a microscopic and the macroscopic domain.[60] It helps to provide some new information about the system, which is not accessible via experiments. It sometimes supports the findings of a theory and makes a bridge between theory and experiments. In all atomistic MD, atoms and molecules interact with their surroundings and even with themselves through a proposed form of potential. Such potential and related parameters describe the intra- and intermolecular interactions between the molecules in the system and can be modeled by empirical forms. The total potential energy ($\mathcal{V}$) can be expressed as below:

$$\mathcal{V} = \mathcal{V}_{bond} + \mathcal{V}_{angle} + \mathcal{V}_{dihedral} + \mathcal{V}_{vdw} + \mathcal{V}_{electrostatic} \tag{9.11}$$

where $\mathcal{V}_{bond}$, $\mathcal{V}_{angle}$, $\mathcal{V}_{dihedral}$, $\mathcal{V}_{vdw}$, $\mathcal{V}_{electrostatic}$ and are bond stretching, angle bending, dihedral, van der Waals, and electrostatic energy components, respectively. The explicit functional forms of each of the energy components of a collection of atoms are given by[60]:

$$\begin{aligned}\mathcal{V} = &\sum_{bonds} \frac{1}{2}\kappa_r(r_{ij} - r_0)^2 + \sum_{angles} \frac{1}{2}\kappa_\theta(\theta_{ijk} - \theta_0)^2 \\ &+ \sum_{i}^{n_{dihedrals}} \sum_{n}^{n_{i,max}} \frac{V_{i,n}}{2}[1 + cos(n\phi_i - \gamma_{i,n})] + \sum_{i<j}^{n_{atoms}} \left(\frac{a_{ij}}{r_{ij}^{12}} - \frac{b_{ij}}{r_{ij}^{6}}\right) \\ &+ \sum_{i<j}^{n_{atoms}} \frac{q_i q_j}{4\pi\epsilon_0 r_{ij}}\end{aligned} \tag{9.12}$$

The first term and second term in eq 9.12 represent the harmonic potential of bond stretching and angle bending when the deviation from the reference value r_0 and θ_0 is small. When the deviation in bond stretching is more, the preferable potential for bond stretching is the Morse potential. The third term in eq 9.12 includes the torsional potential required for maintaining a particular topology of the molecular system. The fourth term describes the famous 12-6 Lennard–Jones potential, while the system-dependent parameters a and b can be found either by the first principle of quantum calculation or from experiments. The fifth term in eq 9.12 represents the electrostatic interaction between the charged particle calculated using Coulomb's law. Thus, the fourth and fifth terms of eq 9.12 describe the sum over of non-bonded interactions between molecules and within molecules. Notably, known functional forms of inter- and intramolecular interactions of a collection of atoms with their respective parameters determine a given configuration's energy. In the computer simulation, such potential energy of a group of atoms refers to a force field. The related parameters are available for various atom types from different force fields such as AMBER[63,64] and CHARMM.[65] Once we have the information about the force fields or the Hamiltonian, we have Newton's equations of motion for the group of atoms of the system. By numerically solving Newton's equations of motion,[60–62] we can record the system's dynamic evolution for a fixed period. These data, sometimes called trajectory data, are then analyzed to obtain various ensemble-averaged (or time-averaged) structural and thermodynamic properties of the system.

9.3.3 EXPERIMENTAL TECHNIQUES

At scales below micrometer, thermal fluctuations and viscous drag play an important role over inertial forces.[37,66] Thus, manipulating and detecting macromolecules at a single-molecule level are always challenging. Over the last three decades, we notice significant progress in single-molecule force spectroscopic techniques. Such techniques and related cutting edge technologies have revolutionized biophysical research and can scan a wide length scale $10^{-10} - 10^{-4}$ m and force $10^{-14} - 10^{-8}$ N.[67,68] Some well-known single-molecule manipulation techniques are atomic force microscopy, optical, and magnetic tweezers.[2,66–69] These methods allow the system to respond under the application of force. The primary outcome is to study the force-extension behavior of the single molecule. Such graphs are nothing but the biomolecule's mechanical fingerprints and provide the molecule's elastic behavior uniquely. Notably, the force-extension behavior for biomolecule is not linear unless we observe it at low forces. Below we briefly explain different single-molecule force spectroscopy techniques (see Fig. 9.5).

Atomic Force Microscope (AFM) is famous for scanning and imaging a material's surface to sub-nanometer level. However, little modification on this instrument provides wide biological applications, such as single-biomolecule manipulation, determining inter- and intramolecular interaction forces with piconewton resolution. AFM consists of a flexible cantilever with a sharp tip. The biomolecule, such as DNA or protein, is tethered between the cantilever's tip and the stage (see Fig. 9.5(a)). The tip of the cantilever deflects while an external force is applied to the biomolecule. One can measure such deflection of the cantilever's tip with an unprecedented resolution by reflecting a laser beam from the cantilever onto the sensitive photodiode. The deflection will be harmonic for a reasonable range, and the force will be proportional to the deflection. Along with the primary function of AFM, we can apply a controlled force to the molecule and study the cantilever's stiffness.[67] Typical stiffness of the AFM cantilever ranges from 1 to 10^5 pN/nm, and one can measure the force that ranges from 5 to 10^3 pN.[70] This technique is widely used to study protein unfolding and receptor–ligand interactions.[71,72]

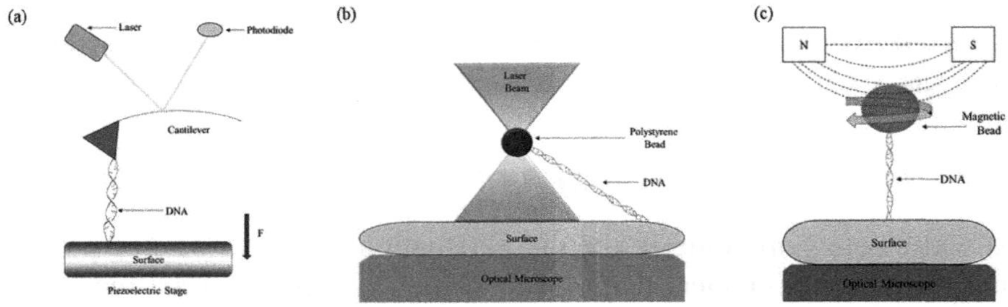

FIGURE 9.5 Different single-molecule force spectroscopic techniques. (a) Atomic force microscope, (b) optical tweezers, and (c) magnetic tweezers.

Optical Tweezers (OT) provide a noninvasive technique providing 3D control over the trapped object (such as a dielectric bead).[67,68] Optical tweezers also offer the ability to apply a controlled force up to 100 pN with simultaneous recording of the trapped particle. A highly focused laser beam can trap the bead in the vicinity of focus (see Fig. 9.5(b)). The bead experiences two types of optical forces: (a) a scattering force, which pushes the bead toward the propagation of the beam, and (b) a gradient force, which attracts the bead along the direction of the electric field gradient. A balance between these two forces will lead to trapping the bead stably. So, for the displacements below ~150 nm, the particle experiences a restoring force. The optical trap acts as a linear spring, and the spring constant depends on the light intensity. One can obtain the trap's stiffness from Hooke's law by calculating the particle's positional variance to the applied force.[67] OT is widely used to study conformational dynamics of a biomolecule, force-velocity behavior of molecular motors, and DNA and RNA elastic properties.[9,10,73,74]

Magnetic Tweezers (MT) are useful and straightforward single-molecule manipulation techniques.[70,75] A paramagnetic bead is placed in a magnetic field gradient (see Fig. 9.5(c)), and the bead experiences a force toward the magnetic field gradient.[67,76] Single-biomolecule is attached to the magnetic bead, and the magnetic bead is kept in between the two permanent magnets. The force applied on the magnetic bead is given by $\boldsymbol{F} = -\frac{1}{2}\nabla(\mu \cdot \boldsymbol{B})$, where $\boldsymbol{B}$ is the applied magnetic field and $\boldsymbol{\mu}$ is the magnetic dipole moment. One can control the bead's stretching force by changing the magnetic field gradient by moving up or down the magnets.[68] Additionally, by rotating the magnets, one can turn the bead ranging from

0.5 to 5 μm and study the force required to twist the biomolecule and how it adapts its structure with the force's application. The stiffness of the MT is smaller in comparison to the OT and AFM. The force range for MT is few femto Newtons (fN) to 10^4pN.[70,76] One uses this technique to study the DNA or hairpin topology.[67,70]

9.4 DNA UNDER THE FORCE

Studying elastic properties of DNA has been a long-time interest. Using magnetic tweezers, Smith et al. studied the force vs. extension behavior of single ds-DNA.[10,74] Discrete data in Fig. 9.6 depict the mechanical fingerprint of λ–phage DNA (approximately 16.5 μm long). It also shows that at low forces, FJC model can capture the behavior of ds-DNA, but at intermediate forces, it cannot. On the other hand, at low and intermediate forces, WLC model explains the force-extension behavior of the DNA well. It indicates that the FJC model's deviation is due to some local curvatures in DNA, which the WLC model captures well. Fitting these data with eq 9.9 up to 10 pN or with eq 9.10 give the persistence length of DNA as 53 nm in 10 mM Na^+ buffer. It is also observed that at forces larger than 65 pN, DNA does an overstretching transition where the WLC model does not work.[2] Thus force-extension curve of ds-DNA has three regions[2]: (a) *entropic region*, (b) *enthalpic region*, and (c) *overstretching transition region*. Below 10 pN forces, DNA remains in the entropic domain. Entropic spring controls the extension of semiflexible DNA. The intermediate force region (above 10 pN but below 65 pN) is known as enthalpic region. Here, force shows a sudden increase due to the extension of flexible DNA. In overstretching transition region (above 65 pN), a little variation of force, length of the DNA gets almost double, a B to S transition,[21,25,77–80] and force-induced DNA melting occurs.[81,82] Thus, on applying large forces, one can expect to see severe deformations of the double helix structure such as unzipping,[83,84] unwinding,[85,86] overstretching,[87,88] and sharp bending of the double helix.[11,17] A coarse-grained description of how DNA behaves in the presence of force is reviewed elsewhere (89). Recent theoretical and experimental studies[13,22,24,90,91] have shown that the short DNA's persistence length varies from its known persistence length ~ 50 nm. These studies indicate that the short DNAs are more flexible than the long DNAs. In the next section, we discuss the ionic effect on DNA elasticity.

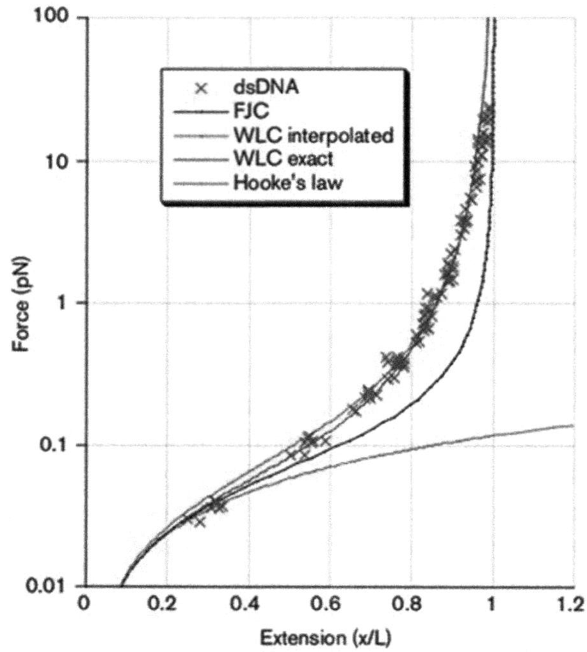

FIGURE 9.6 Force versus extension data (red crosses) for λ–phage ds-DNA obtained from magnetic tweezers experiment in 10 mM Na^+ solution. The data are fit to numerically to WLC (WLC exact), the FJC model, eq 9.10, and Hooke's law.

Source: Reprinted with permission from Ref. [10]. Copyright © 2000, Elsevier

9.5 IONIC EFFECTS ON DNA ELASTICITY

The whole biological functioning of a living being is associated with its coded DNA. So, the decoding process in DNA must take place in a customary manner and standard rate. But, variations in the surroundings of DNA may affect its normal functioning. Even these variations can be a cause of mutation in DNA. Thus, it is essential to study DNA mechanics in varying physiological conditions to evaluate different abnormal corporal behaviors. There are various ions (Na^+, K^+, Ca^{2+}, Mn^{2+}, Mg^{2+}, H^+, and Fe^{2+}) present in living cells that have a significant role in electric communication, photosynthesis, respiration, signaling, muscle action, DNA packaging, chromatin remodeling, and DNA bending. The ion concentration variations can abnormally vary this functioning. So, in this section, we discuss the effect of variation in ion concentration in the mechanical properties of DNA.

9.5.1 PERSISTENCE LENGTH

Experimental studies show that DNA has appreciable conformational flexibility in ionic solutions. The mechanical properties of DNA in ionic solutions are studied within the skeleton of the WLC model, where the single parameter persistence length (ξ) is used. We should consider the electrostatic and elastic interactions to spell out the persistence length of DNA in an ionic solution. Being a negatively charged polymer, the repulsive interactions between different segments of DNA cause an increase in the persistence length, and the elastic interactions come into the picture due to the base pair stacking. Garai et al. performed all atomistic MD simulations on different short DNAs of lengths 12, 24, 38, and 56 base-pairs at various monovalent salt concentrations[24] and calculated elastic properties using Eqs 9.5–9.7 of these DNAs. Figure 9.7(a) depicts the persistence length of different short DNAs in various monovalent salt concentrations and found ξ decreases with the increase of salt concentrations. Baumann et al.[9] observed similar behavior for λ-bacteriophage DNA compared in Figure 9.7(a).

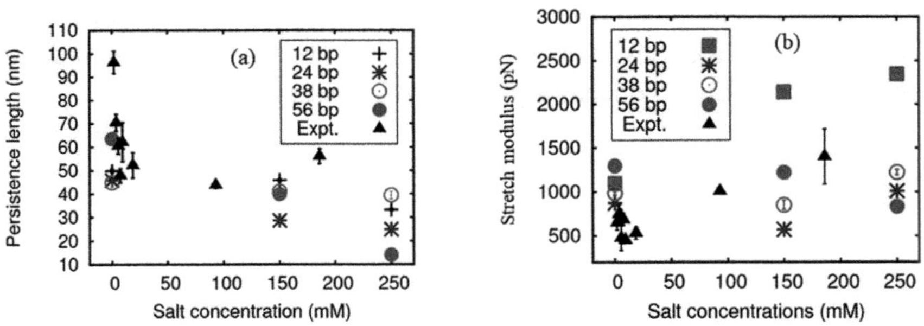

FIGURE 9.7 (a) Persistence length versus salt concentrations, (b) stretch modulus versus salt concentrations for various sizes of short DNA.

Source: Reprinted with permission from Ref. [24]. Copyright © 2015 ACS Publications

The presence of ions in DNA surroundings can modify the electrostatic repulsive interactions, while the elastic interaction is not much affected. Thus, the decrease of persistence length with an increase of salt concentrations is due to the screening of electrostatic repulsion of phosphate groups present on the DNA's backbone. Monovalent and multivalent ions have different effects on the persistence length of DNA. In their experimental study, they found that ξ value becomes lower in the presence of multivalent ions than at the same ionic strength of monovalent salt. Centrally centered charges for multivalent ions (such as Mg^{2+}, $Co(NH_3)_6^{3+}$) yield a lower ξ value of DNA than in linearly distributed charges (such as polyamines putrescine^{2+} and spermidine^{3+}).[9] Baumann et al. further fit their persistence length data with a nonlinear Poisson–Boltzmann theory for uniformly charged cylinders (see eq 9.13) and found the non-electrostatic contribution of the persistence length as 50 nm.

$$\xi = \xi_0 + \xi_e = \xi_0 + \frac{1}{4\kappa_0 l_B} = \xi_0 + 0.324 I^{-1}\,\text{Å} \tag{9.13}$$

where ξ_e and ξ_0 are the electrostatic and non-electrostatic contributions to the persistence length, κ_0 is the Debye-Hückel screening length and l_B as the Bjerrum length. However, a substantial discrepancy was observed when we study the dependence of persistence length on ionic strength (I) obtained from different experimental techniques such as force independent[92–94] or in force-dependent experiments.[9,95] Recently, Guilbaud et al., based on the high-throughput tethered particle,[94,96] determined the persistence length of DNA of length 1201 base-pairs in various monovalent and divalent metallic and alkylammonium ions concentrations ranging from 0.5 mM to 5 M (see Fig. 9.8). They found that Debye-Hückel theories are not sufficient to explain the persistence length behavior for the whole range of *I*. Instead, they found two approximate theories by Netz-Orland[97] and Trizac-Shen[98] to explain the persistence length-ion concentrations behavior better.

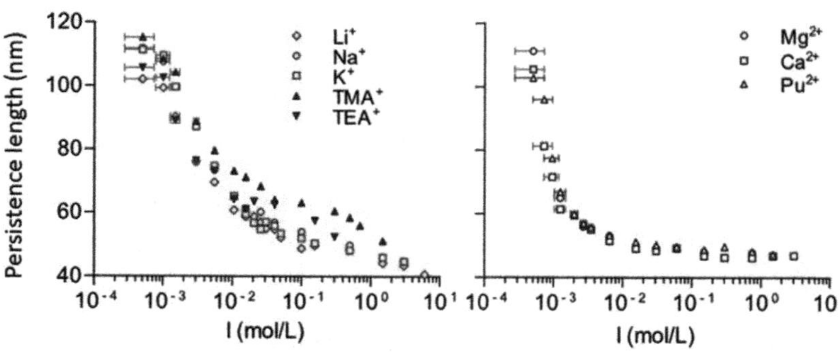

FIGURE 9.8 Persistence length as a function of ionic strengths for different monovalent and divalent ions.

Source: Reprinted with permission from Ref. [94]. Copyright © 2019 American Physical Society.

9.5.2 STRETCH MODULUS

Baumann et al.[9] determined the stretch modulus (γ) of λ–bacteriophage DNA by fitting the force-extension data with the extensible WLC model (see eq. 9.10) in various salt concentrations. They observed that γ decreases as I decrease, a trend opposite to ξ variations. They further identified that macroscopic elastic theory[99] could not explain these opposite trends between γ and ξ with ionic strength. In reality, DNA consists of phosphates on backbone, stacked base-pairs of different sequences, major grooves, and minor grooves that are not identical to an isotropic, homogeneous elastic rod. They predicted that DNA contour length would increase at low I due to enthalpic contribution, increased localized melting of A + T regions, and intramolecular electrostatic repulsion may decrease γ. Garai et al.,[24] from their equilibrium MD studies, have found a similar trend of γ and ξ with monovalent salt concentrations for different short DNAs (see Fig. 9.7b). In their study, they observed that γ increases with salt concentrations are due to reduction of the electrostatic repulsion of the phosphate groups on the DNA backbone.

9.6 PROTEIN-BINDING EFFECTS ON DNA ELASTICITY

Proteins are macromolecules made up of one or more long chains of amino acid residues.[37] There are 20 standard amino acids, and they can arrange themselves in a vast number of ways, and each corresponds to a particular protein with a specific function. So, proteins do a wide variety of tasks such as catalysis, transport of other molecules, response to stimuli, provide immunity, control cell growth, gene regulations, DNA packaging, and DNA sequence processing.[31] The mechanism of protein DNA binding and interaction is difficult to explain because (i) the basic electrostatic (may be attractive or repulsive) and mechanical properties are different for each protein, (ii) the protein binding may be sequence-dependent (sequence-specific proteins, e.g., CENP-B) or independent (nonsequence-specific proteins, e.g., HMGB1, HU, and H-NS), (iii) for nonsequence-specific proteins, the bending stiffness varies corresponding to the different sequences. DNA–protein interactions can be modeled as binary chemical reaction kinetics.[31,100] However, protein binding to DNA depends on various factors such as salt concentration, protein concentration, and binding sites.[31] In bacteriophage capsid, long DNA is packaged in which it is compacted to a substantial length so that it fits in a small volume. In the nucleosome, histone protein facilitates DNA to bend. During chromatin remodeling, DNA is always under force, and its structure may get deformed. The necessary local forces and torques at the deformed DNA in the base-pairs are discussed elsewhere.[16] Deformation of DNA structure due to protein interaction is essential for the various active biological processes, such as replication, transcription, and DNA repair. DNA interacts with multiple proteins during genetic regulation and forms loops with lengths comparable to or smaller than its persistence length.[101] Notably, bending fluctuations of free DNA in solutions are different than when the protein binds the DNA. This is because the former case is more or less isotropic. Still, the latter case is anisotropic due to protein and its binding site's size.[102] Protein can bend the DNA and induce distortions such as untwisting of DNA. Thus binding of a protein to DNA significantly affects the mechanical properties of DNA.[103]

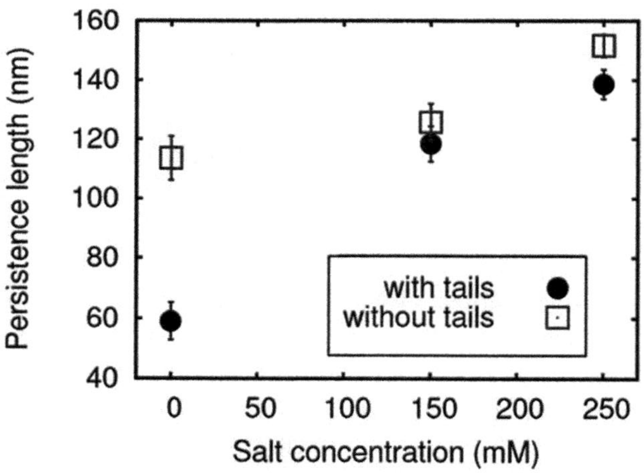

FIGURE 9.9 Persistence length of NCP DNA versus salt concentrations. *Source:* Reprinted with permission from Ref. [24]. Copyright © 2015 ACS

Publications

Garai et al.,[24] in their all atomistic MD studies, have found that the nucleosomal DNA in physiological conditions is stiffer than the bare short DNA. The persistence length of nucleosomal DNA increases with monovalent salt concentrations (see Fig. 9.9). Further, they investigated the role of histone protein tails on nucleosomal DNA's elastic properties. Histone tails are positively charged; thus, they bind to the DNA at low salt concentration due to strong electrostatic attraction. However, the electrostatic interaction reduces with the increase of salt concentration. As a result, the tails become freely suspended, and the existing kinks present in the nucleosome get reduced. Figure 9.9 depicts the histone tails' effects on the persistence length of nucleosomal DNA, and it is more prominent at low salt concentrations. Mogurampelly et al.,[46] in an MD study, have found the binding of positively charged dendrimer on DNA reduces electrostatic repulsion on the DNA backbone; thus, bending flexibility of DNA increases. However, they observed no significant change in the stretch modulus of DNA. High mobility group B (HMGB) structural proteins bind to DNA and play a crucial role in maintaining chromatin structure, transcription, replication, and DNA repair.[104,105] They locally interact with the DNA and bend the DNA. Using single-molecule pulling experiments, McCauley et al. have determined that HMGB protein increases DNA flexibility significantly.[105] Another study by Garai and coworkers has recently found that binding of human TRF1 protein to telomeric DNA makes it stiffer compared to when it is not bound (unpublished data). Like proteins, drugs can also interact with DNA and modify its mechanical properties.[106] In an MD study, Sahoo et al. determined the mechanical properties of DNA in intercalators' presence.[107] They found the stretch modulus of the DNA in intercalators' presence increases significantly compared to the bare ds-DNA, but the persistence length and the bending modulus decrease.

9.7 CONCLUSIONS AND FUTURE PERSPECTIVES

Double-stranded DNA is a negatively charged semiflexible polymer. The structural flexibility of DNA and conformational dynamics are essential for various active biological processes. Additionally, diverse ionic conditions directly impact biomolecular functioning, such as replication, DNA repair, and DNA packaging. The persistence length of DNA determines its stiffness, and for DNA, it is 50 nm. However, it varies depending on the size,[22,24] sequences,[108] and surrounding environments of DNA.[4,92] The bare DNA's persistence length decreases with increased ionic salt concentrations.[9,24,94] Garai et al. have shown recently that ionic liquids (IL) make DNA rigid.[26] IL cations bind to the phosphate backbone, invade into the DNA grooves, and establish electrostatic and hydrophobic interactions with the nitrogen bases and part of the pentose sugar. They further observed that persistence length and stretch modulus increase with the IL strengths, which are different when compared with DNA in monovalent or multivalent salt concentrations. However, a general theory on how persistence length depends on ionic strengths and ionic species is lacking. Notably, ionic conditions and ionic species are not the only factors determining the elastic properties of DNA. The stiffness of DNA also depends on proteins' interactions.[16,24,104,105] The methods described here are applicable to determine structural and mechanical properties of DNA and most DNA–protein complexes.

The study of elastic properties of the DNA in different environments may provide a close connection with health-related diseases. An increase in DNA flexibility causes genetic disorders[109,110] such as fragile X syndrome, myotonic dystrophy, and neurodegenerative disease[111] such as Huntington's disease. Thus, a similar study would further help to identify different communicable and non-communicable diseases. DNA is also a distinct and intelligent self-assembling nanomaterial.[112] Ionic liquids are potential solvents for DNA storage.[113] The study of elastic properties of DNA in various ionic liquids can not only provide the choice of preferable ionic liquids for maintaining DNA stability for long-term storage but also give insights for developing DNA nanomaterials.[26,113] The study of DNA nanomechanics has direct DNA nanotechnology applications, for example, the building of DNA nanostructures (like DNA origami[6,114–116]) and their stability.[117,118] DNA-based nanostructures such as DNA nanorobots represent a promising strategy for drug delivery in cancer therapy.[119] Detailed knowledge of the structural properties of DNA also has wide applications in DNA nanoelectronics.[120–122] Further theoretical, computer simulation and experimental investigation on the structural and mechanical properties of DNA in crowded environments and in confined geometry would provide a better picture of DNA nanomechanics and related applications.[4] Therefore, DNA nanomechanics is in a blooming stage and getting too much attention in recent days due to its broad applicability.

ACKNOWLEDGMENTS

This work was supported by the SERB, DST, Govt. of India for the financial support (Grant No ECR/2017/000683). AG thanks Prof. Prabal K. Maiti and the other collaborators who have contributed to many of the presented results. AG also thanks K. Jose for helping in the development of a few figures.

KEYWORDS

- **DNA**
- **persistence length**
- **stretch modulus**
- **bending modulus**
- **elastic properties**
- **ions**
- **molecular mechanics**

REFERENCES

1. Alberts, B.; Johnson, A.; Lewis, J.; Morgan, D.; Raff, M.; Roberts, K.; Walter, P. Molecular Biology of the Cell. *Mol. Biol. Cell.* **2017.**
2. Bustamante, C.; Bryant, Z.; Smith, S. B. Ten Years of Tension: Single-Molecule DNA Mechanics. *Nature* **2003,** *421*, 423–427.
3. Marko, J. F. DNA Mechanics. In: *Nuclear Architecture and Dynamics*; Elsevier, 2018; pp 3–40.
4. Aggarwal, A.; Naskar, S.; Sahoo, A. K.; Mogurampelly, S.; Garai, A.; Maiti, P. K. Science Direct What Do We Know About DNA Mechanics So Far ? *Curr. Opin. Struct. Biol.* **2020,** *64*, 42–50.
5. Seeman, N. C.; Sleiman, H. F. DNA Nanotechnology. *Nat. Rev. Mater.* **2017,** *3*, 1–23.
6. Seeman, N. C. *Structural DNA Nanotechnology*; Cambridge University Press, 2015.
7. Zahid, M.; Kim, B.; Hussain, R.; Amin, R.; Park, S. DNA Nanotechnology: A Future Perspective. *Nanosc. Res. Lett.* **2013,** *8*, 119.
8. Phillips, R.; Kondev, J.; Theriot, J.; Garcia, H. *Physical Biology of the Cell*; Garland Science, 2012.
9. Baumann, C. G.; Smith, S. B.; Bloomfield, V. A.; Bustamante, C. Ionic Effects on the Elasticity of Single DNA Molecules. *Proc. Natl. Acad. Sci. USA* **1997,** *94*, 6185–6190.
10. Bustamante, C.; Smith, S. B.; Liphardt, J.; Smith, D. Single-Molecule Studies of DNA Mechanics. *Curr. Opin. Struct. Biol.* **2000,** *10*, 279–285.
11. Cloutier, T. E.; Widom, J. Spontaneous Sharp Bending of Double-Stranded DNA. *Mol. Cell.* **2004,** *14*, 355–362.
12. Cloutier, T. E.; Widom, J. DNA Twisting Flexibility and the Formation of Sharply Looped Protein-DNA Complexes. *Proc. Natl. Acad. Sci. USA* **2005,** *102*, 3645–3650.
13. Wiggins, P.; Van Der Heijden, T.; Moreno-Herrero, F.; Spakowitz, A.; Phillips, R.; Widom, J.; Dekker, C.; Nelson, P. C. High Flexibility of DNA on Short Length Scales Probed by Atomic Force Microscopy. *Nat. Nanotechnol.* **2006,** *1*, 137–141.

14. Mathew-Fenn, R. S.; Das, R.; Harbury, P. A. B. Remeasuring the Double Helix. *Science (80-.).* **2008,** *322*, 446–449.
15. Becker, N. B.; Everaers, R. DNA Nanomechanics in the Nucleosome. *Structure* **2009,** *17*, 579–589.
16. Becker, N. B.; Everaers, R. DNA Nanomechanics: How Proteins Deform the Double Helix. *J. Chem. Phys.* **2009,** *130*, 04B602.
17. Vafabakhsh, R.; Ha, T. Extreme Bendability of DNA Less Than 100 Base Pairs Long Revealed by Single-Molecule Cyclization. *Science (80-.).* **2012,** *337*, 1097–1101.
18. Mazur, A. K.; Maaloum, M. DNA Flexibility on Short Length Scales Probed by Atomic Force Microscopy. *Phys. Rev. Lett.* **2014,** *112*, 1–5.
19. Kratky, O.; Porod, G. Röntgenuntersuchung gelöster Fadenmoleküle. *Recl. des Trav. Chim. des Pays-Bas*. **1949,** *68*, 1106–1122.
20. Marko, J. F.; Siggia, E. D. Stretching DNA. *Macromolecules* **1995,** *28*, 8759–8770.
21. Cocco, S.; Yan, J.; Léger, J. F.; Chatenay, D.; Marko, J. F. Overstretching and Force-Driven Strand Separation of Double-Helix DNA. *Phys. Rev. E—Stat. Phys. Plasmas Fluids Relat. Interdiscip. Top.* **2004,** *70*, 18.
22. Yuan, C.; Chen, H.; Lou, X. W.; Archer, L. A. DNA Bending Stiffness on Small Length Scales. *Phys. Rev. Lett.* **2008,** *100*, 1–4.
23. Santosh, M.; Maiti, P. K. Force Induced DNA Melting. *J. Phys. Condens. Matter*. **2009,** 21.
24. Garai, A.; Saurabh, S.; Lansac, Y.; Maiti, P. K. DNA Elasticity from Short DNA to Nucleosomal DNA. *J. Phys. Chem. B*. **2015,** *119*, 11146–11156.
25. Garai, A.; Mogurampelly, S.; Bag, S.; Maiti, P. K. Overstretching of B-DNA with Various Pulling Protocols: Appearance of Structural Polymorphism and S-DNA. *J. Chem. Phys.* **2017,** *147*, 225102.
26. Garai, A.; Ghoshdastidar, D.; Senapati, S.; Maiti, P. K. Ionic Liquids Make DNA Rigid. *J. Chem. Phys.* **2018,** *149*.
27. Travers, A.; Muskhelishvili, G. DNA Structure and Function. *FEBS J. 282*, 2279–2295.
28. http://mcat-review.org. DNA Structure and Function—MCAT Review, 2013.
29. Watson, J. D.; Crick, F. H. C. Molecular Structure of Nucleic Acids: A Structure for Deoxyribose Nucleic Acid. *Nature*. **1953,** *171*, 737–738.
30. Ussery, D. W. DNA Structure: A-, B- and Z-DNA Helix Families. In *Encyclopedia of Life Sciences*, 2002.
31. Marko, J. F. DNA Mechanics. In: *Nuclear Architecture and Dynamics*; Elsevier, 2018; pp 3–40.
32. Ghosh, A.; Bansal, M. A Glossary of DNA Structures from A to Z. *Acta Crystallogr.—Sect. D Biol. Crystallogr.* **2003,** *59*, 620–626.
33. Grandbois, M.; Beyer, M.; Rief, M.; Clausen-Schaumann, H.; Gaub, H. E. How Strong Is a Covalent Bond. *Science (80-.).* **1999,** *283*, 1727–1730.
34. Dickerson, R. E. Definitions and Nomenclature of Nucleic Acid Structure Components. *Nucleic Acids Res.* **1989,** *17*, 1797–1803.
35. Lu, X. J.; Olson, W. K. Resolving the Discrepancies Among Nucleic Acid Conformational Analyses. *J. Mol. Biol.* **1999,** *285*, 1563–1575.
36. Olson, W. K.; Bansal, M.; Burley, S. K.; Dickerson, R. E.; Gerstein, M.; Harvey, S. C.; Heinemann, U.; Lu, X. J.; Neidle, S.; Shakked, Z.; Sklenar, H.; Suzuki, M.; Tung, C. S.; Westhof, E.; Wolberger, C.; Berman, H. M. A Standard Reference Frame for the Description of Nucleic Acid Base-Pair Geometry. *J. Mol. Biol.* **2001,** *313*, 229–237.

37. Nelson, P. *Biological Physics*; WH Freeman: New York, 2004.
38. Kratky, O.; Porod, G. Röntgenuntersuchung gelöster fadenmoleküle. *Recl. des Trav. Chim. des Pays-Bas*. **1949,** *68*, 1106–1122.
39. Padinhateeri, R.; Menon, G. I. Stretching and Bending Fluctuations of Short DNA Molecules. *Biophys. J.* **2013,** *104*, 463–471.
40. Ranjith, P.; Kumar, P. B. S.; Menon, G. I. Distribution Functions, Loop Formation Probabilities, and Force-Extension Relations in a Model for Short Double-Stranded DNA Molecules. *Phys. Rev. Lett.* **2005,** *94*, 1–4.
41. Olson, W. K.; Gorin, A. A.; Lu, X. J.; Hock, L. M.; Zhurkin, V. B. DNA Sequence-Dependent Deformability Deduced from Protein-DNA Crystal Complexes. *Proc. Natl. Acad. Sci. USA* **1998,** *95*, 11163–11168.
42. Lankaš, F.; Šponer, J.; Hobza, P.; Langowski, J. Sequence-Dependent Elastic Properties of DNA. *J. Mol. Biol.* **2000,** *299*, 695–709.
43. Olson, W. K.; Swigon, D.; Coleman, B. D. Implications of the Dependence of the Elastic Properties of DNA on Nucleotide Sequence. *Philos. Trans. R. Soc. A Math. Phys. Eng. Sci.* **2004,** *362*, 1403–1422.
44. Matsumoto, A.; Olson, W. K. Sequence-Dependent Motions of DNA: A Normal Mode Analysis at the Base-Pair Level. *Biophys. J.* **2002,** *83*, 22–41.
45. Mitchell, J. S.; Glowacki, J.; Grandchamp, A. E.; Manning, R. S.; Maddocks, J. H. Sequence-Dependent Persistence Lengths of DNA. *J. Chem. Theory Comput.* **2017,** *13*, 1539–1555.
46. Mogurampelly, S.; Nandy, B.; Netz, R. R.; Maiti, P. K. Elasticity of DNA and the Effect of Dendrimer Binding. *Eur. Phys. J. E.* **2013,** *36*, 1–9.
47. Sahoo, A. K.; Bagchi, B.; Maiti, P. K. Understanding Enhanced Mechanical Stability of DNA in the Presence of Intercalated Anticancer Drug: Implications for DNA Associated Processes. *J. Chem. Phys.* **2019,** *151*, 164902.
48. Michael Rubinstein, R. H. C. *Polymer Physics*; Oxford University Press: New York, 2003.
49. Marin-Gonzalez, A.; Vilhena, J. G.; Moreno, F.-Herrero, and Perez, R. DNA Crookedness Regulates DNA Mechanical Properties at Short Length Scales. *Phys. Rev. Lett.* **2019,** *122*, 48102.
50. Mazur, A. K.; Maaloum, M. DNA Flexibility on Short Length Scales Probed by Atomic Force Microscopy. *Phys. Rev. Lett.* **2014,** 112.
51. Matsumoto, A.; N. Go. Dynamic Properties of Double-Stranded DNA by Normal Mode Analysis. *J. Chem. Phys.* **1999,** *110*, 11070–11075.
52. Mazur, A. K. Wormlike Chain Theory and Bending of Short DNA. *Phys. Rev. Lett.* **2007,** *98*, 1–4.
53. Noy, A.; Golestanian, R. Length Scale Dependence of DNA Mechanical Properties. *Phys. Rev. Lett.* **2012,** *109*, 1–5.
54. Nelson, P. C. Spare the (Elastic) Rod. *Science (80-.).* **2012,** *337*, 1045–1046.
55. Mazur, A. K.; Maaloum, M. Atomic Force Microscopy Study of DNA Flexibility on Short Length Scales: Smooth Bending Versus Kinking. *Nucleic Acids Res.* **2014,** *42*, 14006–14012.
56. Polley, A.; Samuel, J.; Sinha, S. Bending Elasticity of Macromolecules: Analytic Predictions from the Wormlike Chain Model. *Phys. Rev. E—Stat. Nonlinear, Soft Matter Phys.* **2013,** *87*, 1–4.

57. Gross, P.; Laurens, N.; Oddershede, L. B.; Bockelmann, U.; Peterman, G, E. J.; Wuite, L, G. J.. Quantifying How DNA Stretches, Melts and Changes Twist Under Tension. *Nat. Phys.* **2011,** *7*, 731–736.
58. Chou, F. C.; Lipfert, J.; Das, R. Blind Predictions of DNA and RNA Tweezers Experiments with Force and Torque. *PLoS Comput. Biol.* **2014,** *10*, 37–47.
59. Karplus, M.; Lavery, R. Significance of Molecular Dynamics Simulations for Life Sciences. *Isr. J. Chem.* **2014,** *54*, 1042–1051.
60. Allen, M. P.; Tildesley, D. J. *Computer Simulation of Liquids*; Oxford University Press, 2017.
61. Frenkel, D.; Smit, B. *Understanding Molecular Simulation: From Algorithms to Applications*; Elsevier, 2001.
62. Rapaport, D. C. *The Art of Molecular Dynamics Simulation*; Cambridge University Press, 2004.
63. Salomon-Ferrer, R.; Case, D. A.; Walker, R. C. An Overview of the Amber Biomolecular Simulation Package. *Wiley Interdiscip. Rev. Comput. Mol. Sci.* **2013,** *3*, 198–210.
64. Case, D. A.; Cheatham, T. E.; Darden, T.; Gohlke, H.; Luo, R.; Merz, K. M.; Onufriev, A.; Simmerling, C.; Wang, B.; Woods, R. J. The Amber Biomolecular Simulation Programs. *J. Comput. Chem.* **2005,** *26*, 1668–1688.
65. MacKerell, A. D.; Banavali, N. K. All-Atom Empirical Force Field for Nucleic Acids: IApplication, I. to Molecular Dynamics Simulations of DNA and RNA in Solution. *J. Comput. Chem.* **2000,** *21*, 105–120.
66. Riveline, D. "Single Molecule": Theory and Experiments, an Introduction. *J. Nanobiotechnol.* **2013,** *11* (Suppl 1), S1.
67. Neuman, K. C.; Nagy, A. Single-Molecule Force Spectroscopy: Optical Tweezers, Magnetic Tweezers and Atomic Force Microscopy. *Nat. Methods.* **2008,** *5*, 491–505.
68. Zhao, D.; Liu, S.; Gao, Y. Single-Molecule Manipulation and Detection. *Acta Biochim. Biophys. Sin. (Shanghai).* **2018,** *50*, 231–237.
69. Wang, M. D.; Yin, H.; Landick, R.; Gelles, J.; Block, S. M. Stretching DNA with Optical Tweezers. *Biophys. J.* **1997,** *72*, 1335–1346.
70. Neuman, K. C.; Lionnet, T.; Allemand, J.-F. Single-Molecule Micromanipulation Techniques. *Annu. Rev. Mater. Res.* **2007,** *37*, 33–67.
71. Rief, M.; Gautel, M.; Oesterhelt, F.; Fernandez, J. M.; Gaub, H. E. Reversible Unfolding of Individual Titin Immunoglobulin Domains by AFM. *Science (80-.).* **1997,** *276*, 1109–1112.
72. Marshall, B. T.; Long, M.; Piper, J. W.; Yago, T.; McEver, R. P.; Zhu, C. Direct Observation of Catch Bonds Involving Cell-Adhesion Molecules. *Nature* **2003,** *423*, 190–193.
73. Fazal, F. M.; Block, S. M. Optical Tweezers Study Life Under Tension. *Nat. Photonics.* **2011,** *5*, 318–321.
74. Smith, S. B.; Finzi, L.; Bustamante, C. Direct Mechanical Measurements of the Elasticity of Single DNA Molecules by Using Magnetic Beads. *Science (80-.).* **1992,** *258*, 1122–1126.
75. De Vlaminck, I.; Dekker, C. Recent Advances in Magnetic Tweezers. *Annu. Rev. Biophys.* **2012,** *41*, 453–472.
76. Tanase, M.; Biais, N.; Sheetz, M. Magnetic Tweezers in Cell Biology. *Methods Cell Biol.* **2007,** *83*, 473–493.

77. Danilowicz, C.; Limouse, C.; Hatch, K.; Conover, A.; Coljee, V. W.; Kleckner, N.; Prentiss, M. The Structure of DNA Overstretched from the 5′5′ Ends Differs from the Structure of DNA Overstretched from the 3′3′ Ends. *Proc. Natl. Acad. Sci. USA* **2009,** *106*, 13196–13201.
78. Double-Stranded, I.; Smith, S. B.; Cui, Y., Bustamante, C. Overstretching **1996,** 271.
79. Cluzel, P.; Lebrun, A.; Heller, C.; Lavery, R.; Viovy, J. L.; Chatenay, D.; Caron, F. DNA: An Extensible Molecule. *Science (80-.).* **1996,** *271*, 792–794.
80. Bosaeus, N.; El, A. H.-Sagheer, Brown, T.; Smith, S. B.; Åkerman, B.; Bustamante, C.; Norde, B.ń. Tension Induces a Base-Paired Overstretched DNA Conformation. *Proc. Natl. Acad. Sci. USA* **2012,** *109*, 15179–15184.
81. Rouzina, I.; Bloomfield, V. A. Force-Induced Melting of the DNA Double Helix 1: Thermodynamic Analysis. *Biophys. J.* **2001,** *80*, 882–893.
82. Van Mameren, J.; Gross, P.; Farge, G.; Hooijman, P.; Modesti, M.; Falkenberg, M.; Wuite, L, G. J.; Peterman, E. J. G. Unraveling the Structure of DNA During Overstretching by Using Multicolor, Single-Molecule Fluorescence Imaging. *Proc. Natl. Acad. Sci. USA* **2009,** *106*, 18231–18236.
83. Cocco, S.; Monasson, R.; Marko, J. F. Unzipping Dynamics of Long DNAs. *Phys. Rev. E—Stat. Physics, Plasmas, Fluids, Relat. Interdiscip. Top.* **2002,** *66*, 11.
84. Danilowicz, C.; Coljee, V. W.; Bouzigues, C.; Lubensky, D. K.; Nelson, D. R.; Prentiss, M. DNA Unzipped Under a Constant Force Exhibits Multiple Metastable Intermediates. *Proc. Natl. Acad. Sci. USA* **2003,** *100*, 1694–1699.
85. Marko, J. F.; Neukirch, S. Global Force-Torque Phase Diagram for the DNA Double Helix: Structural Transitions, Triple Points, and Collapsed Plectonemes. *Phys. Rev. E—Stat. Nonlinear Soft Matter Phys.* **2013,** *88*, 062722.
86. Sheinin, M. Y.; Forth, S.; Marko, J. F.; Wang, M. D. Underwound DNA Under Tension: Structure, Elasticity, and Sequence-Dependent Behaviors. *Phys. Rev. Lett.* **2011,** *107*, 108102.
87. Zhang, X.; Chen, H.; H. Fu, Doyle, P. S.; Yan, J. Two Distinct Overstretched DNA Structures Revealed by Single-Molecule Thermodynamics Measurements. *Proc. Natl. Acad. Sci. USA* **2012,** *109*, 8103–8108.
88. Zhang, X.; Chen, H.; S. Le, Rouzina, I.; Doyle, P. S.; Yan, J. Revealing the Competition Between Peeled ssDNA, Melting Bubbles, and S-DNA During DNA Overstretching by Single-Molecule Calorimetry. *Proc. Natl. Acad. Sci. USA* **2013,** *110*, 3865–3870.
89. Kumar, S.; Li, M. S. Biomolecules Under Mechanical Force. *Phys. Rep.* **2010,** *486*, 1–74.
90. Wu, Y. Y.; Bao, L.; Zhang, X.; Tan, Z. J. Flexibility of Short DNA Helices with Finite-Length Effect: From Base Pairs to Tens of Base Pairs. *J. Chem. Phys.* **2015,** *142*.
91. Vafabakhsh, R.; Ha, T. Extreme Bendability of DNA Less. *Science* **a***(80-.). 1097*, 1–6.
92. Brunet, A.; Tardin, C.; Salom, L.é, Rousseau, P.; Destainville, N.; Manghi, M. Dependence of DNA Persistence Length on Ionic Strength of Solutions with Monovalent and Divalent Salts: A Joint Theory-Experiment Study. *Macromolecules* **2015,** *48*, 3641–3652.
93. Savelyev, A. Do Monovalent Mobile Ions Affect DNA's Flexibility at High Salt Content? *Phys. Chem. Chem. Phys.* **2012,** *14*, 2250–2254.
94. Guilbaud, S.; Salomé, L.; Destainville, N.; Manghi, M.; Tardin, C. Dependence of DNA Persistence Length on Ionic Strength and Ion Type. *Phys. Rev. Lett.* **2019,** *122*, 1–6.

95. Saleh, O. A.; McIntosh, D. B.; Pincus, P.; Ribeck, N. Nonlinear Low-Force Elasticity of Single-Stranded DNA Molecules. *Phys. Rev. Lett.* **2009,** *102*, 1–4.
96. Plénat, T.; Tardin, C.; Rousseau, P.; Salomé, L. High-Throughput Single-Molecule Analysis of DNA-Protein Interactions by Tethered Particle Motion. *Nucleic Acids Res.* **2012,** *40*.
97. Netz, H. H.; Orland, H. Variational Charge Renormalization in Charged Systems. *Eur. Phys. J. E.* **2003,** *11*, 301–311.
98. Trizac, E.; Shen, T. Bending Stiff Charged Polymers: The Electrostatic Persistence Length. *EPL*. **2016,** *116*, 18007.
99. Landau, L. D.; Lifshitz, E. M. *Course of Theoretical Physics, Theory of Elasticity*, Vol. 7, 1986.
100. Halford, S. E.; Marko, J. F. How Do Site-Specific DNA-Binding Proteins Find Their Targets? *Nucleic Acids Res.* **2004,** *32*, 3040–3052.
101. Garcia, H. G.; Grayson, P.; Han, L.; Inamdar, M.; Kondev, J.; Nelson, P. C.; Phillips, R.; Widom, J.; Wiggins, P. A. Biological Consequences of Tightly Bent DNA: The Other Life of a Macromolecular Celebrity. *Biopolymers* **2007,** *85*, 115–130.
102. Travers, A. A.; Thompson, J. M. T. An Introduction to the Mechanics of DNA. *Philos. Trans. R. Soc. A Math. Phys. Eng. Sci.* **2004,** *362*, 1265–1279.
103. Yan, J.; Marko, J. F. Effects of DNA-Distorting Proteins on DNA Elastic Response. *Phys. Rev. E—Stat. Physics, Plasmas, Fluids, Relat. Interdiscip. Top.* **2003,** *68*, 12.
104. Murugesapillai, D.; McCauley, M. J.; Maher, L. J.; Williams, M. C. Single-Molecule Studies of High-Mobility Group B Architectural DNA Bending Proteins. *Biophys. Rev.* **2017,** *9*, 17–40.
105. McCauley, M. J.; Rueter, E. M.; Rouzina, I.; Maher, L. J.; Williams, M. C. Single-Molecule Kinetics Reveal Microscopic Mechanism by Which High-Mobility Group B Proteins Alter DNA Flexibility. *Nucleic Acids Res.* **2013,** *41*, 167–181.
106. Rocha, M. S. Extracting Physical Chemistry from Mechanics: A New Approach to Investigate DNA Interactions with Drugs and Proteins in Single Molecule Experiments. *Integr. Biol. (UK)*. **2015,** *7*, 967–986.
107. Sahoo, A. K.; Bagchi, B.; Maiti, P. K. Understanding Enhanced Mechanical Stability of DNA in the Presence of Intercalated Anticancer Drug: Implications for DNA Associated Processes. *J. Chem. Phys.* **2019,** *151*.
108. Coleman, B. D.; Olson, W. K.; Swigon, D. Theory of Sequence-Dependent DNA Elasticity. *J. Chem. Phys.* **2003,** *118*, 7127–7140.
109. Bacolla, A.; Gellibolian, R.; Shimizu, M.; Amirhaeri, S.; Kang, S.; Ohshima, K.; Larson, J. E.; Harvey, S. C.; Stollar, B. D.; Wells, R. D. Flexible DNA: Genetically Unstable CTG·CAG and CGG·CCG from Human Hereditary Neuromuscular Disease Genes. *J. Biol. Chem.* **1997,** *272*, 16783–16792.
110. Wang, Y. H.; Griffith, J. Expanded CTG Triplet Blocks from the Myotonic Dystrophy Gene Create the Strongest Known Natural Nucleosome Positioning Elements. *Genomics*. **1995,** *25*, 570–573.
111. Godde, J. S.; Wolffe, A. P. Nucleosome Assembly on CTG Triplet Repeats. *J. Biol. Chem.* **1996,** *271*, 15222–15229.
112. Penth, M.; Schellnhuber, K.; Bennewitz, R.; Blass, J. Nanomechanics of Self-Assembled DNA Building Blocks. **2021,** *13*, 9371.
113. Tateishi-Karimata, H.; Sugimoto, N. Structure, Stability and Behaviour of Nucleic Acids in Ionic Liquids. *Nucleic Acids Res.* **2014,** *42*, 8831–8844.

114. Fischer, S.; Hartl, C.; Frank, K.; Rädler, J. O.; Liedl, T.; Nickel, B. Shape and Interhelical Spacing of DNA Origami Nanostructures Studied by Small-Angle X-ray Scattering. *Nano Lett.* **2016,** *16*, 4282–4287.
115. Hong, F.; Zhang, F.; Liu, Y.; Yan, H. DNA Origami: Scaffolds for Creating Higher Order Structures. *Chem. Rev.* **2017,** *117*, 12584–12640.
116. Ji, J.; Karna, D.; Mao, H. DNA Origami Nano-Mechanics. *Chem. Soc. Rev.* **2021**.
117. Joshi, H.; Kaushik, A.; Seeman, N. C.; Maiti, P. K. Nanoscale Structure and Elasticity of Pillared DNA Nanotubes. *ACS Nano*. **2016,** *10*, 7780–7791.
118. Naskar, S.; Gosika, M.; Joshi, H.; Maiti, P. K. Tuning the Stability of DNA Nanotubes with Salt. *J. Phys. Chem. C.* **2019,** *123*, 9461–9470.
119. Li, S.; Jiang, Q.; Liu, S.; Zhang, Y.; Tian, Y.; Song, C.; Wang, J.; Zou, Y.; Anderson, G. J.; Han, J. Y.; Chang, Y.; Liu, Y.; Zhang, C.; Chen, L.; Zhou, G.; Nie, G.; Yan, H.; Ding, B.; Zhao, Y. A DNA Nanorobot Functions as a Cancer Therapeutic in Response to a Molecular Trigger In Vivo. *Nat. Biotechnol.* **2018,** *36*, 258–264.
120. Kelley, S. O.; Barton, J. K. Electron Transfer Between Bases in Double Helical DNA. *Science (80-.).* **1999,** *283*, 375–381.
121. Bag, S.; Mogurampelly, S.; W.A.G. III, Maiti, P. K. Dramatic Changes in DNA Conductance with Stretching: Structural Polymorphism at A Critical Extension. *Nanoscale*. **2016,** *8*, 16044–16052.
122. Aggarwal, A.; Bag, S.; Maiti, P. K. Remarkable Similarity of Force Induced dsRNA Conformational Changes to Stretched dsDNA and Their Detection Using Electrical Measurements. *Phys. Chem. Chem. Phys.* **2018,** *20*, 28920–28928.
123. Brett, A.M.O.; Serrano S.H.P.; Piedade A. J. P. Chapter 3 - Electrochemistry of DNA, Comprehensive Chemical Kinetics, 1999, 37, 91–119.
124. Lu, J.; Olson, W.K. 3DNA: a software package for analyzing, rebuilding, and visualization of three-dimensional nucleic acid structures, Nucleic Acids Research, 2003, 31, 1213–1227.

CHAPTER 10

The Ultrasonic-Assisted Synthesis of Locust Gum/Peg-Silver Nanoparticles and Its Mathematical Modeling of Rheological Parameters

SELCAN KARAKUŞ[1], MERT AKIN INSEL[2], INCI ALBAYRAK[3], and NEVIN TAŞALTIN[4,5]

[1]*Department of Chemistry, Faculty of Engineering, Istanbul University-Cerrahpaşa, Turkey*

[2]*Department of Chemical Engineering, Yıldız Technical University, Istanbul, Turkey*

[3]*Department of Mathematical Engineering, Yıldız Technical University, Istanbul; Turkey*

[4]*Department of Renewable Energy Tech., Maltepe University, Istanbul, Turkey*

[5]*Department of Electrical & Electronics Engineering, Maltepe University, Istanbul, Turkey*

ABSTRACT

In this study, the novel locust gum/polyethylene glycol-silver nanoparticles (LG/PEG-Ag NPs) were prepared using a green and facile ultrasonic method. Also, this study highlighted the effect of ultrasonic cavitation on morphological and structural properties of the biopolymer blend-based

Foundation and Growth of Macromolecular Science: Advances in Research for Sustainable Development. Meegle S. Mathew, PhD, Józef T. Haponiuk, PhD & Sabu Thomas, PhD, DSc (Eds.)

nanostructure. The ultrasonic-dependent rheological properties of LG/PEG-Ag NPs were investigated, and mathematical models were constructed with the results of viscosity studies using the dilute solution viscometer (DSV) method. The LG/PEG-Ag NPs were characterized by different techniques such as Fourier transform-infrared spectroscopy (FTIR), scanning electron microscopy/energy dispersive X-ray spectroscopy (SEM/EDX), and Brunauer–Emmett–Teller (BET). According to the results, LG/PEG-Ag NPs exhibited a uniform spherical shape with a diameter of 320 nm while the surface area was 130.533 m^2/g. The influences of the operating parameters such as the sonication time, amplitude, salt, and temperature on the intrinsic viscosity of fabricated LG/PEG-Ag NPs were examined in different experimental conditions to optimize the system. Additionally, the optimization on the effect of the rheological parameters of the LG/PEG-Ag NPs with the highest correlation constant was determined using the Huggins model. The experimental results provided a mathematical model to describe the effect of the LG/PEG-Ag NPs on the intrinsic viscosity for nanofluids. Finally, comparative error analysis models were conducted to verify the validity of the obtained relationships.

10.1 NTRODUCTION

Recently, innovative engineered nanomaterials have attracted the attention of scientists in interdisciplinary science of applications such as textile, environment, food, sensor, energy, agriculture, and drug due to the quantum effects, larger surface areas, and self-assembly properties.[1–5] Generally, nanostructures can be classified into different types based on their morphologies (fibers, rings, tubes, spheres, and planes), dimensions (zero-dimensional (0-D), one-dimensional (1-D), two-dimensional (2-D), and three-dimensional (3-D), and chemical composition (fullerenes, carbon nanotubes, metal nanoparticles (NPs), semiconductor NPs, ceramic NPs, polymeric NPs, and lipid-based NPs) at nanoscale.[6–13] The dimension-based nanostructures are 0-D (nanoparticles), 1-D (nanotubes, nanorods, and nanowires), 2-D (nanofilms, nanolayers, and nano-coatings), and 3-D (bulk powders and multi-nanolayers) nanomaterials.[14,15] Especially, it was noticed that the recent developments have been remarkable for the fabrication, characterization, and experimental design of inorganic (gold, silver, titanium, and copper)-based, known as 0-D nanomaterials, with green chemistry principles in the literature.[16–20] Accordingly, among these nanoparticles, biopolymer matrix-based silver nanoparticles have biological, chemical, thermal, optical, mechanical, morphological, and economical properties.[21–25]

The synthesis of silver nanoparticles by a large number of different methods has been reported such as microwave, laser ablation, hydrothermal, mechanical, chemical reduction, electrochemical, ultraviolet-visible light, and sonication method.[26–30] While producing silver nanoparticles, it attracts great attention due to some advantages such as using nontoxic solvent in the sonication method, and not having costly requirements such as high temperature and pressure to obtain narrow-sized nanoparticles in a short time.[31,32]

According to the literature based on the silver nanostructures with different biopolymer matrices, it was noticed that there were several studies. Munir et al. reported the preparation of the eucalyptus camaldulensis gum-based zinc and silver nanoparticles and investigated the antimicrobial properties of the nanostructures.[33] Talodthaisong et al. reported the fabrication of the composite guar gum-silver nanoparticle hydrogels and showed the self-healing, injectable, and antibacterial properties of the nanoparticles.[34] Zhai et al. investigated the determination of the hydrogen sulfide using a colorimetric sensor based on gellan gum-silver nanoparticles bio-nanocomposite in intelligent packaging applications.[35] However, it was found that a LG/PEG-Ag NPs was not synthesized. In this study, novel locust gum/polyethylene glycol-silver nanoparticles (LG/PEG-Ag NPs) were prepared using a green and facile sonication method based on the formation of OH* radicals from ultrasonic cavitation in aqueous solutions. The case that the study seeks to illuminate is the effect of ultrasonic cavitation on the morphology and structural properties of LG/PEG-based Ag nanostructures. Also, the highlight of this study is the role of ultrasonic cavitation on the rheological properties of LG/PEG-Ag NPs. The mathematical models that experimentally proved these were constructed with the results of viscosity studies using the dilute solution viscometer (DSV) method due to the operating parameters, such as the sonication time, amplitude, salt, and temperature. Also, the LG/PEG-Ag NPs were characterized using various advance techniques, such as Fourier transform-infrared spectroscopy (FTIR), scanning electron microscopy / energy dispersive X-ray spectroscopy (SEM/EDX), and Brunauer–Emmett–Teller (BET). According to the SEM and BET results, it was observed that LG/PEG-Ag NPs exhibited a uniform spherical shape with a diameter of 320 nm while the surface area was 130.533 m^2/g. Finally, comparative error analysis models were applied to verify the validity of the obtained relationships. Considering the results of optimization and error analysis, it was seen that the new nano-formulation can be possible to fabricate with the green chemistry approach and the homogeneous and nano-distributed structure was obtained according to the characterization results. The synthesized LG/PEG-Ag NPs have a major potential, especially in biotechnological studies.

10.2 EXPERIMENTAL PART

10.2.1 MATERIALS

LG (from ceratonia siliqua seeds, molecular weight: 310 kDa), PEG 400 (molecular weight: 380–420 Da), and silver nitrate (AgNO3) (≥99.0%) were purchased from Sigma Aldrich Company. Sodium hydroxide (NaOH) was purchased from Merck Company. The sterile syringe filters were purchased from Whatman (6900-2502 GD/X 25). All chemicals and reagents were used without further purification.

10.2.2 PREPARATION OF THE LG/PEG

LG weighing 2.5 g was mixed in 500 mL distilled water at 25°C under magnetic stirring for 1 h. It was kept unmixed for five days in the dark. Then, 0.5 g of PEG was added in the solution of LG and sonicated at 25°C for 10 min. Finally, the mixture was filtered through a 0.22 μm sterile syringe filter polytetrafluoroethylene (PTFE) membrane.

10.2.3 THE ULTRASONIC FABRICATION OF THE LG/PEG-AG NPS

The ultrasonic process was carried out in a Bandelin Sonopuls (HD 3100) homogenizer with integrated amplitude control 10–100%. 2.5 g/500 mL of LG solution was mixed with 0.5 g of PEG at 25°C under magnetic stirring for 10 min. 0.84 g of AgNO3 was mixed in distilled water (500 mL) for 10 min at room temperature using a magnetic stirrer. 0.4 g of NaOH was mixed in distilled water (250 mL) for 5 min at room temperature using a magnetic stirrer. Then, the solution of LG/PEG was mixed with the AgNO3 solution and sonicated for 10 min at 25°C. 50 mL of NaOH solution (0.02 g/50 mL) was added.

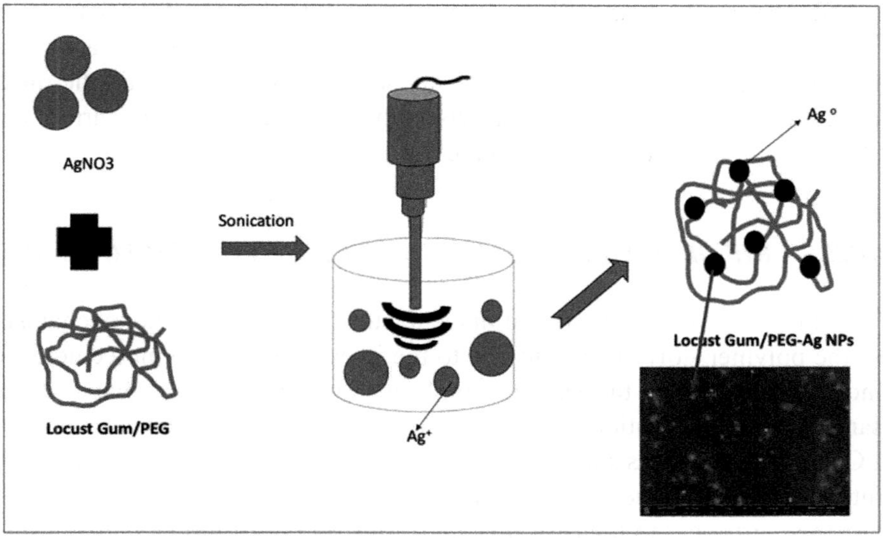

FIGURE 10.1 Schematic diagram of the preparation of the LG/PEG-Ag NPs.

It is observed that the formation of LG/PEG-Ag NPs which was visually approved by color change from colorless to brownish black color (Fig. 10.1). This change of color was a signal that LG/PEG-Ag NPs was obtained (pH = 8.1). Finally, the solution was filtered through a 0.22 μm syringe filter PTFE membrane and stored in sterile polypropylene containers until use at room temperature.

10.2.4 CHARACTERIZATION OF THE LG/PEG-AG NPS

In this study, the LG/PEG and LG/PEG-Ag NPs were characterized by different techniques, such as SEM, FTIR, BET, DLS, and XRD. The morphologies of the samples were determined by a SEM (FEI QUANTA) (TedPella, double-coated, and 8 mmW × 20 mL) with a 6–10 mm working distance, 0–130 Pa pressure and voltage of 7–10 kV under low vacuum medium. The functional groups of the samples were determined by a FTIR (Perkin Elmer Spectrum Two) (KBr powder) in the 4000–400 cm^{-1} frequency range with a resolution of 4 cm^{-1} and eight scans. The specific surface area of the sample was determined by a BET (ASAP 2020 model). DLS measurements of the samples were performed using a Malvern Zetasizer Nano ZS® instrument with 12 runs at 25°C. The viscosities of the samples were performed using a AND viscometer to examine operating parameters, such as the sonication time, amplitude, salt, and temperature on the intrinsic viscosity of fabricated LG/PEG-Ag NPs. The WinCT-Viscosity software was used to aid the rheological calculations. Experimental viscosity measurements were carried out at constant temperature in special measuring vessels.

10.2.5 RHEOLOGICAL CHARACTERIZATION OF THE LG/PEG-AG NPS

The intrinsic viscosity ([n]) is a major factor to understand the mechanism of the polymer–solvent system due to the interactions of the molecules, the molecular structure, the size, the molecular weight, and the shape of the samples in dilute solution.[36–38] In this study, the rheological characterization of LG/PEG-Ag NPs was studied on the change in the experimentally measured intrinsic viscosity due to sonication time, amplitude, salt, and temperature. For this purpose, the focus of this study was on optimization studies using rheological measurements of LG/PEG-Ag NPs in a dilute solution.

The intrinsic viscosity ([η]) was calculated from the specific viscosity (η_{sp}) eq 10.1, using Huggins eq 10.2 as follows[39–40]:

Specific viscosity:

$$\eta_{sp} = \frac{t}{t_0} - 1 \tag{10.1}$$

$$\eta_{sp}/C = [\eta] + k_1 [\eta]^2 C \tag{10.2}$$

where, C: concentration, t0: the pure solvent flow time, t: the polymer solution flow time, and k_1: Huggins constant.

10.2.6 MATHEMATICAL MODELING OF INTRINSIC VISCOSITY

In this study, in order to obtain the intrinsic viscosity as a function of the operating variables: concentration (C), sonication time (t_s), amplitude of sonication (AMP), salt concentration (M_{NaOH}), and temperature (T), a mathematical model was constructed as follows:

Step 1: $\eta_c(C)$ is obtained by using linear regression on the intrinsic viscosity data corresponding to concentration, while the other variables kept constant at fixed operating parameters.

Step 2: Exponential regressions are applied to the intrinsic viscosity data corresponding to sonication time and amplitude of sonication, and linear regressions are applied to the intrinsic viscosity data corresponding to salt concentration and temperature.

Step 3: The latterly obtained functions are normalized with respect to the constant values mentioned before.

Step 4: All normalized functions and $\eta_c(C)$ are multiplied.

10.2.7 STATISTICAL ERROR ANALYSIS

In this study, different statistical error analysis methods were applied to identify the mathematical models, such as the hybrid fractional error function (HYBRID), the sum of squares of errors (SSE), and the average relative error (ARE). All expressions for the experimental rheological results of the nanostructures and these methods were shown in eqs 10.3–10.5, respectively.[41–44]

$$\frac{100}{n-p}\sum_{i=1}^{n}\left[\frac{\left(\eta_e-\eta_{cal}\right)^2_i}{\eta_e}\right] \tag{10.3}$$

$$\sum_{i=1}^{n}\left(\eta_{cal}-\eta_e\right)^2_i \tag{10.4}$$

$$\frac{100}{n-p}\sum_{i=1}^{n}\left|\frac{\eta_e-\eta_{cal}}{\eta_e}\right|_i \tag{10.5}$$

Where, n: number of data points, p: number of parameters, η_e: experimental viscosity, and η_{cal}: calculated viscosity.

10.3 RESULTS AND DISCUSSIONS

10.3.1 RHEOLOGICAL CALCULATIONS OF THE LG/PEG- AG NPS

Rheological experiments have a significant function in control during fabricating, design, and characterization of the nanostructures. According to this situation, [η] was calculated to determine the rheological behaviors of the uniform nano-formulations. By applying rheology measurements, effects of experimental conditions were compared, which have an important function in the design of LG/PEG-Ag NPs. Therefore, the effects of the experimental conditions, such as the sonication time, amplitude, salt, and temperature on the intrinsic viscosity of fabricated LG/PEG-Ag NPs were experimentally determined for optimizing the system and these results are shown in Figures 10.2–10.5, respectively.

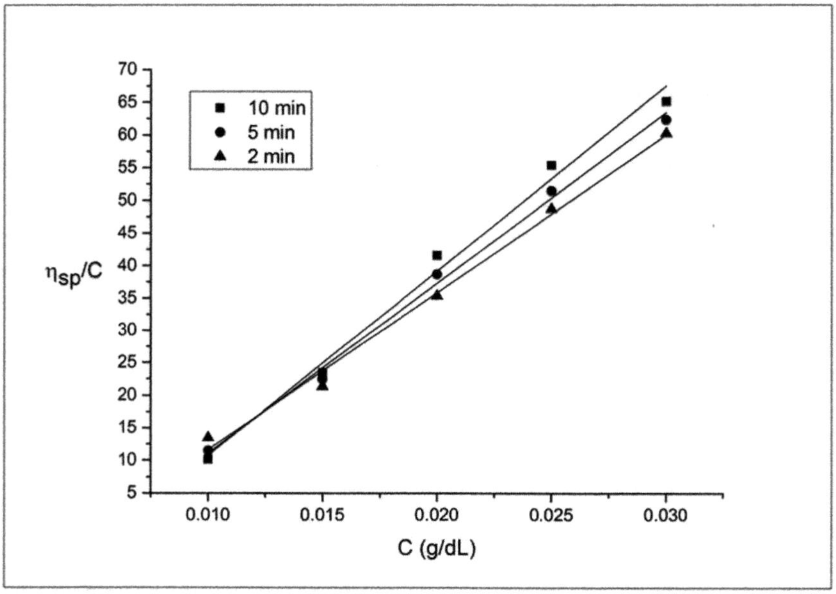

FIGURE 10.2 The effects of the sonication time (2–10 min) on [η] of the LG/PEG-Ag NPs.

The intrinsic viscosities of the LG/PEG-Ag NPs from 60.23 to 65.12 dL/g with changing the sonication time of the experimental process from 2 to 10 min were calculated and the results were given for the LG/PEG-Ag NPs in Figure 10.2. According to the results, it is found that the intrinsic viscosities of NPs changed with the sonication time and had an effective signal on the control of our process.

The intrinsic viscosities of the LG/PEG-Ag NPs from 75.43 to 65.12 dL/g with changing amount of the amplitude of LG/PEG-Ag NPs from 20 to 40% were calculated by Huggins equation (Fig. 10.3). According to the rheological experimental results, it is observed that the intrinsic viscosities of the LG/PEG-Ag NPs increased with the increase of the concentration of the nanostructure. It was observed that the intrinsic viscosities changed at different experimental conditions.[45]

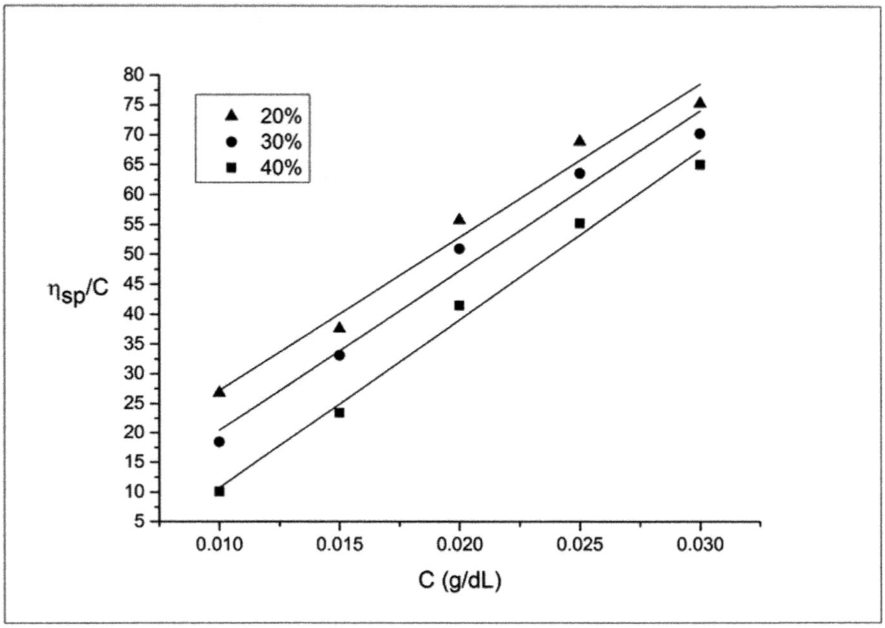

FIGURE 10.3 The effects of the amplitude (20–40%) on [η] of the LG/PEG-Ag NPs.

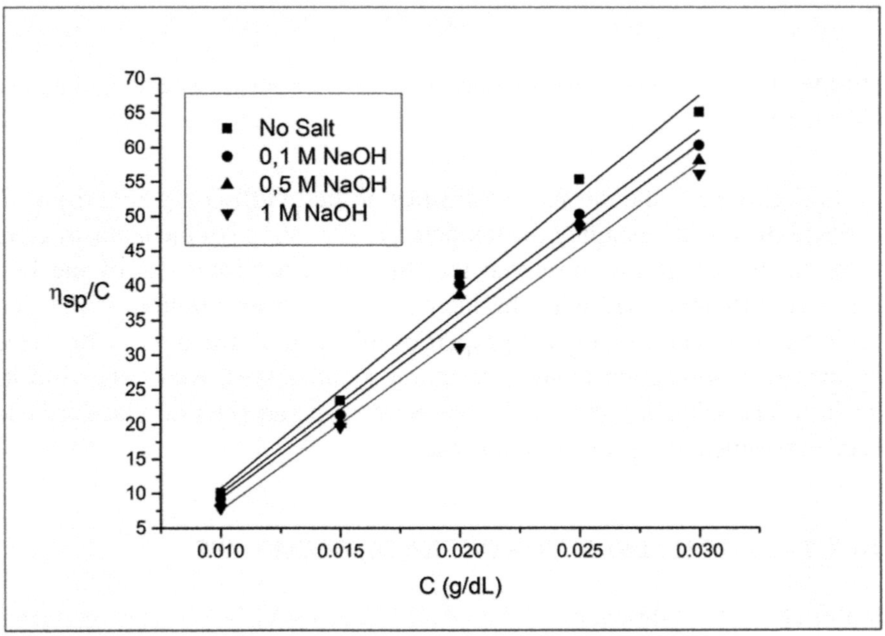

FIGURE 10.4 The effects of the salt on [η] of the LG/PEG-Ag NPs.

Figure 10.4 shows the changes in viscosity representing the liquid rheological property by adding different amounts of NaOH (0–1 M). The intrinsic viscosities of the LG/PEG-Ag NPs from 65.12 to 56.14 dL/g with changing amount of the NaOH of LG/PEG-Ag NPs from 0 to 1 M were calculated (Fig. 10.4).[46]

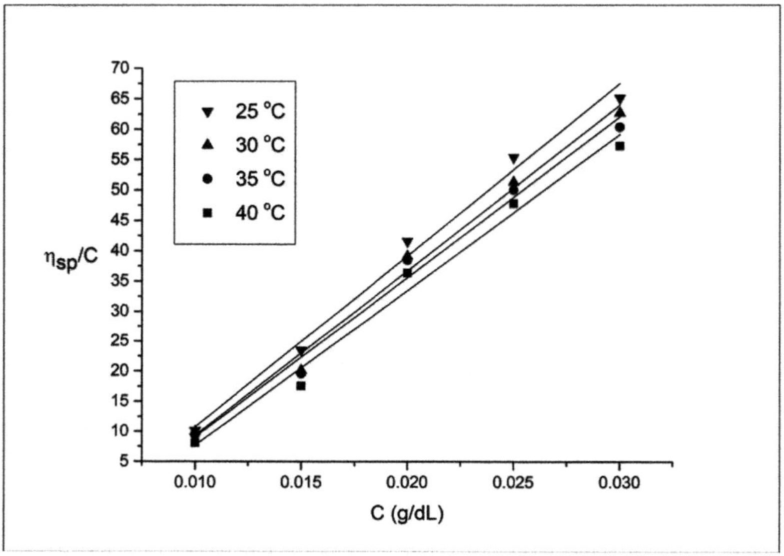

FIGURE 10.5 The effects of the temperature (25–40°C) on intrinsic viscosities of the LG/PEG-Ag NPs.

In Figure 10.5, the intrinsic viscosities of the LG/PEG-Ag NPs from 57. to 65.12 dL/g with changing the temperature (25–40°C) of the nanostructure were calculated. It was observed that the intrinsic viscosities of the LG/PEG-Ag NPs increased with the increase of the concentration of the LG/PEG-Ag NPs. The rheological properties of the nanostructure, synthesized by the green sonication method based on sound waves, were related to its intrinsic viscosity, and the effects of operating parameters on viscosity had been experimentally proven in this study.[47]

10.3.2 CHARACTERIZATION OF THE LG/PEG-AG NPS

In this study, the fabrication of the novel LG/PEG-Ag NPs was investigated under the sonic waves from Ag+ ions to Ag NPs using the green and facile ultrasonic method. The morphology of synthesized LG/PEG-Ag NPs was performed by SEM technique (Fig. 10.6). The high-resolution SEM image of the LG showed that the surface of LG was very flat and had a highly nonporous morphology (Fig. 10.6a). However, after the sonication process, LG/PEG blend mainly acted as reducing and stabilizing agents to prevent the aggregation of Ag NPs. According to the SEM results, it was found that the LG/PEG-Ag NPs exhibited a uniform spherical shape with a diameter of 320 nm (Fig. 10.6b). BET results of the LG/PEG-Ag NPs showed that the surface area of the nanostructure was 130.533 m^2/g.

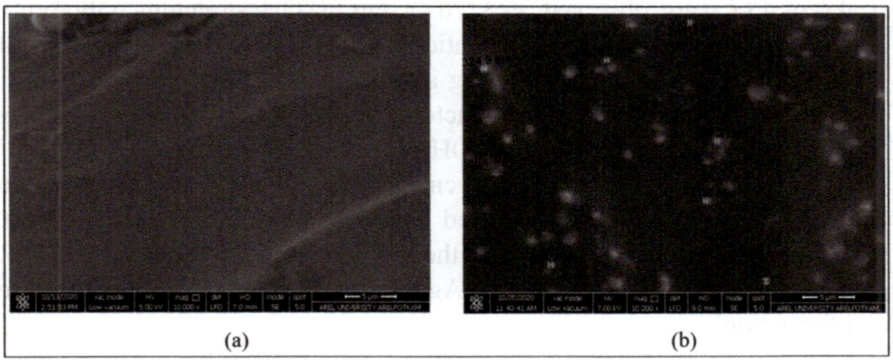

FIGURE 10.6 The SEM images of (a) LG and (b) LG/PEG-Ag NPs.

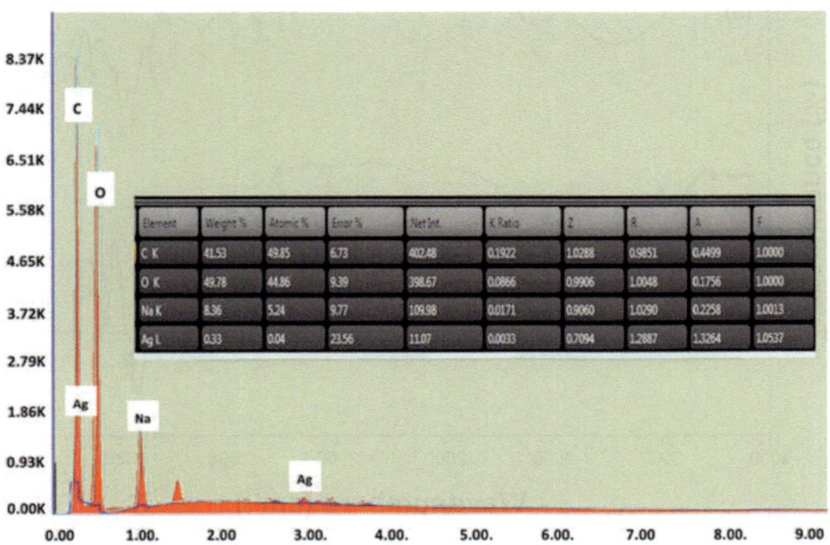

Element	Weight %	Atomic %	Error %	Net Int.	K Ratio	Z	R	A	F
C K	41.53	49.85	6.73	402.48	0.1922	1.0288	0.9851	0.4499	1.0000
O K	49.78	44.86	9.39	398.67	0.0866	0.9906	1.0048	0.1756	1.0000
Na K	8.36	5.24	9.77	109.98	0.0171	0.9060	1.0290	0.2258	1.0013
Ag L	0.33	0.04	23.56	11.07	0.0033	0.7094	1.2887	1.3264	1.0537

FIGURE 10.7 The SEM/EDX result of the LG/PEG-Ag NPs.

The elemental composition for the LG/PEG-Ag NPs was determined by SEM/EDX (Fig. 10.7) as carbon, oxygen, sodium, and silver. The SEM/EDX results of the LG/PEG-Ag NPs demonstrated that it proved that Ag^+ ions to Ag NPs were reduced by the green LG/PEG matrix. The morphological results and our assumptions were compatible with obtained experimental results.

The FTIR graphs of LG and LG/PEG-Ag NPs were used to investigate functional groups in the chemical structure and the interaction between Ag NPs and polymer matrix. FTIR results of (a) LG and (b) LG/PEG-Ag NPs were given in Figure 10.8. According to the FTIR results, the characteristic peaks of LG were observed at 3276.47 cm^{-1} (–OH stretching vibration), 2923.56 cm^{-1} (C-H stretching vibration), 1639.2 cm^{-1} (C-O stretching of ketone), 1375.96 cm^{-1} (C-H bonding in alkane), and 1011.48 cm^{-1} (C-O-C vibrations).[48] Furthermore, the characteristic peaks of the LG/PEG-Ag NPs were observed at 3393.14 cm^{-1} (–OH stretching vibration), 2870.52 cm^{-1} (C-H stretching vibration), 1571.70 cm^{-1} (C=O, carbonyl groups), 1348.96 cm^{-1} (C–H wagging vibrations), and 1085.73 cm^{-1} (amine groups N–H). Small peaks at 517.93 cm^{-1} due to the Ag NPs were also observed. FTIR data proved that the polymer matrix Ag NPs are involved in the stabilization and formation.

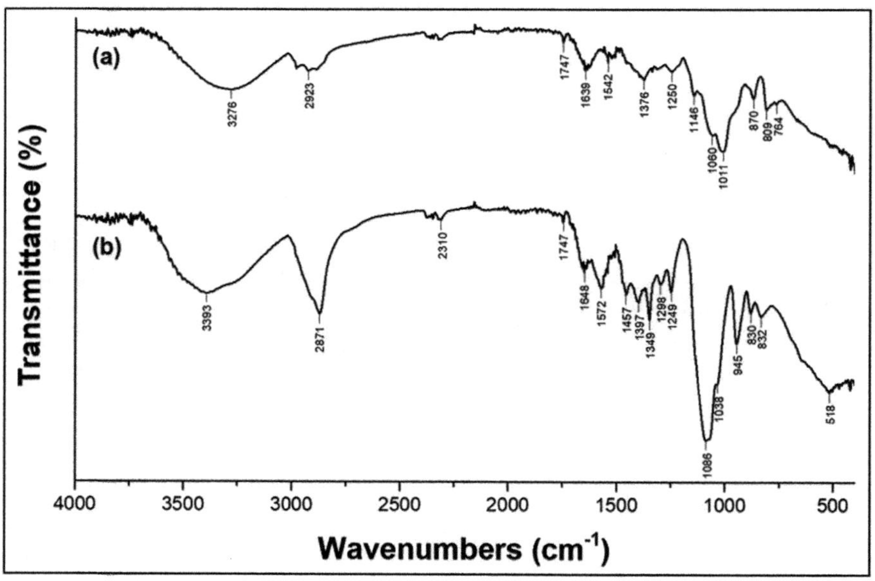

FIGURE 10.8 FTIR graphs of (a) LG and (b) LG/PEG-Ag NPs.

10.3.3 MATHEMATICAL MODELING OF INTRINSIC VISCOSITY OF LG/PEG-AG NPS

In this study, the intrinsic viscosity as a function of operating variables concentration (C), sonication time (t_s), amplitude of sonication (AMP), salt concentration (M_{NaOH}), and temperature (T) was obtained. First, a linear regression is applied to the intrinsic viscosity data corresponding to concentration (R^2=0.988) while the other variables are kept constant at values t_s=10 min, AMP=40%, M=0 (no salt), T=25°C. Thus, $\eta_C(C)$ is obtained as in eq 10.6, and its illustration is presented in Figure 10.2.

$$\eta_C(C) = {\eta_{sp}}/{C} = -17.638 + 2837.8\,C \qquad (10.6)$$

Then, exponential and linear regressions are applied to the remaining operating variables at fixed operating parameters, and the equations presented in Table 10.1 are obtained. $\eta_{ts}(t_s)$, $\eta_{AMP}(AMP)$, $\eta_M(M_{NaOH})$, and $\eta_T(T)$ are illustrated in Figure 10.9.

TABLE 10.1 Equations Via Experimental Fixed Operating Parameters of LG/PEG-Ag NPs.

Equations	Eq. #	Fixed Operating Parameters				
		C (g/dL)	t_s (min)	AMP	M_{NaOH} (M)	T (°C)
$\eta_{ts}(t_s) = {\eta_{sp}}/{C} = 43.95792 - 11.866\,e^{-t_s/6.2537}$	(10.7)	0.02	–	40%	0	25
$\eta_{AMP}(AMP) = {\eta_{sp}}/{C} = 60.934 - 1.3445\,e^{\frac{AMP}{14.993}}$	(10.8)	0.02	10	–	0	25
$\eta_M(M_{NaOH}) = {\eta_{sp}}/{C} = 41.859 - 9.9419\,M_{NaOH}$	(10.9)	0.02	10	40%	–	25
$\eta_T(T) = {\eta_{sp}}/{C} = 49.859 - 0.32680\,T$	(10.10)	0.02	10	40%	0	–

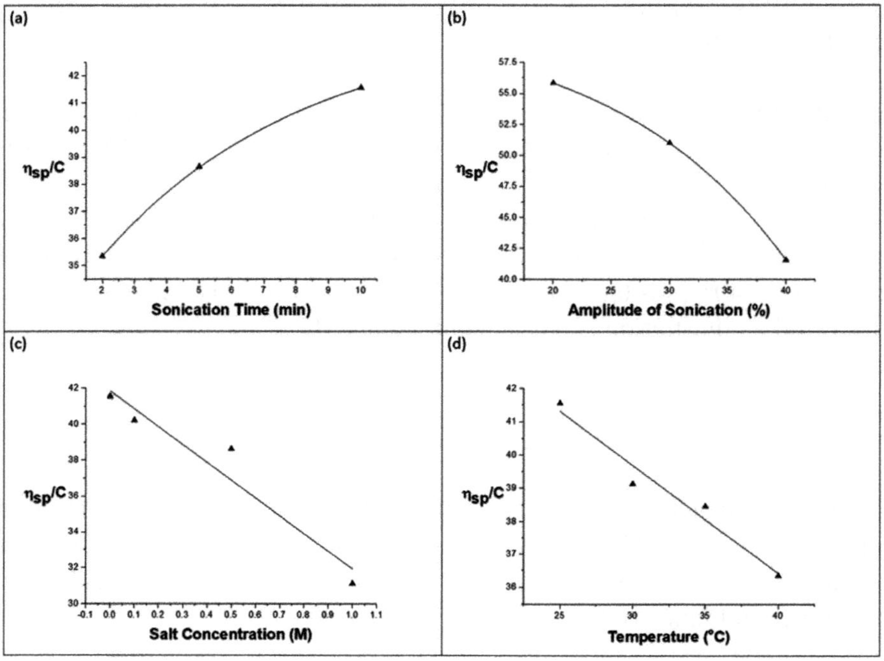

FIGURE 10.9 The effects of sonication time (a), amplitude of sonication (b), salt concentration (c), and temperature (d) on the intrinsic viscosity of LG/PEG-Ag NPs.

The functions $\eta_{ts}(t_s), \eta_{AMP}(AMP), \eta_M(M_{NaOH})$, and $\eta_T(T)$ are normalized as in eqs 10.11–10.14, respectively. The normalized functions yield 1 at fixed operating parameters t_s=10 min, AMP=40%, M=0, T=25°C.

$$\eta_{tsN}(t_s) = \frac{\eta_{ts}(t_s)}{\eta_{ts}(10)} = 1.0091 - 0.27241\, e^{-t_s/6.2537} \tag{10.11}$$

$$\eta_{AMPN}(AMP) = \frac{\eta_{AMP}(AMP)}{\eta_{AMP}(40)} = 1.4662 - 3.2351 \bullet 10^{-2}\, e^{\frac{AMP}{14.993}} \tag{10.12}$$

$$\eta_{MN}(M_{NaOH}) = \frac{\eta_M(M_{NaOH})}{\eta_M(0)} = 1.0000 - 0.23751\, m_{NaOH} \tag{10.13}$$

$$\eta_{TN}(T) = \frac{\eta_T(T)}{\eta_T(25)} = 1.1960 - 7.8390 \cdot 10^{-3}\, T \tag{10.14}$$

Finally, all normalized functions and $\gamma_C(C)$ are multiplied, thus the intrinsic viscosity is obtained as a function of all operating variables (eq 10.15).

$$\eta_{spC}\left(C, t_s, AMP, T, m_{NaOH}\right) = \eta_{sp}/C \\ = \eta_{tsN}\left(t_s\right)\ \eta_{AMPN}\left(AMP\right)\ \eta_{TN}\left(T\right)\ \eta_{MN}\left(M_{NaOH}\right)\ \eta_C\left(C\right) \tag{10.15}$$

10.3.4 STATISTICAL ERROR ANALYSIS

In this study, the error analysis results of the LG/PEG-Ag NPs were compared with different statistical error methods, such as SSE, HYBRID, and ARE due to the rheological measurements. The comparison of the statistical error analysis results of the synthesized LG/PEG-Ag NPs were given in Table 1.2. The results showed the accuracy of rheological measurements with the minimum error results using eqs. 10.3–10.5.

TABLE 10.2 Statistical Error Analysis Results for the Synthesized LG/PEG-Ag NPs.

Methods	Results
HYBRID	53.421
SSE	1011.6
ARE	10.327

It can also be seen from Figure 10.10 that eq. 10.15 provides a sufficient description for the determination of intrinsic viscosity of LG/PEG-Ag NPs.

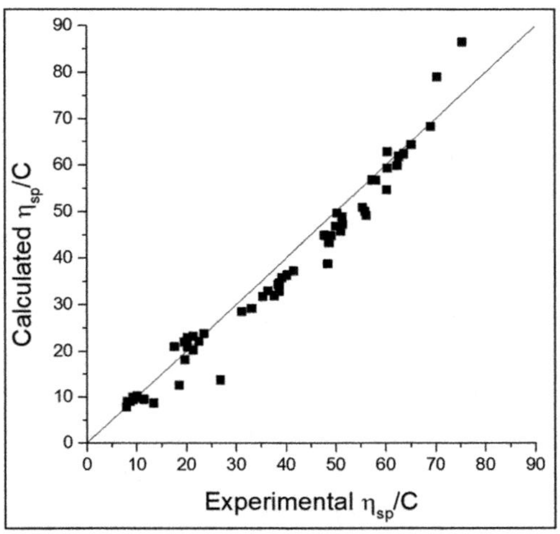

FIGURE 10.10 Comparison of experimental values with calculated values.

According to our experimental results, the estimated formation of the LG/PEG-Ag NPs mechanism was described in eqs. 10.16–10.19.

$$H_2O \rightarrow {}^*OH + {}^*H \quad (10.16)$$

$${}^*H + {}^*H \rightarrow H_2 \quad (10.17)$$

$${}^*OH + {}^*OH \leftrightarrow H_2O_2 \quad (10.18)$$

$$LG/PEG\text{-}Ag^+ + H_2O_2 \leftrightarrow LG/PEG\text{-}Ag^o + H^+ + HOO^* \quad (10.19)$$

The comparison of our results of the synthesized Ag NPs with the results of previous studies in the literature was represented in Table 10.3.

TABLE 10.3 The Comparison of Preparation Methods and Morphological Properties of the Various Green Ag NPs.

Sample	Method	Morphological properties	Refs.
PVP-coated Ag NPs	Ultrasonication	70 nm	[49]
Ag NPs	Solution method	Spherical, 50 ± 20 nm	[50]
Plant leaf extracts-Ag NPs	Phytosynthesis	15–500 nm	[51]
Fungus Aspergillus flavus-Ag NPs	Biological synthesis	8.92 ± 1.61 nm	[52]
Ag NPs	Biological synthesis	18.2 ± 3.6 nm	[53]
LG/PEG-Ag NPs	Ultrasonication	Spherical, 320 nm	This study

10.4 CONCLUSIONS

The novel LG/PEG-Ag NPs were successfully prepared by the green ultrasonic method and the synthesized LG/PEG-Ag NPs displayed rheological sensitivity to the sonication time, amplitude, salt, and temperature due the intrinsic viscosity. These effective experimental factors were calculated to determine the rheological behavior of the Ag NPs using Huggins model. The morphological results showed that the LG/PEG-Ag NPs had the uniform spherical shaped structure with large surface area. Also, a mathematical model is obtained for the determination of intrinsic viscosity of LG/PEG-Ag NPs considering the operating variables. Error analysis is performed, and the experimental values are compared with the calculated values. Thus, the obtained model is a valid description for the intrinsic viscosity.

KEYWORDS

- **biopolymer**
- **intrinsic viscosity**
- **silver nanoparticle**
- **mathematical model**
- **rheological parameter**
- **optimization**
- **nanostructure**
- **locust gum**

REFERENCES

1. Tarafdar, J. C.; Sharma, S.; Raliya, R. Nanotechnology: Interdisciplinary Science of Applications. *Afr. J. Biotechnol.* **2013,** *12* (3).
2. Onyia, A. I.; Ikeri, H. I.; Chima, A. I. Surface and Quantum Effects in Nanosized Semiconductor. *Am. J. Nano Res. App.* **2020,** *8* (3), 35–41.
3. Shende, P.; Wakade, V. S. Biointerface: A Nano-Modulated Way for Biological Transportation. *J. Drug Target* **2020,** *28* (5), 456–467.
4. Hamad, H. T.; Al-Sharify, Z. T.; Al-Najjar, S. Z.; Gadooa, Z. A. A Review on Nanotechnology and Its Applications on Fluid Flow in Agriculture and Water Recourses. *IOP Conf. Ser. Mater. Sci. Eng.* June **2020,** *870* (1), *012038*.
5. Shaheen, N.; Yousuf, M. A.; Shakir, I.; Zulfiqar, S.; Agboola, P. O.; Warsi, M. F. Wet Chemical Route Synthesis of Spinel Oxide Nano-Catalysts for Photocatalytic Applications. *Physica B Condensed Matter* **2020,** *580*, 411820.
6. Xue, Y.; Chen, S.; Yu, J.; Bunes, B. R.; Xue, Z.; Xu, J.; Lu, B.; Zang, L. Nanostructured Conducting Polymers and Their Composites: Synthesis Methodologies, Morphologies and Applications. *J. Mater. Chemistry C* **2020,** *8* (30), 10136–10159.
7. Kamal, N.; Zaki, A. H.; El-Shahawy, A. A.; Sayed, O. M.; El-Dek, S. I. Changing the Morphology of One-Dimensional Titanate Nanostructures Affects Its Tissue Distribution and Toxicity. *Toxicol. Ind. Health* **2020,** *36* (4), 272–286.
8. Seifi, H.; Gholami, T.; Seifi, S.; Ghoreishi, S. M.; Salavati-Niasari, M. A Review on Current Trends in Thermal Analysis and Hyphenated Techniques in the Investigation of Physical, Mechanical and Chemical Properties of Nanomaterials. *J. Analyt. Appl. Pyrolysis* **2020,** 104840.
9. Saleh, T. A. Nanomaterials: Classification, Properties, and Environmental Toxicities. Environ. Technol. Innov. **2020,** 101067.
10. Khan, F. A. Nanomaterials: Types, Classifications, and Sources. In *Applications of Nanomaterials in Human Health*; Springer: Singapore, 2020; pp 1–13.
11. Radhapyari, K.; Datta, S.; Dutta, S.; Jadon, N.; Khan, R. Graphene-Based Nanostructures for Biomedical Applications. In *Two-Dimensional Nanostructures for Biomedical Technology*; Elsevier, 2020; pp 101–135.

12. Thangadurai, T. D.; Manjubaashini, N.; Thomas, S.; Maria, H. J. Fundamentals of Nanostructures. In *Nanostructured Materials*; Springer: Cham, 2020; pp 29–45.
13. Gatoo, M. A.; Naseem, S.; Arfat, M. Y.; Mahmood Dar, A.; Qasim, K.; Zubair, S. Physicochemical Properties of Nanomaterials: Implication in Associated Toxic Manifestations. *BioMed Res. Int.* **2014,** 2014.
14. Siddiqui, H.; Parra, M. R.; Pandey, P.; Qureshi, M. S.; Haque, F. Z. Performance Evaluation of Optimized Leaf-Shaped Two-Dimension (2D) Potassium Doped CuO Nanostructures with Enhanced Structural, Optical and Electronic Properties. *Ceram. Int.* 2020.
15. Khan, I.; Saeed, K.; Khan, I. Nanoparticles: Properties, Applications and Toxicities. *Arab. J. Chem.* **2019,** *12* (7), 908–931.
16. Din, M. I.; Rehan, R. Synthesis, Characterization, and Applications of Copper Nanoparticles. *Analyt. Lett.* **2017,** *50* (1), 50–62.
17. Cometa, S.; Iatta, R.; Ricci, M. A.; Ferretti, C.; De Giglio, E. Analytical Characterization and Antimicrobial Properties of Novel Copper Nanoparticle–Loaded Electrosynthesized Hydrogel Coatings. *J. Bioactive Compatible Polym.* **2013,** *28* (5), 508–522.
18. Albalawi, F.; Hussein, M. Z.; Fakurazi, S.; Masarudin, M. J. Engineered Nanomaterials: The Challenges and Opportunities for Nanomedicines. *Int. J. Nanomed.* **2021,** *16,* 161–184.
19. Nahar, K. N.; Rahaman, M.; Khan, G. M.; Islam, M.; Al-Reza, S. M. Green Synthesis of Silver Nanoparticles from Citrus Sinensis Peel Extract and Its Antibacterial Potential. *Asian J. Green Chem.* **2021,** *5* (1), 135–150.
20. Zhang, H.; Chen, S.; Jia, X.; Huang, Y.; Ji, R.; Zhao, L. Comparation of the Phytotoxicity Between Chemically and Green Synthesized Silver Nanoparticles. *Sci. Total Environ.* **2021,** *752*, 142264.
21. Roy, S.; Kim, H. C.; Kim, J. W.; Zhai, L.; Zhu, Q. Y.; Kim, J. Incorporation of Melanin Nanoparticles Improves UV-Shielding, Mechanical and Antioxidant Properties of Cellulose Nanofiber Based Nanocomposite Films. *Mater. Today Commun.* **2020,** *24,* 100984.
22. Iravani, S.; Korbekandi, H.; Mirmohammadi, S. V.; Zolfaghari, B. Synthesis of Silver Nanoparticles: Chemical, Physical and Biological Methods. *Res. Pharma. Sci.* **2014,** *9* (6), 385.
23. Castro-Mayorga, J. L.; Randazzo, W.; Fabra, M. J.; Lagaron, J. M.; Aznar, R.; Sánchez, G. Antiviral Properties of Silver Nanoparticles Against Norovirus Surrogates and Their Efficacy in Coated Polyhydroxyalkanoates Systems. *LWT-Food Sci. Technol.* **2017,** *79,* 503–510.
24. Youssef, A. M.; El-Sayed, S. M. Bionanocomposites Materials for Food Packaging Applications: Concepts and Future Outlook. *Carbohyd. Polym.* **2018,** *193*, 19–27.
25. Roy, S.; Rhim, J. W. Effect of CuS Reinforcement on the Mechanical, Water Vapor Barrier, UV-Light Barrier, and Antibacterial Properties of Alginate-Based Composite Films. *Int. J. Biol. Macromol.* **2020,** *164*, 37–44.
26. Pryshchepa, O.; Pomastowski, P.; Buszewski, B. Silver Nanoparticles: Synthesis, Investigation Techniques, and Properties. *Adv. Colloid Interface Sci.* **2020,** 102246.
27. Owaid, M. N. Green Synthesis of Silver Nanoparticles by Pleurotus (Oyster Mushroom) and Their Bioactivity. *Environ. Nanotechnol. Monitor. Manage.* **2019,** *12*, 100256.
28. Bhattarai, B.; Zaker, Y.; Bigioni, T. P. Green Synthesis of Gold and Silver Nanoparticles: Challenges and Opportunities. *Curr. Opin. Green Sustain. Chem.* **2018,** *12*, 91–100.

29. Beyene, H. D.; Werkneh, A. A.; Bezabh, H. K.; Ambaye, T. G. Synthesis Paradigm and Applications of Silver Nanoparticles (AgNPs), a Review. *Sustain. Mater. Technol.* **2017,** *13*, 18–23.
30. Khodashenas, B.; Ghorbani, H. R. Synthesis of Silver Nanoparticles with Different Shapes. *Arab. J. Chem.* **2019,** *12* (8), 1823–1838.
31. Konwarh, R.; Karak, N.; Sawian, C. E.; Baruah, S.; Mandal, M. Effect of Sonication and Aging on the Templating Attribute of Starch for "Green" Silver Nanoparticles and Their Interactions at Bio-Interface. *Carbohyd. Polym.* **2011,** *83* (3), 1245–1252.
32. Lei, Z.; Zhang, L.; Wei, X. One-Step Synthesis of Silver Nanoparticles by Sonication or Heating Using Amphiphilic Block Copolymer as Templates. *J. Colloid Interface Sci.* **2008,** *324* (1–2), 216–219.
33. Munir, H.; Mumtaz, A.; Rashid, R.; Najeeb, J.; Zubair, M. T.; Munir, S.; Bilal, M.; Cheng, H. Eucalyptus Camaldulensis Gum as a Green Matrix to Fabrication of Zinc and Silver Nanoparticles: Characterization and Novel Prospects as Antimicrobial and Dye-Degrading Agents. *J. Mater. Res. Technol.* **2020,** *9* (6), 15513–15524.
34. Talodthaisong, C.; Boonta, W.; Thammawithan, S.; Patramanon, R.; Kamonsutthipaijit, N.; Hutchison, J. A.; Kulchat, S. Composite Guar Gum-Silver Nanoparticle Hydrogels as Self-Healing, Injectable, and Antibacterial Biomaterials. *Mater. Today Commun.* **2020,** *24*, 100992.
35. Zhai, X.; Li, Z.; Shi, J.; Huang, X.; Sun, Z.; Zhang, D.; Zou, x.; Sun, Y.; Zhang, J.; Holmes, M.; Povey, M.; Wang, S. A Colorimetric Hydrogen Sulfide Sensor Based on Gellan Gum-Silver Nanoparticles Bionanocomposite for Monitoring of Meat Spoilage in Intelligent Packaging. *Food Chem.* **2019,** *290*, 135–143.
36. Pyun, C. W.; Fixman, M. Intrinsic Viscosity of Polymer Chains. *J. Chem. Phys.* **1965,** *42* (11), 3838–3844.
37. Krigbaum, W. R.; Flory, P. J. Molecular Weight Dependence of the Intrinsic Viscosity of Polymer Solutions II. *J. Polym. Sci.* **1953,** *11* (1), 37–51.
38. Debye, P. The Intrinsic Viscosity of Polymer Solutions. *J. Chem. Phys.* **1946,** *14* (10), 636–639.
39. Huggins, M. L. The Viscosity of Dilute Solutions of Long-Chain Molecules IV: Dependence on Concentration. *J. Am. Chem. Soc.* **1942,** *64* (11), 2716–2718.
40. Ma, X.; Pawlik, M. Intrinsic Viscosities and Huggins Constants of Guar Gum in Alkali Metal Chloride Solutions. *Carbohyd. Polym.* **2007,** *70* (1), 15–24.
41. Kausar, A.; Shahzad, R.; Iqbal, J.; Muhammad, N.; Ibrahim, S. M.; Iqbal, M. Development of New Organic-Inorganic, Hybrid Bionanocomposite from Cellulose and Clay for Enhanced Removal of Drimarine Yellow HF-3GL Dye. *Int. J. Biol. Macromol.* **2020,** *149*, 1059–1071.
42. Alkurdi, S. S.; Al-Juboori, R. A.; Bundschuh, J.; Bowtell, L.; Marchuk, A. Inorganic Arsenic Species Removal from Water Using Bone Char: A Detailed Study on Adsorption Kinetic and Isotherm Models Using Error Functions Analysis. *J. Hazard. Mater.* **2020,** 124112.
43. Kabuba, J.; Banza, M. Ion-Exchange Process for the Removal of Ni (II) and Co (II) from Wastewater Using Modified Clinoptilolite: Modeling by Response Surface Methodology and Artificial Neural Network. *Results Eng.* **2020,** *8*, 100189.
44. Alrobei, H.; Prashanth, M. K.; Manjunatha, C. R.; Kumar, C. P.; Chitrabanu, C. P.; Shivaramu, P. D. et al. Adsorption of Anionic Dye on Eco-Friendly Synthesised

Reduced Graphene Oxide Anchored with Lanthanum Aluminate: Isotherms, Kinetics and Statistical Error Analysis. *Ceram. Int.* **2020.**

45. Kasaai, M. R.; Arul, J.; Charlet, G. Intrinsic Viscosity–Molecular Weight Relationship for Chitosan. *J. Polym. Sci. B Polym. Phys.* **2000,** *38* (19), 2591–2598.
46. Brunchi, C. E.; Morariu, S.; Bercea, M. Intrinsic Viscosity and Conformational Parameters of Xanthan in Aqueous Solutions: Salt Addition Effect. *Colloids Surfaces B Biointerfaces* **2014,** *122*, 512–519.
47. Majzoobi, M.; Seifzadeh, N.; Farahnaky, A.; Mesbahi, G. Effects of Sonication on Physical Properties of Native and Cross-Linked Wheat Starches. *J. Texture Stud.* **2015,** *46* (2), 105–112.
48. Bello, O. S.; Adegoke, K. A.; Sarumi, O. O.; Lameed, O. S. Functionalized Locust Bean Pod (Parkia biglobosa) Activated Carbon for Rhodamine B Dye Removal. *Heliyon* **2019,** *5* (8), e02323.
49. Loza, K.; Diendorf, J.; Sengstock, C.; Ruiz-Gonzalez, L.; Gonzalez-Calbet, J. M.; Vallet-Regi, M.; et al. The Dissolution and Biological Effects of Silver Nanoparticles in Biological Media. *J. Mater. Chem. B* **2014,** *2* (12), 1634–1643.
50. Kittler, S.; Greulich, C.; Gebauer, J. S.; Diendorf, J.; Treuel, L.; Ruiz, L.; Calbet, J.M.G.; Regi, M.V.; Köller, M. & Epple, M. The Influence of Proteins on the Dispersability and Cell-Biological Activity of Silver Nanoparticles. *J. Mater. Chem.* **2010,** *20* (3), 512–518.
51. Song, J. Y.; Kim, B. S. Rapid Biological Synthesis of Silver Nanoparticles Using Plant Leaf Extracts. *Bioprocess Biosyst. Eng.* **2009,** *32* (1), 79–84.
52. Vigneshwaran, N.; Ashtaputre, N. M.; Varadarajan, P. V.; Nachane, R. P.; Paralikar, K. M.; Balasubramanya, R. H. Biological Synthesis of Silver Nanoparticles Using the Fungus *Aspergillus flavus*. *Mater. Lett.* **2007,** *61* (6), 1413–1418.
53. Aygün, A.; Gülbağça, F.; Nas, M. S.; Alma, M. H.; Çalımlı, M. H.; Ustaoglu, B.; Altunoğlu, Y.Ç.; Baloğlu, M.C.; Cellat, K.; Şen, F. Biological Synthesis of Silver Nanoparticles Using Rheum Ribes and Evaluation of Their Anticarcinogenic and Antimicrobial Potential: A Novel Approach in Phytonanotechnology. *J. Pharma. Biomed. Analy.* **2020,** *179*, 113012.

CHAPTER 11

Macromolecular-Enzyme-Triggered Electrochemical Biosensing

S. K. SUJA and G. JAYANTHI KALAIVANI

Department of Chemistry, Lady Doak College, Madurai, Tamil Nadu, India

ABSTRACT

Enzymes are complex biological macromolecules that play a vital role in the speeding up of several biological reactions. Without enzymes, the essential biological reactions will take some billion years, and the existence of life will be questioned. The knowledge about different enzymes is therefore very much required. Enzymes fall under different classifications viz., oxidoreductases, hydrolases, transferases, lyases, ligases, etc. These enzymes trigger the catalytic conversion of substrate molecules into the products. They tend to exhibit high specificity, for example, glucose oxidase is highly specific toward β-D glucose and non-specific toward other monosaccharides, pyranose oxidase for catalyzing oxidation of molecules with pyranose ring structure, alcohol oxidase for alcohols, choline oxidase for choline, etc. This kind of specific action finds significance in clinical analysis and biosensing devices that operate by measuring a particular substrate molecule in biological samples (blood or urine). This chapter highlights the different classification of enzymes, structure, and function of the enzymes, mechanism of action, and their utility in the electrochemical biosensing studies.

Foundation and Growth of Macromolecular Science: Advances in Research for Sustainable Development. Meegle S. Mathew, PhD, Józef T. Haponiuk, PhD & Sabu Thomas, PhD, DSc (Eds.)

11.1 INTRODUCTION

The origin of enzymes dates back to 1833 when the first enzyme diastase was discovered and was followed by the discovery of other hydrolytic enzymes viz., invertase and pepsin.[1,2] Frederick Wilhelm Kühne, a German physiologist, was the first to coin the word enzyme in the year 1877, while demonstrating the production of alcohols from sugars by yeast.[3] Enzymes are a kind of proteins that are responsible for catalyzing several important biochemical reactions that support life. They are biocatalysts that are required in small quantities to perform various biochemical reactions in biological species. They exhibit a significant role in the electrochemical biosensing of clinically relevant biomolecules. They are huge macromolecules that are composed of polymers made of amino acids, interconnected via amide bonds, whose molecular weight vary from kilodaltons as in the case of insulin to megadaltons as in the case of ribosomes.[4] They consist of more than one active site, in which the substrate molecules bind thereby catalyzing a specific reaction. An active site consists of a few amino acid residues, and the remaining part of the enzyme comprises of protein, which offers stability to the 3D network of the enzyme. The active region present in the enzyme macromolecules are often deeply buried inside the hydrophobic pockets that determines the specificity of their respective substrate. Each and every enzyme bonds with a substrate of stereochemical specificity, and it catalyzes one conformer of the substrate and not the other.

Enzymes trigger multitude of reactions and find enormous applications in industrial production and electrochemical sensing. Enzyme-based electrochemical biosensing is a novel technique applied in several fields ranging from environment, food industries, and biomedical fields. Enzyme-based biosensors are used to measure analytes in biological fluids for clinical diagnostics that help in early diagnosis of diseases. They also serve as a useful technique to know the levels of important analytes such as glucose, alcohol, etc., in food samples. Several enzymes demonstrate pivotal roles in the sensitive and selective analysis of specific analytes present in biological fluids, food samples, etc. The factors that affect the performance of enzymes in electrochemical biosensors include temperature, pH, choice of buffer solution, enzyme loading, choice of the support matrix for enzyme immobilization, etc. The characteristic features of the electrode system used in the biosensor are sensitivity, selectivity, storage stability, reproducibility, and reusability. The enzyme needs to be suitably incorporated in the electrode thereby satisfying the basic requirements of the biosensors. The storage stability of enzymes tends to vary depending on the storage conditions (room temperature, refrigeration) and also the immobilizing matrix. Nanomaterials can serve as stable support for the immobilization of enzyme, which makes the enzyme retain the original conformation of the enzyme thereby preventing its denaturation, improving the storage stability, and hence satisfying all the requirements for the design of a stable electrode system for biosensing applications.[5]

This chapter aims at providing an outline of the different enzymes, their classifications, oxidoreductases, GMC oxidases, and their structure, function, mechanism of action, and their applications in the field of electrochemical biosensing.

11.2 CLASSIFICATION OF ENZYMES

Basically, there are six different classifications of enzymes as prepared in 1964 by *International Union of Biochemistry (IUB),* which is now referred as *International Union of Biochemistry and Molecular Biology (IUBMB).* They include oxidoreductases, transferases, hydrolases, lyases, isomerases, and ligases. A new major classification of enzyme has been included in the list by IUBMB which is translocases.[6] The description of the various enzymes is given in Table 11.1.

TABLE 11.1 Description of the Various Types of Enzymes.

Types of enzymes	General reaction catalyzed	Enzyme commission number (EC No.)
Oxidoreductases	Oxidation-reduction reactions $\mathbf{X_{ox} + Y_{red}}$ $\mathbf{X_{red} + Y_{ox}}$	EC 1
Transferases	Transfer of atoms or groups **X-Y+ A** **X + Y-A**	EC 2
Hydrolases	Hydrolysis $\mathbf{X\text{-}Y + H_2O}$ **X-H + Y-OH**	EC 3
Lyases	Removal of groups **X=Y + A-B** **X(A)-Y(B)**	EC 4
Isomerases	Isomerization **X-Y-Z** **X-Z-Y**	EC 5
Ligases	Joining of two molecules to break pyrophosphate bond **X + Y + ATP** **X-Y + ADP + Pi**	EC 6

TABLE 11.1 *(Continued)*

Types of enzymes	General reaction catalyzed	Enzyme commission number (EC No.)
Translocases	Movement of ions or molecules across membranes **X-Y + Z || X + Z + || Y** **(Side 1) (Side 2)**	EC 7

Among the various classes of enzymes, oxidoreductases are utilized in the redox reactions in the biological organisms. They find extensive applications in disease diagnosis, prognosis, and treatment.[7]

11.2.1 CLASSIFICATION OF OXIDOREDUCTASES

Oxidoreductases are those classifications of enzymes, which are involved in catalyzing oxidation-reduction processes. They promote the transference of electrons from oxidant species to reductant species. They contain different cofactors viz., heme, flavin, and metal ions in order to promote the redox processes.[8] Heme-containing peroxidases and peroxygenases are activated by H_2O_2 (electron acceptor); flavin-containing oxidases, and dehydrogenases are activated by O_2 and other oxidants, such as Fe^{3+} and quinones, respectively, while copper-containing oxidases and mono-oxygenases are activated by O_2 by a more complicated mechanism of activation.[9]

Oxidoreductases can be broadly categorized as oxidases and dehydrogenases. Oxidases are those that utilize O_2 molecule as the hydrogen or electron acceptor, while dehydrogenases oxidize substrate molecules by the transfer of hydrogen to an electron-accepting species, which can be either NAD^+/$NADP^+$ or flavin moiety. The remaining varieties of oxidoreductases include peroxidases, hydroxylases, oxygenases, and reductases.[10] Peroxidase catalyzes the reduction of hydrogen peroxide (H_2O_2), hydroxylase adds hydroxyl groups to the substrates, oxygenase adds oxygen into organic substrates, and reductase catalyzes reductions (Fig. 11.1).[11]

Among the various oxidoreductases, the carbohydrate oxidases play a significant role in day today life and they find extensive biological applications.

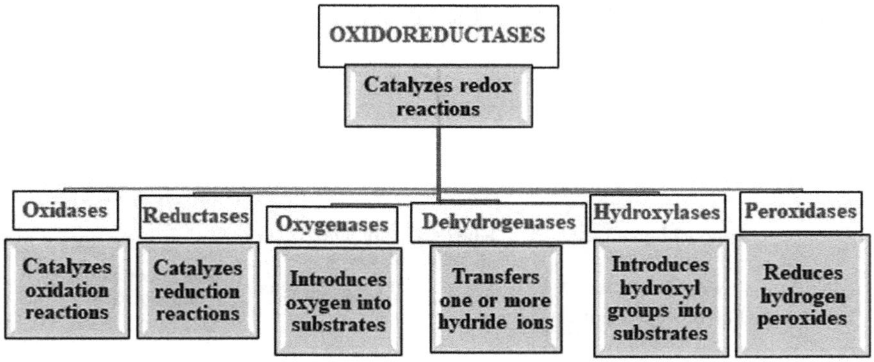

FIGURE 11.1 Classification and function of oxidoreductases.

11.2.2 CLASSIFICATION OF CARBOHYDRATE OXIDASES

Carbohydrate oxidase is used to describe an enzyme that acts on the same class of carbohydrates and use molecular oxygen for the oxidation process.[12] There is an increased number of studies on this classification of enzyme in the last few decades, and the same is shown in Figure 11.2. The cofactor present in these enzymes varies in order to perform specific oxidation reactions. They possess either copper or flavin as the redox cofactor. Copper-dependent oxidases[13], for example, galactose oxidase uses one copper ion that is bound to the protein core, and the surrounding amino acid residues hold the ion in its place. The copper-containing carbohydrate oxidases share a common structure. The flavin-dependent carbohydrate oxidases consist of covalently bonded flavin adenine dinucleotide (FAD) cofactor and are classified as glucose-methanol-choline oxidase (GMC) family[14] and the vanillyl-alcohol oxidase (VAO) family[15] based on their structures. The GMC oxidases superfamily include aryl-alcohol oxidase, methanol oxidase, glucose oxidase, and pyranose oxidase, while VAO family includes gluco-oligosaccharide oxidase,[16] chito-olgosaccharide oxidase,[17] and xylo oligo-saccharide oxidase.[18] They are distinguished on the basis of the presence of covalent bonds and number of covalent bonds between FAD and the protein portion of the enzyme.[19] The members of the VAO family possess a common structure, and is composed of a conserved FAD-binding domain which binds the adenine part of the FAD and a variable cap domain, which covers the isoalloxazine ring of the cofactor thereby forming a substrate binding pocket.[20] The majority of VAO-type oxidases involve in the oxidation of alcohols (EC 1.1.3.x) leading to the formation of aldehydes, ketones, or lactones.[21]

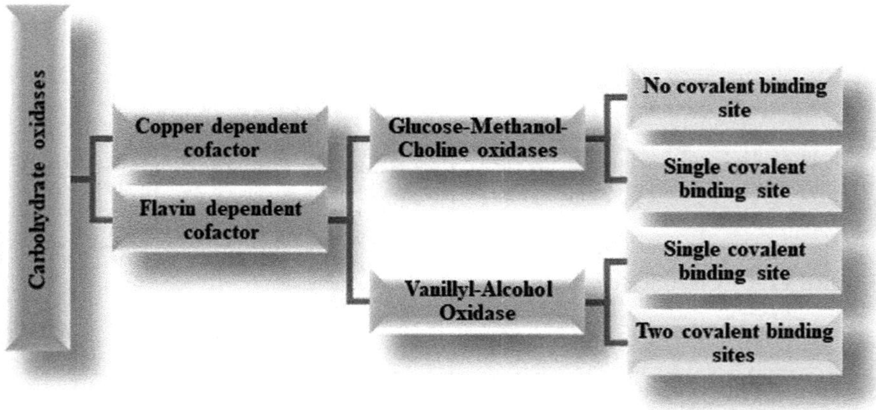

FIGURE 11.2 Classification of carbohydrate oxidases.

11.2.3 CLASSIFICATION OF GMC OXIDASES

In 1992, Cavener identified the GMC oxidases family based on the resemblances in the sequence of glucose dehydrogenase from *Drosophila melanogaster*, choline dehydrogenase from *Escherichia coli*, glucose oxidase from *Aspergillus niger*, and methanol (alcohol) oxidase from *Hansenula polymorpha.*[22] All of them have the same fold and contain covalently or noncovalently bound FAD cofactor. The associates of GMC oxidases contain FAD-binding and substrate-binding domains.

The GMC members include flavin-containing (1) oxidases, such as aryl-alcohol oxidase (AAOx, EC 1.1.3.7), methanol oxidase (MOx, also known as alcohol oxidase, EC 1.1.3.13), pyranose 2-oxidase (POx, EC 1.1.3.10), glucose oxidase (GOx, EC1.1.3.7), and choline oxidase (EC 1.1.3.17), and [23] (2) dehydrogenases, such as cellobiose dehydrogenase (CDH, EC 1.1.99.18) that contain flavin and heme separate domains, and glucose dehydrogenase (GDH, 1.1.99.35), pyranose dehydrogenase (PDH, EC 1.1.99.29) (Fig. 11.3). The members of this superfamily tend to gain two electrons from alcohol such that the reduced flavin passes to O_2 forming H_2O_2 (oxidases) or to other oxidizing substrates (dehydrogenases). The GMC family of enzymes are found in bacteria, yeasts, fungi, insects,[22,24] and also in plants.[25,26] Among the GMC oxidases, those that are derived from fungal sources are more attractive as they find applications in the field of electrochemical biosensing and food analysis.[27,28]

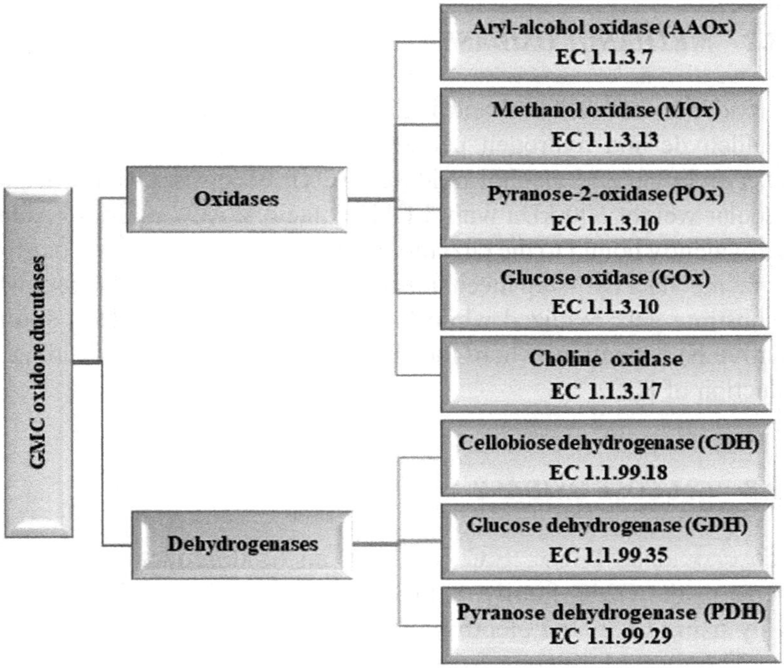

FIGURE 11.3 Classification of GMC oxidoreductases.

11.3 STRUCTURE AND FUNCTION OF GMC-BASED FLAVIN-DEPENDENT CARBOHYDRATE OXIDASES

11.3.1 ARYL-ALCOHOL OXIDASE

Aryl-alcohol oxidase (AAOx, alcohol oxidoreductase, EC 1.1.3.7) is a monomeric flavoenzyme having molecular mass ranging from approximately 70 to 80 kDa. They are bound to a FAD redox cofactor that promotes the transfer of two hydrogen atoms and two electrons from aryl alcohol substrates to O_2 molecule thereby forming aldehyde. The first step of the catalytic process is the reduction of the cofactor (FAD to $FADH_2$) followed by oxidation to the native form. The oxidation of alcohols by AAOx is not reversible due to the strong oxidizing character of O_2. The standard pathways aimed at monitoring the reaction that is catalyzed by the oxidase enzyme either by evaluating the consumption of O_2 or by the production of H_2O_2.[29]

11.3.2 METHANOL OXIDASE

The oxidation of methanol and higher alcohols, with O_2 molecule to give formaldehyde and hydrogen peroxide, is catalyzed by methanol oxidase (MOx) or in general alcohol oxidase (AOx), which is a homo-octamer of molecular weight 600-kDa with 8 FAD cofactors. AOx utilizes FAD that is non-covalently bound to the enzyme. Small primary alcohols are oxidized by the enzyme via a two-step mechanism: (1) a reductive half-reaction, wherein the substrate gets oxidized while FAD gets reduced to $FADH_2$ and (2) an oxidative half-reaction, where the flavin is reoxidized by O_2 leading to the production of H_2O_2.[30]

11.3.3 GLUCOSE OXIDASE

Glucose oxidase (GOx; E.C. 1.1.3.4), an oxidoreductase enzyme that belongs to the GMC family. It promotes the catalysis of glucose oxidation by using O_2 as the electron acceptor to form glucono-δ-lactone and H_2O_2. The formed glucono-δ-lactone gets hydrolyzed into gluconic acid, while the generated H_2O_2 is broken down into oxygen and water. GOx, being a holoenzyme, consists of two identical subunits (80 kDa) at the active site containing FAD cofactor. In the catalytic process, these identical subunits act as redox carriers. GOx is highly specific, and hence it is more preferred than the other enzymes that oxidizes glucose. GOx is obtained from a variety of sources that include plants, animals, red algae, bacteria, and fungi, but the one that is obtained from fungi is preferred for industrial applications since its discovery. Fungal strains from *Aspergillus* sp. and *Penicillium* sp. are widely used in the industrial production of biotechnology-based products. Therefore, majority of research worldwide are focused on glucose oxidase originated from fungal strains because of the enormous utility in the commercial field, such as production of gluconic acid as it acts as a stabilizing agent, as an antioxidant, and as a chelating agent during food processing. Also, the hydrogen peroxide formed during the oxidation of glucose improves the shelf-life of beverages. GOx is used in the construction of biosensors that monitor the level of glucose content in biological samples and food samples.[31]

11.3.4 PYRANOSE OXIDASE

Pyranose oxidase (POx; pyranose 2-oxidase, pyranose:oxygen 2-oxido-reductase, EC 1.1.3.10) is an FAD-dependent enzyme that promotes the oxidation of D-glucose, D-galactose, and D-xylose thereby reducing O_2 to H_2O_2. Their properties (especially catalytic efficiency (k_{cat}/k_m)) are superior to GOx for oxidizing D-glucose and has a lower k_m value than GOx Their electrochemical biosensing applications are limited to its origin from fungal sources. Recently, they are discovered from bacterial sources (*Proteobacteria, bacilli*) wherein FAD is non-covalently bound to the enzyme, and it paves the way for further application in the field of electrochemical biosensing. These enzymes are first reported in the year 1968 and found in wood degrading fungi. They have two domains viz., flavin-binding and substrate-binding domain. Flavin-binding domain is a Rossmannfold or βαβ-mononucleotide-binding motif. The substrate-binding domain has more variations. Fungal-based POx consists of four monomers in the sequence that are identical, and each of them has FAD in the catalytic site which is covalently bonded to a histidine residue. It contains a central void and is deeply buried with restricted access to the substrate that limits its activity to monosaccharide molecules.[32]

11.3.5 CHOLINE OXIDASE

Choline oxidase (E.C. 1.1.3.17, choline-oxygen 1-oxidoreductase) favours the flavin-mediated oxidation of choline to glycine betaine. It has both fungal as well as bacterial origin. Choline oxidation was first reported in the year 1937, and the enzyme choline oxidase was isolated from rat liver in the year to monitor the effect of nitrogen mustards (cytotoxic organic compound with choloroethylamine functional group) on enzymes and tissue in solution exists as a homodimer of 120 kDa, as predicted by size exclusion chromatography and non-denaturing polyacrylamide gel electrophoresis.[33]

The active site structural details of GMC-based flavin-dependent carbohydrate oxidases are shown in Table 11.2.

TABLE 11.2 Structures of GMC-Based Flavin-Dependent Carbohydrate Oxidases.

Enzyme /source	Pdb id	Active site structures	Reference
Aryl-Alcohol oxidase from *Pleurotus eryngii*	3FIM 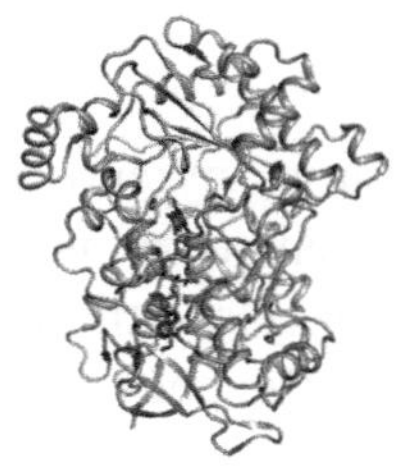	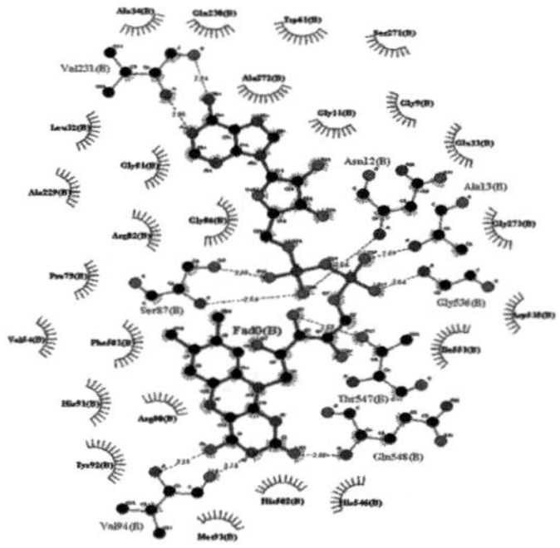	[68]
Alcohol oxidase from *Pichia Pastoris*	5HSA 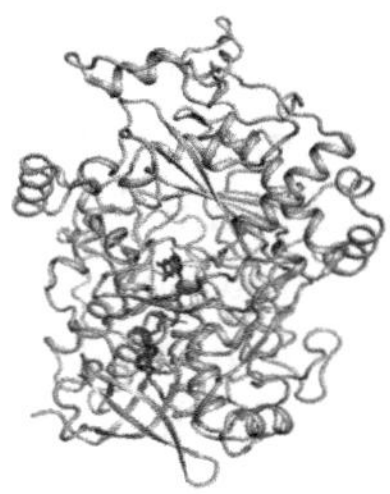	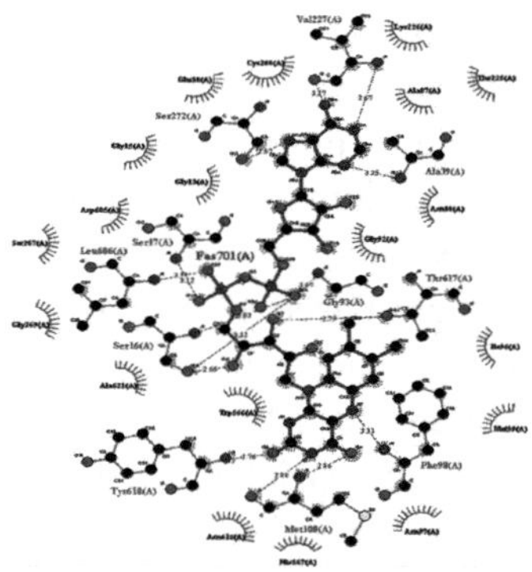	[69]
Glucose oxidase form *Aspergillus Niger*	1CF3 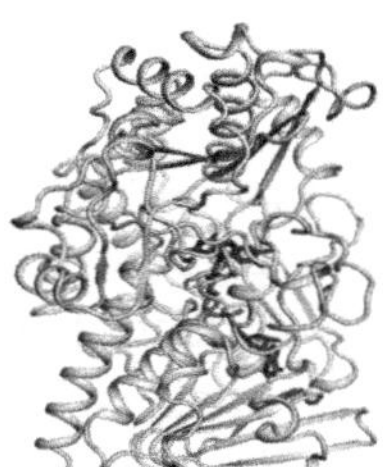	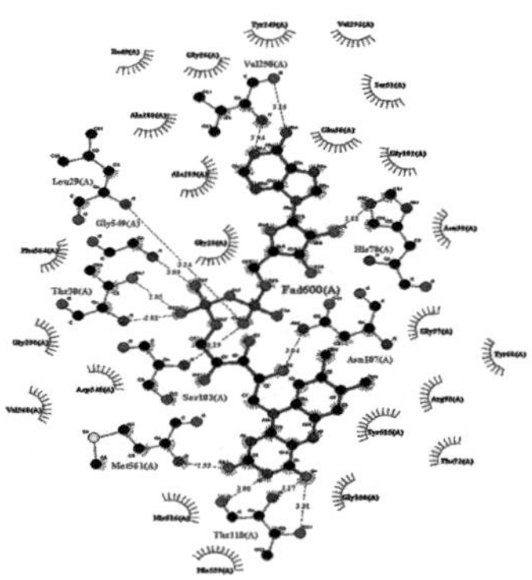	[70]

TABLE 11.2 *(Continued)*

Enzyme /source	Pdb id	Active site structures	Reference
Pyranose oxidase from *Peniophora sp. SG*	1TZL 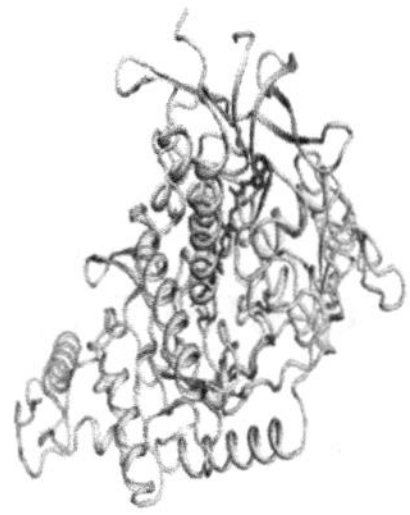	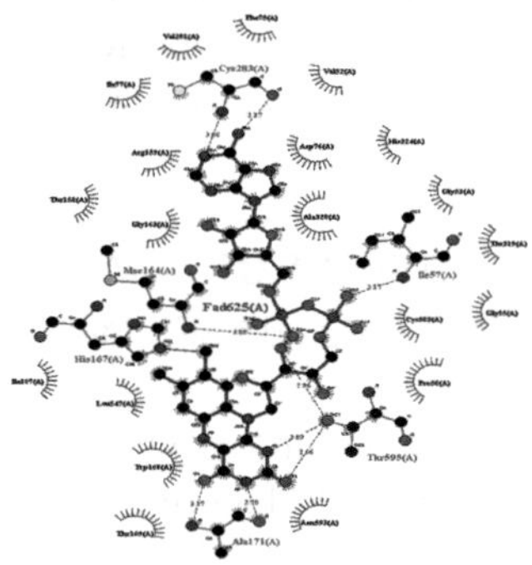	[71]
Choline oxidase from *Arthrobacter globiformis*	4MJW 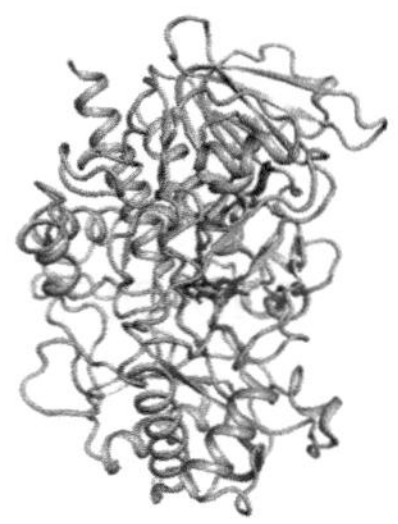	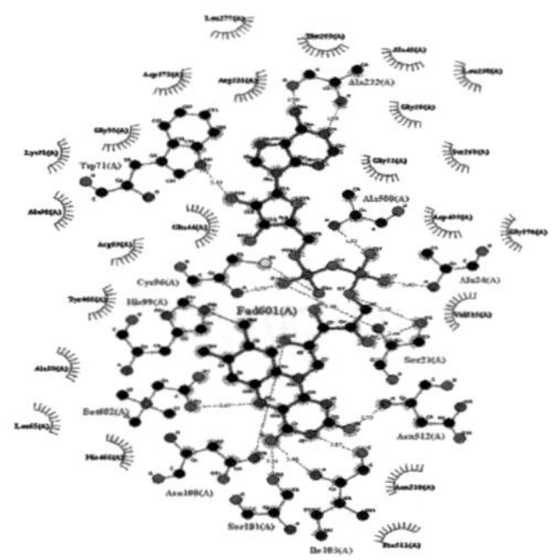	[72]

11.4 GENERAL MECHANISM

Although the different GMC oxidases act on various substrates, such as sugars, alcohols, etc., they share a general mechanism of action of FAD-dependent oxidoreductases. The mechanism includes two reactions viz., oxidative half-reaction and reductive half-reaction. In the reductive half-reaction, reduction of FAD with the concurrent oxidation of the electron donating substrate occurs. In the oxidative half-reaction, reoxidation of $FADH_2$ back to FAD takes place. Overall, the mechanism relies on His/His or His/Asn base pairs, which exist in the catalytic site of GMC oxidoreductases.[14]

The electron-donating substrates of GMC oxidoreductases include different sugars, alcohols, and choline, but in general the mechanism of action is analogous to those of FAD-dependent oxidoreductases. The overall mechanism is considered as a ping-pong Bi-Bi mechanism. The mechanistic pathway involves catalytic His/His or His/Asn pair in the reactive site.[34] The electron accepting species of GMC oxidoreductases family is oxygen or other alternative species such as quinones, phenol radicals, or metal ions. Based on the choices toward electron acceptors, they are classified into two broad categories viz., oxidases (O_2 as electron acceptor), and dehydrogenases (with insignificant or low reactivity with O_2).[14]

11.5 APPLICATIONS OF GMC-BASED FLAVIN-DEPENDENT CARBOHYDRATE OXIDASES IN ELECTROCHEMICAL BIOSENSING

The common cofactor in oxidase enzymes is FAD, which is not covalently bound to the enzyme.[35] The oxidase-based biosensors either detect H_2O_2 at the anodic potential of +0.7 V versus Ag/AgCl or detect the consumption of O_2 at the cathodic potential of −0.7 V versus Ag/AgCl. Peroxidase enzyme can be incorporated in the design of the biosensor in order to detect H_2O_2 at a low reduction potential.[36]

11.5.1 ELECTROCHEMICAL SENSING OF CLINICALLY SIGNIFICANT ANALYTES

11.5.1.1 GLUCOSE SENSING

Diabetes mellitus is referred to as a “silent epidemic”, which is a metabolic disorder caused when there is insufficiency of insulin in the body resulting elevated glucose concentration (hyperglycemia), or reduced glucose concentration in blood (hypoglycemia). The term “diabetes” was derived from the Greek word (diabainein), which means “pass through” or “siphon”, and the term “mellitus” was derived from Latin word which means “sweetened with honey.”

Insulin is a hormone that is synthesized and secreted from the pancreas to mediate metabolic reactions that involve glucose. Hence, it demonstrates a significant role in the control of glucose concentration in blood (Fig. 11.4). It initiates glucose uptake by liver, muscle, and fat cells for energy thereby controlling glucose concentration in blood. If the body has enough energy, insulin will signal the liver to uptake glucose and pile up ~5% of its mass as glycogen. Insufficient amount of insulin, or the inability to respond to insulin, leads to diabetes.

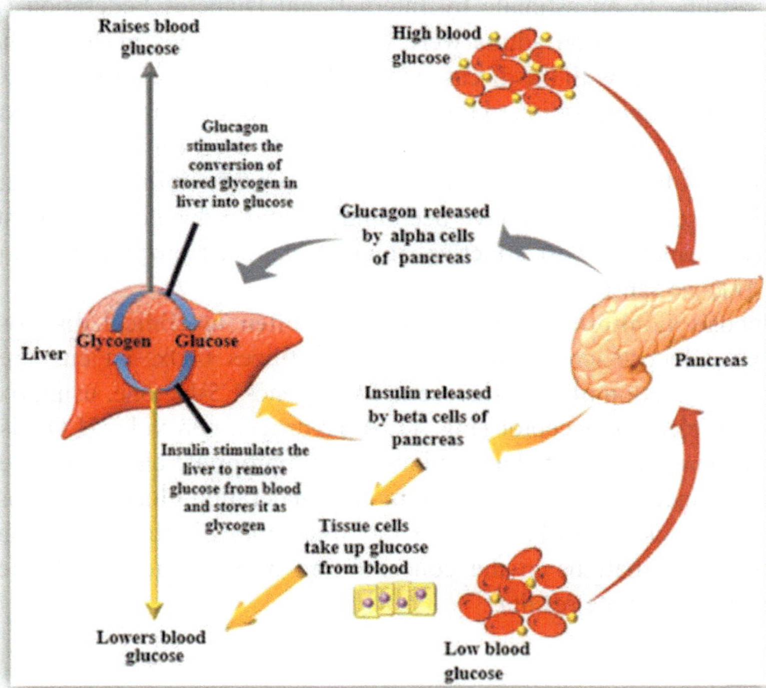

FIGURE 11.4 Metabolic reactions of Insulin in maintaining blood glucose concentration.

Diabetes has been divided into three main types. Type I diabetes was initially referred to as insulin-dependent diabetes mellitus or juvenile diabetes. It is primarily triggered by decreased production of insulin present in the body. Thus, majority of the cells in the body are unable to absorb glucose from blood, and hence uses other forms of energy. Ketones produced by the liver are used as another form of energy; however, elevated levels of the ketones lead to ketoacidosis. This condition is manifested typically among the youngsters. Regular supplement of insulin is therefore essential for people suffering from such type of diabetes. Type II diabetes is noninsulin-dependent, and can be regarded as adult-onset diabetes. It arises from body's resistance to insulin, wherein the target organs fail to respond to insulin. Majority (> 95%) of the cases belong to this type. The next type is referred as gestational diabetes which occurs when pregnant women, normally without a prehistory of diabetes, develop a high concentration of glucose in blood. Women who suffer from gestational diabetes typically have a greater risk of eventually developing of type II diabetes.[37]

According to an estimate by International Diabetes Federation (IDF), more than 387 million adults were living with diabetes in 2014, and the number of people with the disease may afflict beyond 592 million in 2035.[38] Consequently, quantization of the glucose (diabetes biomarker) is of extreme importance. Most of the conventional methods used for measuring blood glucose involve complex procedures, low reproducibility, and stability. Therefore, several studies have been undertaken to develop simple, stable, selective, sensitive, cost-effective, reliable, and fast enzymatic/non-enzymatic glucose sensors.

Glucose monitoring is performed very often in clinical diagnosis of diabetes. Glucose-sensing device usually utilizes GOx, an enzyme that promotes oxidation of β-D-glucose to D-glucono-1,5-lactone, using oxygen (O_2) as an electron acceptor (Fig. 11.5). Hydrogen peroxide (H_2O_2) is then electrochemically detected at a suitable and appropriate electrode potential. This is the mechanism involved in the first-generation glucose biosensor. A major drawback of this biosensor is the fluctuations in the oxygen tension, leading to fluctuations in the concentration of H_2O_2 generated; thereby, producing inaccurate results.[37]

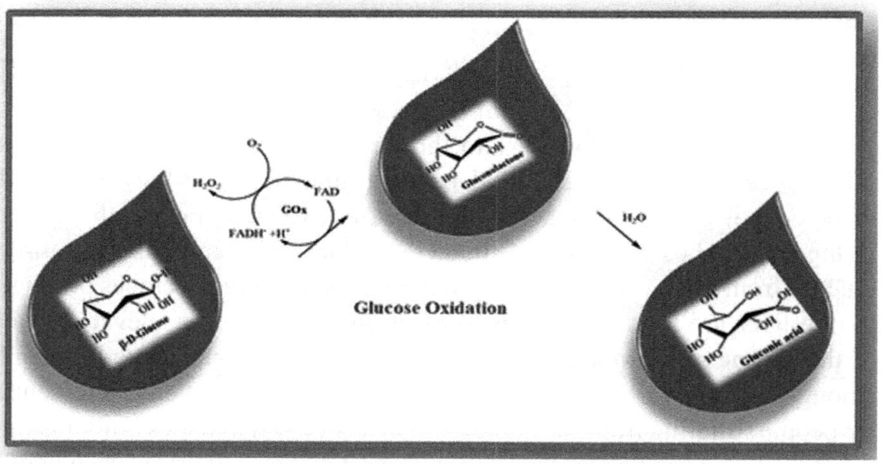

FIGURE 11.5 Mechanism of oxidation of glucose by glucose oxidase.

Source: Adapted from Ref. [37]

An electroactive bio-nanohybrid film based on polypyrrole-nafion-functionalized multi-walled carbon nanotubes nanocomposite was deposited on the glassy carbon electrode by a single-step electropolymerization, with the subsequent immobilization of chitosan-glucose oxidase on the surface to fabricate a sensitive glucose biosensor. The prepared nanohybrid composite provided high surface area and enhanced GOx loading efficiency. The electrochemical studies offered a good catalytic property of the film to detect glucose having a sensitivity of 2860.3 μA mM^{-1} cm^{-2} with the linear range up to 4.7 mM, and a detection limit as low as 5 μM. The developed biosensor exhibited good selectivity, storage stability, repeatability, and reproducibility in real serum samples.[39] The preparation of a bio-nanohybrid material by doping functionalized MWCNTs in nafion (Nf) with polypyrrole followed by electropolymerization in addition to GOx immobilization on a platinum (Pt) electrode was reported. The electrode exhibited excellent electrocatalytic activity to monitor glucose with a good sensitivity of 54.2 μA mM^{-1} cm^{-2} and a linear response of 4.1 mM, with a low detection limit of 5 μM along with good selectivity, stability, and practical applicability.[40] The process of monitoring and controlling the fermentation process of glucose remains a greatest challenge in the lab-scale and industrial-scale experiments. The reliability in the quantification of glucose is the key step in on-line mode, which promotes the fermentation process easy with standardized reactor efficiency, maximum productivity, and minimum waste. In spite of the existence of commercial biosensing platforms for continuous glucose monitoring, on-line monitoring of glucose in fermentation processes was not much available. The study of a commercial flow-through-cell integrated with electrochemical glucose biosensors was reported for the first time, and the sensor detected glucose up to 150 mM, which is four to sixfold more than that found in the literature. The automated biosensor platform demonstrated a good mechanical stability and considered as an inexpensive tool to monitor glucose despite the presence of other electroactive species.[41] The recent development in the noninvasive continuous glucose monitoring (CGM) in the alternative body fluids apart from blood serum that offers real-time application in determining the interstitial glucose levels was discussed. The challenges toward the development of biocompatible membrane coatings that prevent the surface fouling of electrochemical glucose sensors were focussed. The analytical applications in the field of self-powered glucose sensors, paper-based glucose sensing, and multiplexed detection of diabetes-related biomarkers were also highlighted.[42]

An RBC membrane-coated enzymatic glucose sensor in order to mimic the amount of glucose uptake was developed. The optimized sensor monitored glucose from 1 to 15 mM with a low limit of detection as 0.66 mM, and has a sensitivity of 2.978 μA mM^{-1}. The developed sensor sensed glucose without interference, such as galactose, ascorbic acid, uric acid, etc., with an accuracy ranging from interfering molecule were 0.8–2.3%, which was approximately 4.8–14.2 times more accurate than the unmodified one. A sensor was applied similar to the result verified for human serum sample that contains numerous interfering molecules. The performance suggested that the sensor has greater ability to sense the samples after 4 weeks of storage.[43] A novel enzymatic electrochemical glucose biosensor possessing antimicrobial properties were obtained, using TiO_2 nanotubes modified by AgO and Prussian blue nanoparticles was developed. An electrochemical anodization was employed for the deposition of GOx onto Prussian blue-modified TiO_2 nanotube arrays that were functionalized by using Au and AgO NPs. The AgO-deposited TiO_2 nanotubes were synthesized by using an electrochemical anodization process following which Ag was deposited by electroplating process using the same electrolyte. Cyclic voltammetry and chronoamperometric studies revealed that the developed electrochemical biosensors exhibited excellent properties such as fast transfer of electrons, high stability, and low detection limit for glucose level determination. Under the optimized conditions, the amperometric response exhibited a linear dependence of the biosensor for glucose with a detection limit of 4.91 μM having sensitivity of 185.1 mA M^{-1} cm^{-2} in Au-modified biosensor and a detection limit of 58.7 μM with 29.1 mA M^{-1} cm^{-2} sensitivity, which was obtained using AgO-modified biosensor.[44]

POx-based biosensors also find applications in the electrochemical sensing of glucose in food and beverage samples. Pyranose is a general term used for saccharides with a structure that has a six-membered ring made of five carbon atoms and one oxygen atom. It can oxidize D-glucose (in both the α- and β-anomeric forms) leading to the formation of 2-keto D-glucose and also other monosaccharides at the C2 position while reducing oxygen molecule to hydrogen peroxide. The reactivity of POx in contrast to GOx, with both the anomers of D-glucose and also with other sugar substrates such as D-xylose, D-galactose, L-sorbose, and L-arabinose, is more frequently used in biosensors as it reacts with the β-anomer only. The other monosaccharides in their aldopyranose form can also be oxidized at the C2 position to the corresponding keto-sugars by POx. The K_m (Michaelis constant) for D-glucose of different POx ranges from 0.74 to 5.0 mM, which is lesser than the values reported for glucose oxidase isolated from *Aspergillus niger*, 22–28 mM or 33–248 mM. Turnover numbers for D-glucose (k_{cat}) of POx from different fungal sources ranged from 1.48 to 111 s^{-1}.[32]

11.5.1.2 ALCOHOL SENSING

Over the decades, the detection of ethanol is required in a variety of fields like forensic science, food industries, and biomedical analysis. Alcohol is considered as one among the common poisonous substances consumed by human beings, and is the major factor responsible for several accidents. It is being analyzed in biological samples like urine, blood serum, etc. As the quantity of ethanol rises to a maximum, leading to alcohol poisoning, and hence quantification of ethanol is much needed to check the quality of products in industries like food, beverages, etc., and additionally in biomedical field for determining the toxicological effects on human beings.[45–48] The quantification of alcohol plays an important role in determining the different factors that affect the health of the individuals like liver cirrhosis, cardiovascular ailments as well as strokes. A variety of nanomaterials that have enzyme-like characteristics are widely utilized in the field of electrochemical alcohol biosensing.

AOx biosensors utilize optical as well as electrochemical methods[49] thereby relying on the formation of H_2O_2. In electrochemical method, amperometric electrodes are used to analyze the production of H_2O_2 by AOx, due to the chemical reaction involving oxidation or reduction of H_2O_2 at the working electrode. Alcohols were determined from various sources such as fruit juices,[50] forensic samples,[51] etc. A wearable noninvasive alcohol biosensor was fabricated by integrating alcohol oxidase enzyme and Prussian blue for sensing alcohol in blood stream. It is a transdermal alcohol biosensor that has the ability to quantify the level of alcohol that diffuse through skin in the bloodstream. The drawback with the earlier biosensors includes surface fouling, fluctuations in the readings of the sensor over time. Hence, a wearable sensor, which overcomes the drawback of surface fouling by enabling in vivo fluctuation-free signals for up to 24 h and an exchangeable cartridge connection, which promoted successive monitoring the level of alcohol. The laboratory experiments using the sensor demonstrated that the electron transfer process is limited to mass-transfer by the diffusion-controlled membrane, which has lesser permeability than human skin, and exhibited a linear concentration range between 0 mM and 50 mM.[52]

A suitable detection method combining Au@PtRu nanozymes and AOx was reported. The Au@PtRu nanorods demonstrated peroxidase-like bioactivity, which was larger than that exhibited by Au and Au@Pt nanorods. The alcohol sensor was developed based on the nanozyme catalyzed production of hydroxyl radicals in the presence of H_2O_2 by detecting the H_2O_2 produced during the oxidation of alcohol to acetaldehyde by AOx. The limit of detection of alcohol when the nanozyme coupled with AOx was found to be 23.8 μM in buffer solution that was used for clinical assays. The real time practical applicability of the developed sensor for the detection of alcohol was checked in mouse blood samples. The obtained results were compared with that of commercial alcohol meters. The ability of the Au@PtRu nanorods with peroxidase-like activity for alcohol quantification has opened a novel avenue for the preparation of nanozymes and its utility in biomedical applications.[53] A light, easy to wear, cost-effective, wrist-worn alcohol biosensor, which are applicable for biomedical research was fabricated. Three pilot studies that monitor the relation between drinking attitude and transdermal alcohol concentration were described. The first study compared *skin*-derived transdermal alcohol concentration with that of alcohol sensor worn in ankle and alcohol concentration in breath as analyzed in the laboratory. The next report compared *skin*-derived transdermal alcohol concentration with that of breathe alcohol concentration during a naturalistic drinking experience. The final report used the *skin alcohol concentration* to detect alcohol continuously for 2 weeks. On the whole, the results revealed the transdermal detection of alcohol in skin was reliable with further refinements such as analytical software that promotes passive detection of alcohol with accuracy.[54]

11.5.1.3 CHOLINE SENSING

Choline which is referred to as vitamin B_4, a quaternary ammonium compound, is a vital nutrient for maintaining proper body function. Choline can be found as free choline, phosphocholine, glycerophosphocholine, sphingomyelin or phosphatidylcholine. Choline is an important nutrient that exhibits a significant role for proper cell functioning, and the daily intake of it is prescribed by the U.S. National Academy of Sciences. Choline and their derivatives tend to play a significant role in the synthesis of the neurotransmitter acetylcholine, signaling of the cell membrane, and the proper functioning of liver and muscles.[55] The deficiency of choline was found to have an impact on some diseases viz., Parkinson's disease, Alzheimer's disease, and in multiple cancer ailments.[56] Hence, it is important that the level of choline need to be monitored. The metabolism of choline makes it a probable biomarker for premedical diagnosis of diseases. Compared to the time-consuming, expensive, and tedious conventional methods, biosensor-based detection of choline offers a cost-effective, easy to operate, sensitive, selective, and specific way of determining choline concentration. Although, several methods such as liquid chromatography (LC), liquid chromatography/mass spectrometry (LC/MS), gas chromatography, ion chromatography, UV-vis spectroscopy, and capillary zone electrophoresis are available for detecting choline, majority of them are not cost-effective and also time consuming. Choline oxidase-based biosensors are found to be most promising owing to their efficient way of determining the concentration of choline in various biological samples.[55]

In the choline oxidase-based biosensing, the first step involves the oxidation of choline to betaine, and the reduction of O_2 to H_2O_2. In the second step, the enzyme-associated betaine is hydrated to form gem-diol-choline, which is subsequently oxidized to glycine. The primary objective of bienzyme biosensors based on acetylcholinesterase and choline oxidase was to examine acetylcholine in clinical samples. The redox potential of acetyl choline is quite large for the direct determination. Hence, biosensors employing acetylcholinesterase (AChE) and choline oxidase were deposited on the electrode to achieve electrochemical response as per the following reactions.

$$\underset{\text{Acetylcholine}}{(CH_3)_3\overset{+}{N}\text{-}CH_2CH_2COOCH_3} + H_2O \xrightarrow{\text{AChE}} \underset{\text{Choline}}{(CH_3)_3\overset{+}{N}\text{-}CH_2CH_2OH} + \underset{\text{Acetic Acid}}{CH_3COOH}$$

$$\underset{\text{Choline}}{CH_3)_3\overset{+}{N}\text{-}CH_2CH_2OH} + O_2 + H_2O \xrightarrow{\text{Choline Oxidase}} \underset{\text{Betaine}}{(CH_3)_3\overset{+}{N}\text{-}CH_2COOH} + \underset{\text{Hydrogen peroxide}}{H_2O_2}$$

The biocatalytic reactions of AChE–choline oxidase biosensor consisted of AChE-catalyzed hydrolysis of ACh into choline and acetic acid succeeded by the choline oxidase catalyzed oxidation of choline. The subsequent reaction produces H_2O that was oxidized on the electrode surface thereby generate an output signal proportional to AChE concentration. An oxygen electrode can also be used in place of metal or carbon electrode as choline oxidase consumes dissolved O_2 for choline oxidation. Hence, the consumption of O_2 can be detected by employing commercial O_2 electrode. A novel and highly sensitive amperometric choline biosensor has been fabricated by immobilizing choline oxidase on the surface of multiwalled carbon nanotubes (MWCNTs) modified platinum electrode employing sol-gel method. The modified electrode showed that the MWCNTs improved the current response and decreased the potential toward the detection of H_2O_2. The fabricated biosensor exhibited sensitive detection of choline between the concentration linear 5.0 and 100 μM at 0.15 V detection potential. The sensitivity of the biosensor was found to be 9.48 μA mM^{-1} with 0.5 μM as the detection limit. It also exhibited anti-interference property and good stability by retaining 85% of the initial current response for a month. The developed biosensor was applied for the quantification of choline in human serum samples.[57]

Two enzymes viz., acetylcholinesterase and choline oxidase were coimmobilized with chitosan/gold-coated ferric oxide nanoparticles, and then deposited over gold electrode. Amperometric analysis of acetylcholine biosensor revealed the efficient determination of acetylcholine within the linear range 0.005–400 μM, with 0.005 μM as the limit of detection. The developed biosensor was used to measure acetylcholine in plasma obtained from normal persons and Alzheimer patients. The biosensor was found to be stable for 3 months when stored at 4°C.[58] An amperometric acetylcholine biosensor was prepared by embedding acetylcholinesterase and choline oxidase on to electrochemically polymerized polyaniline-polyvinyl sulphonate film (by cross linking with glutaraldehyde) and bovine serum albumin. Determination of ACh was carried out by oxidation of enzymatically produced H_2O_2 at +0.4 V versus Ag/AgCl. The pH and temperature were standardized to be 8.0°C and 60°C, respectively. The linear range for the detection of acetylcholine was 0.1–0.6 μM.[59] A biocompatible biosensor prepared using nafion/choline oxidase/manganese dioxide composite film, which was deposited over GCE, was sensitive toward the detection of choline in milk, milk powder, and feedstuff samples. The modified electrode system provided high specific surface area, hydrophilicity, permeability, and biocompatibility. Cyclic voltammetric studies revealed that the linear range of the sensor under optimized conditions was found to be 8.0–1.0 mM with a low detection limit of 5.0 μM.[60] An amperometric biosensor for the determination of choline by utilizing the electrocatalytic properties of MWCNTs, gold nanoparticles and chitosan, which were coated over GCE along with the immobilization of choline oxidase via glutaraldehyde crosslinking. The fabricated biosensor demonstrated a linear concentration range from 3 to 120 μM with a limit of detection and sensitivity of 0.6 μM and 204 μA cm^{-2} mM^{-1}, respectively. The first study on the determination of choline by electrochemical impedance spectroscopic (EIS) on choline quantification by EIS. The detection of choline present in milk samples was reported.[61] A portable sensor was developed for the estimation of choline. The determination of choline and its derivatives, such as acetylcholine, benzoylcholine, and atropine in samples using silica isoporous membrane were investigated. The sensor was utilized for the estimation of choline and its derivatives in biological samples viz., urine and blood samples. The limit of detection values were found to be 1.12, 1.30, 1.08, and 0.99 μM, for choline, acetyl choline, atropine, and benzoylcholine present in urine samples, respectively, and the values obtained in blood serum samples were found to be 3.61, 3.38, 2.32, and 1.81 μM, respectively.[62] An amperometric biosensor was developed by immobilizing choline oxidase onto a membrane that was antifouling and anti-interferant, and was applied for the determination of choline in biological samples of patients who suffer from kidney-related diseases and undergoing repeated hemodialysis. The fabricated biosensor demonstrated a linear response for the choline level up to 100 μM with a sensitivity of 156 nA mM^{-1} mm^{-2} and a detection limit of 1 μM. Preliminary analysis resulted in choline levels that are in good agreement with the expected values.[63] Immobilization of acetylcholinesterase and choline oxidase over gold nanoparticles coated graphene oxide-based nanocomposite followed by depositing over ITO glass plate was effected and the developed sensor was utilized for the detection of ultra-small concentration of acetylcholine. The synergetic properties of gold nanoparticles and grapheme oxide resulted in efficient sensing of the analyte by the biosensor. The electrochemical measurements revealed that the detection of acetylcholine was effected within the concentration range from 100 pM to 1000 nM having a detection limit of less than 100 pM. The real-time utility of the sensor was determined by applying the developed sensor for detecting acetylcholine present in serum samples.[64] A stable amperometric biosensor, which was fabricated by the suitable incorporation of acetylcholinesterase and choline oxidase enzymes for the determination of acetylcholine, using flow injection analysis was reported for the first time. Several factors like the quantity of immobilized enzymes, pH, detection potential, flow rate, and amount of acetyl choline were standardized. The detection limit was found to be 4.1 μM. The biosensor exhibited anti interference property for the detection of 0.1 mM acetyl choline coexisted with glucose, uric acid, and ascorbic acid. The biosensor exhibited a long-term stability of over 100 days after carrying out > 400 measurements.[65]

A dual enzymatic biosensor based on acetylcholine esterase and choline oxidase was fabricated using carbon paste electrode modified with carbon nano dot-(3-aminopropyl) triethoxysilane. The developed biosensor was utilized for the monitoring of acetylcholine through the electro oxidation of H_2O_2 at 0.4 V (vs Ag/AgCl).[66] Electroenzymatic microsensor was fabricated for the theoretical determination of choline by immobilizing choline oxidase over the surface of Pt, and the developed sensor was guided by simulations performed by using detailed mathematical model. It is an implantable microsensor with an array of electroenzymatic sensing sites to determine the changes in the concentration of choline (precursor of the neurotransmitter acetylcholine) owing to rapid turnover in the brain. The simulated model suggested that the function of the sensor could be improved by standardizing the thickness of the immobilized choline oxidase (~5 μm) and polymer coatings (200 nm), which exhibited a higher sensitivity of 660 nA μM^{-1} cm^{-2}. This provided a better option in the fabrication of microelectrode array probes for sensing the signals of acetylcholine by correlating the physiological functions with the electrochemical signals in order to improve the study of brain activity.[67]

The schematic representation of the mechanism of action of GMC-based oxidases is shown in Figure 11.6.

11.6 FUTURE PERSPECTIVES

Enzyme-based electrochemical biosensors are of huge significance in the areas of food, environmental pollutants, and clinical analyses. It offers great challenge to develop, nondestructive, noninvasive, economically viable, and wearable biosensors that detect environmentally and clinically significant analytes in food, biological, and environmental samples. Real time analysis in other biological fluids apart from blood viz., saliva or sweat is of great importance in the field of clinical diagnostics. Designing portable device with microfabricated noninvasive sensing system that can continuously monitor, quantify, and store the data of more than one analyte is the greatest challenge as it will pave the way for early diagnosis and prevent diseases. A knowledge on the fabrication of biocompatible nanocomposites-based enzyme biosensors for the determination of choline and its metabolites leads to further advancement and commercialization.

11.7 CONCLUSIONS

Sensors incorporated with enzymes have become integral part of everyone's life. The specificity, selectivity, and sensitivity make enzyme-based electrochemical biosensing as a more reliable technique compared to other techniques. Since the discovery of enzymes' many research work concentrated in the industrial and biomedical fields. Knowledge about the different types of enzymes, their structure, function, mechanism of action, methods of immobilization, and their utilization in the various field is very much needed. Majority of the reactions that occur in the body and in the nature

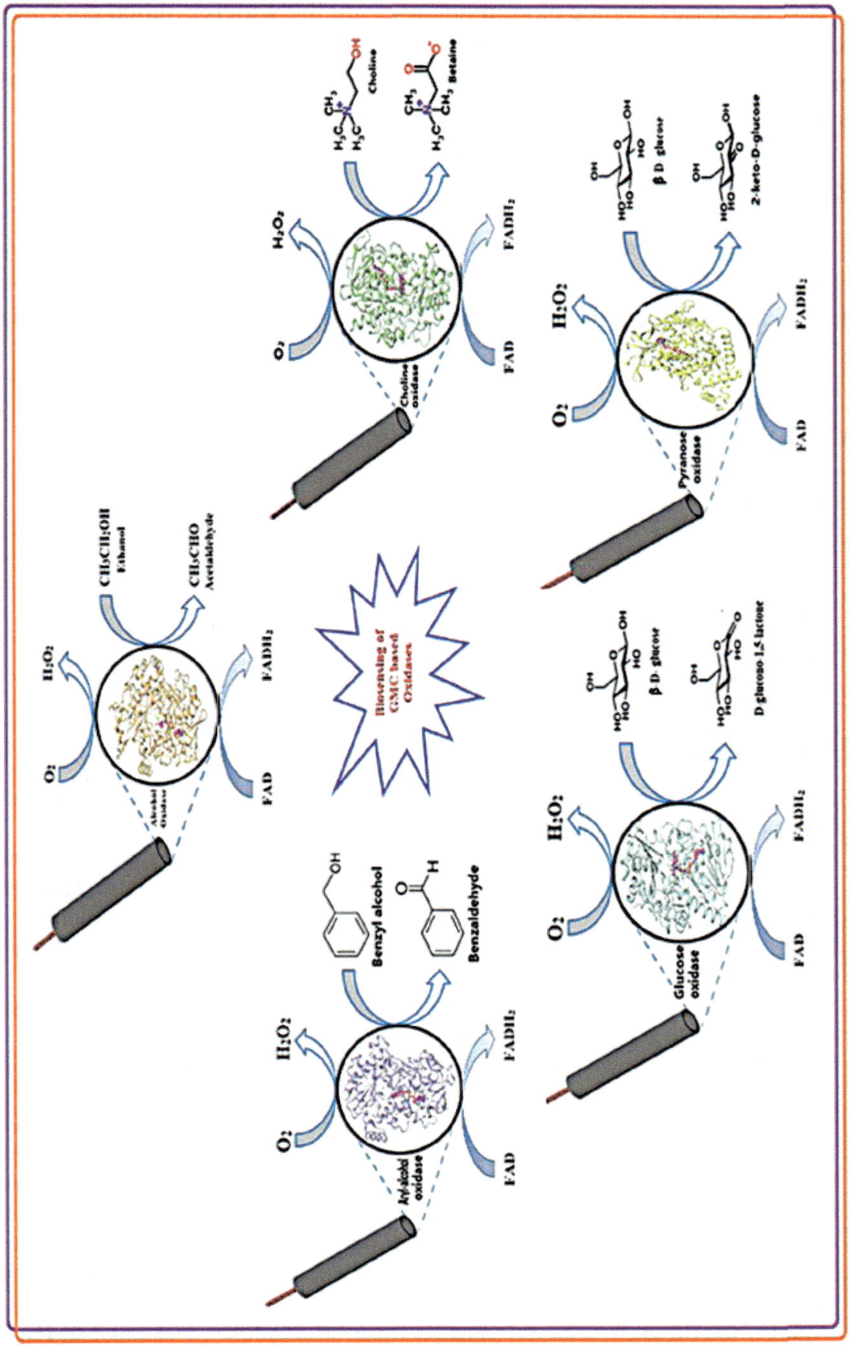

FIGURE 11.6 Schematic representation of mechanism of action of GMC based oxidases.

are redox processes. Hence oxidoreductases are the widely studied enzymes among the other enzymes. This chapter is written to provide an overview of oxidoreductases, and more specially oxidases and their electrochemical sensing applications.

ACKNOWLEDGMENT

The authors greatly acknowledge the Principal and Management of Lady Doak College, Madurai, for their constant support and encouragement. The contributions made by the researchers in the studies based on enzymatic sensors are highly acknowledged.

KEYWORDS

- **enzymes**
- **oxidases**
- **glucose**
- **electrochemical**
- **sensing**

REFERENCES

1. Cuesta, S. M.; Rahman, S. A.; Furnham, N.; Thornton, J. M. The Classification and Evolution of Enzyme Function. *Biophys. J.* **2015,** *109*, 1082–1086.
2. Heckmann, C. M.; Paradisi, F. Looking Back: A Short History of the Discovery of Enzymes and How They Became Powerful Chemical Tools. *Chemcatchem* **2020,** *12* (24), 6082–6102.
3. Robinson, P. K. Enzymes: Principles and Biotechnological Applications. *Essays Biochem.* **2015,** *59*, 1–41.
4. Singh, R.; Kumar, M.; Mittal, A.; Mehta P. K. Microbial Enzymes: Industrial Progress in 21st Century. *3 Biotech.* **2016,** *6* (2) 174.
5. Navaee, A.; Salimi, A. Enzyme-Based Electrochemical Biosensors. *Electrochem. Biosens.* **2019,** 167–211.
6. Vandenberghe, L. P. S.; Karp, S. G.; Pagnoncelli, M. G. B.; Tavares, M. L.; Libardi Junior, N.; Diestra, K. V.; Viesser, J. A.; Soccol, C. R. Chapter 2 - Classification of Enzymes and Catalytic Properties. In *Biomass Biofuels Biochemicals: Advances in Enzyme Catalysis and Technologies*, 2019; pp 11–30.
7. Berry, C. E.; Hare, J. M. Xanthine Oxidoreductase and Cardiovascular Disease: Molecular Mechanisms and Pathophysiological Implications. *J. Physiol.* **2004,** *555* (3), 589–606.
8. Gygli G.; van Berkel W. J. H. Oxizymes for Biotechnology. *Curr. Biotechnol.* **2015,** *4*, 100–110.

9. Martínez, A. T.; Ruiz-Dueñas, F. J. ; Camarero, S.; Serrano, A.; Linde, D.; Lund, H.; et al. Oxidoreductases on their Way to Industrial Biotransformations. *Biotechnol. Adv.* **2017,** *35*, 815–831.
10. Younus H. Oxidoreductases: Overview and Practical Applications. In *Biocatalysis*; Husain Q., Ullah, M., Eds.; Springer: Cham, 2019.
11. Trisolini, L.; Gambacorta, N.; Gorgoglione, R.; Montaruli, M.; Laera, L.; Colella, F.; Volpicella, M.; De Grassi A.; Pierri, C. L. FAD/NADH Dependent Oxidoreductases: From Different Amino Acid Sequences to Similar Protein Shapes for Playing an Ancient Function. *J. Clin. Med.* **2019,** *8* (12), 2117.
12. Savino, S.; Fraaije, M. W. The Vast Repertoire of Carbohydrate Oxidases: An Overview. *Biotechnol. Adv.* **2020,** 107634.
13. Daou, M.; Faulds, C. B. Glyoxal Oxidases: Their Nature and Properties. *World J. Microbiol. Biotechnol.* **2017,** *33* (5), 87.
14. Sützl, L.; Foley, G.; Gillam, E. M. J.; Bodén, M..; Haltrich, D. The GMC Superfamily of Oxidoreductases Revisited: Analysis and Evolution of Fungal GMC Oxidoreductases. *Biotechnol. Biofuels* **2019,** *12*, 118.
15. Robert, H. H.; van den, H.; Fraaije, M. W.; Mattevi, A.; Laane, C.; van Berkel, W. J. H. Vanillyl-Alcohol Oxidase, a Tasteful Biocatalyst. *J. Mol. Catal. B Enzym.* **2001,** *11*, 185–188.
16. Lin, S.; Yang, T.; Inukai, T.; Yamasaki, M.; Tsai, Y. Purification and Characterization of a Novel Glucooligosaccharide Oxidase from *Acremonium stricture* T1. *Biochim. Biophys. Acta* **1991,** *18*, 41–47.
17. Heuts, D. P. H. M.; Janssen, D. B.; Fraaije, M. W. Changing the Substrate Specificity of a Chitooligosaccharide Oxidase from *Fusarium graminearum* by Model-Inspired Site-Directed Mutagenesis. *FEBS Lett.* **2007,** *581*, 4905–4909.
18. Ferrari, A. R.; Rozeboom, H. J.; Dobruchowska, J. M.; Van Leeuwen, S. S.; Vugts, A. S. C.; Koetsier, M. J.; Visser, J.; Fraaije, M. W.; Hart, G. Discovery of a Xylooligosaccharide Oxidase from *Myceliophthora thermophila* C1. *J. Biol. Chem.* **2016,** *291*, 23709–23718.
19. Heuts, D. P. H. M.; Scrutton, N. S.; McIntire, W. S.; Fraaije, M. W. What's in a Covalent Bond? On the Role and Formation of Covalently Bound Flavin Cofactors. *FEBS J.* **2009,** *276* (13), 3405–3427.
20. Xu, F.; Golightly, E. J.; Fuglsang, C. C.; Schneider, P.; Duke, K. R.; Lam, L.; Christensen, S.; Brown, K. M.; Jørgensen, C. T.; Brown, S. H. A Novel Carbohydrate: Acceptor Oxidoreductase from *Microdochium nivale*. *Eur. J. Biochem.* **2001,** *268*, 1136–1142.
21. Alessandro, R. F.; Henriëtte, J. R.; Aniek, S. C. V.; Martijn, J. K.; Robert, F.; Marco, W. F. Characterization of Two VAO-Type Flavoprotein Oxidases from *Myceliophthora thermophile*. *Molecules* **2018,** *23* (1), 111.
22. Cavener, D. R. GMC Oxidoreductases. A Newly Defined Family of Homologous Proteins with Diverse Catalytic Activities. *J. Mol. Biol.* **1992,** *223*, 811–814.
23. Fan, F.; Mahmoud, G.; Giovanni, G. Cloning, Sequence Analysis, and Purification of Choline Oxidase from *Arthrobacter globiformis*: A Bacterial Enzyme Involved in Osmotic Stress Tolerance. *Arch. Biochem. Biophys.* **2004,** *421* (1), 149–158.
24. Iida, K.; Cox-Foster, D. L.; Yang, X.; Ko, W. Y.; Cavener, D. R. Expansion and Evolution of Insect GMC Oxidoreductases. *BMC Evol. Biol.* **2007,** *7*, 75.
25. Dreveny, I.; Gruber, K.; Glieder, A.; Thompson, A.; Kratky, C. The hydroxynitrile Lyase from Almond: A Lyase that Looks Like an Oxidoreductase. *Structure* **2001,** *9*, 803–815.
26. Dreveny, I.; Andryushkova, A. S.; Glieder, A.; Gruber, K.; Kratky, C. Substrate Binding in the FAD-Dependent Hydroxynitrile Lyase from Almond Provides Insight into the

Mechanism of Cyanohydrin Formation and Explains the Absence of Dehydrogenation Activity. *Biochemistry* **2009,** *48*, 3370–3377.
27. Wong, C. M.; Wong, K. H.; Chen, X. D. Glucose Oxidase: Natural Occurrence, Function, Properties and Industrial Applications. *Appl. Microbiol. Biotechnol.* **2008,** *78*, 927–938.
28. Ludwig, R.; Ortiz, R.; Schulz, C.; Harreither, W.; Sygmund, C.; Gorton, L. Cellobiose Dehydrogenase Modified Electrodes: Advances by Materials Science and Biochemical Engineering. *Anal. Bioanal. Chem.* **2013,** *405*, 3637–3658.
29. Munteanu, F. D.; Ferreira, P.; Ruiz-Dueñas, F. J.; Martínez, A. T.; Cavaco-Paulo, A. Bioelectrochemical Investigations of Aryl-alcohol Oxidase from Pleurotus eryngii. *J. Electroanal. Chem.* **2008,** *618*, 83–86.
30. Vonck, J.; Parcej, D. N.; Mills, D. J. Structure of Alcohol Oxidase from *Pichia pastoris* by Cryo-Electron Microscopy. *PLoS One* **2016,** *11* (7), e0159476.
31. Singh, R. S. ; Singh, T.; Pandey, A. Chapter 1 - Microbial Enzymes-An Overview. In *Biomass, Biofuels, Biochemicals, Advances in Enzyme Technology*; Singh, R. S., Singhania, R. R., Pandey, A., Larroche, C., Eds.; 2019; pp 1–40.
32. Abrera, A. T.; Sützl, L.; Haltrich, D. Pyranose Oxidase: A Versatile Sugar Oxidoreductase for Bioelectrochemical Applications. *Bioelectrochemistry* **2020,** *132*, 107409.
33. Gadda, G. Chapter Six - Choline Oxidases. In *The Enzymes*; Chaiyen, P., Tamanoi, F., Eds.; Academic Press, 2020; vol 47, pp 137–166.
34. Hernandez-Ortega, A.; Ferreira, P.; Martinez, A. T. Fungal Aryl-alcohol Oxidase: A Peroxide-Producing Flavoenzyme Involved in Lignin Degradation. *Appl. Microbiol. Biotechnol.* **2012,** *93*, 1395–410.
35. Harper, A.; Anderson, M. R. Electrochemical Glucose Sensors-Developments Using Electrostatic Assembly and Carbon Nanotubes for Biosensor Construction. *Sensors* **2010,** *10*, 8248–8274.
36. Rocchitta, G.; Spanu, A.; Babudieri, S.; Latte, G.; Madeddu, G.; Galleri, G.; Nuvoli, S.; Bagella, P.; Demartis, M. I.; Fiore, V.; Manetti, R.; Serra, P. A. Enzyme Biosensors for Biomedical Applications: Strategies for Safeguarding Analytical Performances in Biological Fluids. *Sensors (Basel)* **2016,** *16* (6), 780.
37. Wang, H. C.; Lee, A. R. Recent Developments in Blood Glucose Sensors. *J. Food Drug Anal.* **2015,** *23*, 191–200.
38. Joao da, R. F.; Ogurtsova, K.; Linnenkamp, U.; Guariguata, L.; Seuring, T. ; Zhang, P.; Cavan, D.; Makaroff, L. E. IDF Diabetes Atlas Estimates of 2014 Global Health Expenditures on Diabetes. *Diabetes Res. Clin. Pract.* **2016,** *117*, 48–54.
39. Shrestha, B. K.; Ahmad, R., Mousa , H. M.; Kim , I. G.; Kim , J. I.; Neupane , M. P.; Park , C. H.; Kim, C. S. High-Performance Glucose Biosensor Based on Chitosan-Glucose Oxidase Immobilized Polypyrrole/Nafion/Functionalized Multi-Walled Carbon Nanotubes Bio-nanohybrid Film. *J. Coll. Interface Sci.* **2016,** *482*, 39–47.
40. Shrestha, B. K.; Ahmad, R; Shrestha , S.; Park , C. H.; Kim, C. S. Globular Shaped Polypyrrole Doped Well-Dispersed Functionalized Multiwall Carbon Nanotubes/Nafion Composite for Enzymatic Glucose Biosensor Application. *Sci. Rep.* **2017,** *7* (1), 16191.
41. Katrin, P.; Daria, S.; Yuliya, S. E.; Krist, V. G.; Helena, J. Automated Electrochemical Glucose Biosensor Platform as an Efficient Tool Toward On-Line Fermentation Monitoring: Novel Application Approaches and Insights. *Front. Bioeng. Biotechnol.* **2020,** *8* 436.
42. Teymourian, H.; Barfidokht, A.; Wang, J. Electrochemical Glucose Sensors in Diabetes Management: An Updated Review (2010–2020). *Chem. Soc. Rev.* **2020, 49,** 7671–7709.

43. Kim, I.; Kim, C.; Lee, D.; Lee, S. W.; Gyudo Lee, G.; Yoon, D. S. A Bio-Inspired Highly Selective Enzymatic Glucose Sensor Using a Red Blood Cell Membrane. **Analyst 2020, 145,** 2125–2132.
44. Farajpour, N.; Deivanayagam, R.; Phakatkar, A.; Narayananl, S.; Shahbazian-Yassar, R.; Shokuhfar, T. A Novel Antimicrobial Electrochemical Glucose Biosensor Based on Silver–Prussian Blue-Modified TiO_2 Nanotube Arrays. *Wiley Online Library* **2020,** *3* (2), e10061.
45. Santos, A. S.; Freire, R. S; Kubota, L. T. Highly Stable Amperometric Biosensor for Ethanol Based on Meldola's Blue Adsorbed on Silica Gel Modified with Niobium Oxide. *J. Electroanal. Chem.* **2003,** *547*,135–142.
46. Santos, A. S.; Pereira, A. C.; Durán, N.; et al. Amperometric Biosensor for Ethanol Based on Co-immobilization of Alcohol Dehydrogenase and Meldola's Blue on Multi-Wall Carbon Nanotube. *Electrochim Acta* **2006,** *52*, 215–220.
47. Niculescu, M.; Erichsen, T.; Sukharev, V., et al. Quinohemoprotein Alcohol Dehydrogenase-Based Reagentless Amperometric Biosensor for Ethanol Monitoring During Wine Fermentation. *Anal. Chim. Acta* **2010,** *463*, 39–51.
48. Hooda, V.; Kumar, V.; Gahlaut, A.; Hooda, V. Alcohol Quantification: Recent Insights into Amperometric Enzyme Biosensors. *Artif. Cells Nanomed. Biotechnol.* **2018,** *46* (2), 398–410.
49. Tanriverdi, S.; Tuncagil, S.; Toppare, L. A New Amperometric Alcohol Oxidase Biosensor Based on Conducting Polymer of (4, 7-dithien- 2-yl-2, 1, 3-benzothiadiazole). *J. Macromol. Sci.* **2012,** *49*, 185–190.
50. Hammerle, M.; Hilgert, K.; Horn, M. A.; et al. Analysis of Volatile Alcohols in Apple Juices by an Electrochemical Biosensor Measuring in the Headspace Above the Liquid. *Sens. Actuatators B Chem.* **2011,** *158*, 313–318.
51. Gahlaut, A.; Dahiya, M.; Dhull, R.; et al. Fabrication of Nafion/HRPSWCNT/MWCNT-ZnO Based Alcohol Biosensor: Diagnostics for Forensic Analysis. *Int. J. Chem. Anal. Sci.* **2014,** *5*, 15–20.
52. Lansdorp, B.; Ramsay, W.; Hamid, R.; Strenk, E. Wearable Enzymatic Alcohol Biosensor. *Sensors* **2019,** *19* (10), 2380.
53. Lv, F.; Gong, Y.; Cao, Y.; Deng, Y.; Liang, S.; Tian, X.; Gu, H.; Yin, J. J. A Convenient Detection System Consisting of Efficient Au@PtRu Nanozymes and Alcohol Oxidase for Highly Sensitive Alcohol Biosensing. *Nanoscale Adv.* **2020, 2,** 1583–1589.
54. Wang, Y.; Fridberg, D. J.; Shortell, D. D.; Leeman, R. F.; Barnett, N. P.; Cook, R. L.; Porges, E. C. Wrist-Worn Alcohol Biosensors: Applications and Usability in Behavioral Research. *Alcohol* **2021,** *92*, 25–34.
55. Rahimi, P.; Joseph, Y. Enzyme-Based Biosensors for Choline Analysis: A Review. *Trends Analyt. Chem.* **2019,** *110*, 367–374.
56. Lopez, M. S. A. P.; Hervas Perez, J. P.; Lopez-Cabarcos, E.; Opez-Ruiz, B. L. Amperometric Biosensors Based on Choline Oxidase Entrapped in Olyacrylamide Microgels. *Electroanalysis* **2007,** *19*, 370–378.
57. Song, Z.; Zhao, Z.; Qin, X.; et al. Highly Sensitive Choline Biosensor Based on Carbon Nanotube-Modified Pt Electrode Combined with Sol-gel Immobilization. *Front. Chem. China* **2007,** *2*, 146–150.

58. Chauhan, N.; Pundir, C. S. Amperometric Determination of Acetylcholine-A Neurotransmitter, by Chitosan/gold-Coated Ferric Oxide Nanoparticles Modified Gold Electrode. *Biosens. Bioelectron.* **2014,** *61*, 1–8.
59. Tunc¸A. T.; Aynacı Koyuncu, E.; Arslan, F. Development of an Acetylcholinesterasecholine Oxidase Based Biosensor for Acetylcholine Determination. *Artif. Cells Nanomed. Biotechnol.* **2016,** *44*, 1659–1664.
60. Yu, G.; Zhao , Q.; Wu , W.; Wei, X., Lu, Q. A Facile and Practical Biosensor for Choline Based on Manganese Dioxide Nanoparticles Synthesized In-Situ at the Surface of Electrode by One-Step Electrodeposition. *Talanta* **2016,** *146*, 707–713.
61. Magar, H. S.; Ghica, M. E.; Abbas, M. N.; Brett C. M. A. A Novel Sensitive Amperometric Choline Biosensor Based on Multiwalled Carbon Nanotubes and Gold Nanoparticles. *Talanta* **2017,** *167*, 462–469.
62. Xie, L.; Huang, X.; Su, B. Portable Sensor for the Detection of Choline and Its Derivatives Based on Silica Isoporous Membrane and Gellified Nanointerfaces. *ACS Sens.* **2017,** *2* (6), 803–809.
63. Guerrieri, A.; Ciriello, R.; Crispo, F.; Bianco, G. Detection of Choline in Biological Fluids from Patients on Haemodialysis by an Amperometric Biosensor Based on a Novel Anti-Interference Bilayer. *Bioelectrochemistry* **2019,** *129*,135–143.
64. Tyagi, C.; Chauhan, N.; Tripathi, A.; Jain, U.; Avasthi, D. K. Voltammetric Measurements of Neurotransmitter-Acetylcholine Through Metallic Nanoparticles Embedded 2-D Material. *Int. J. Biol. Macromol.* **2019,** *140*, 415–422.
65. Tvorynska, S.; Barek, J.; Josypčuk, B. Acetylcholinesterase-choline Oxidase-Based Mini-Reactors Coupled with Silver Solid Amalgam Electrode for Amperometric Detection of Acetylcholine in Flow Injection Analysis. *J. Electroanal. Chem.* **2020,** *860*, 113883.
66. Bodur, O. C.; Dinc¸ S.; Ozmen, M.; Arslan, F. A Sensitive Amperometric Detection of Neurotransmitter Acetylcholine Using Carbon Dot-Modified Carbon Paste Electrode. *Biotechnol. Appl. Biochem.* **2020,** *66* (22), 20–29.
67. Huang, I. W.; Clay, M.; Cao, Y.; Nie, J.; Guo, Y.; Monbouguette, H. G. Electroenzymatic Choline Sensing at Near the Theoretical Performance Limit. **Analyst 2021, 146,** 1040–1047.
68. Fernandez, I. S.; Ruiz-Duenas, F. J.; Santillana, E.; Ferreira, P.; Martinez, M. J.; Martinez, A. T.; Romero, A. Novel Structural Features in the GMC Family of Oxidoreductases Revealed by the Crystal Structure of Fungal Aryl-alcohol Oxidase. *Acta Crystallogr. D Biol. Crystallogr.* **2009,** *65*, 1196–1205.
69. Koch, C.; Neumann, P.; Valerius, O.; Feussner, I.; Ficner, R. Crystal Structure of Alcohol Oxidase from *Pichia pastoris*. *PLoS One* **2016,** *11*, e0149846–e0149846.
70. Wohlfahrt, G.; Witt, S.; Hendle, J.; Schomburg, D.; Kalisz, H. M.; Hecht, H. J. 1.8 and 1.9 Å Resolution Structures of the *Penicillium amagasakiense* and *Aspergillus niger* Glucose Oxidases as a Basis for Modelling Substrate Complexes. *Acta Crystallogr. D Biol. Crystallogr.* **1999,** *55*, 969–977.
71. Bannwarth, M.; Bastian, S.; Heckmann-Pohl, D.; Giffhorn, F.; Schulz, G. E. Crystal Structure of Pyranose 2-Oxidase from the White-Rot Fungus Peniophora sp. *Biochemistry* **2004, 43**, 11683–11690.
72. Salvi, F.; Wang, Y. F.; Weber, I. T.; Gadda, G. Structure of Choline Oxidase in Complex with the Reaction Product Glycine Betaine. *Acta Crystallogr. D Biol. Crystallogr.* **2014,** *70*, 405–413.

CHAPTER 12

Supramolecular Gel-Based Materials for Sensing Environmentally Sensitive Molecules

S. K. SUJA and S. MATHIYA

Department of Chemistry, Lady Doak College, Madurai, Tamil Nadu, India

ABSTRACT

Supramolecular gels are at the cutting edge of material science. The interactions which constitute the solid-like network are non-covalent interactions like hydrogen bonds, van der Waals forces, π–π stacking, dipole–dipole, and coordination interactions. The integrant includes small molecules which assemble into the polymeric framework. These "paramount gels" with diversity in their structure exhibit manifold applications. Also, the possibility of manipulating the gel with entities sensitive to external light or chemical molecules makes them good candidate materials for sensing applications. The weak non-covalent reversible interactions contribute to the stimuli–response behavior. Accurate, selective, simple, and inexpensive detection method for sensing environmentally sensitive molecules is the primary requisite in the fast-growing technology world. The more complex nature of the environmentally polluting samples imposes a great challenge to the researchers to develop selective and accurate sensors. Gels are competent sensor materials owing to their macroscopic physical responses and their potential to act as optical and electrochemical sensors. The structure of these gels can be tuned to accomplish the detection of broad spectrum of pollutants. The selectivity in the detection of pollutants is enhanced by hybrid gels and suitably engineered gel nanocomposites. Gel-based materials exhibit improved analytical

Foundation and Growth of Macromolecular Science: Advances in Research for Sustainable Development. Meegle S. Mathew, PhD, Józef T. Haponiuk, PhD & Sabu Thomas, PhD, DSc (Eds.)

characteristics of sensors. Therefore, this chapter attempts to present an overview of the current state of research on supramolecular gel-based materials as sensors to sense environmentally sensitive molecules.

12.1 INTRODUCTION

12.1.1 HISTORY

Gels are 3D soft materials that exhibit properties between those of liquids and solids.[1] They are industrially significant as they possess intriguing physical properties.[2] The first gel which was reported in the literature by Lipowitz in 1841 was prepared from lithium urate. The sugar-based gelator which placed its record in the scientific literature is 1,3:2,4-di-O-benzylidene-d-sorbitol in the year 1891.[3] Gels are formed when hierarchically ordered 3D microstructures entrap solvents, thereby controlling their mobility in fluids.[4] The solvents are trapped in high quantities within the crosslinks created.[5] Paul Flory established four canonical categories of gels[6]:

(i) High-order lamellar gels (e.g., phospholipids)
(ii) Disordered covalent networks (e.g., vulcanized rubber)
(iii) Semiordered physical networks (e.g., gelatin)
(iv) Disordered particulate gels (e.g., reticular fiber networks).

The definition of gel given by the International Union of Pure and Applied Chemistry (IUPAC) is, "nonfluid colloidal network or polymer network that is expanded throughout its whole volume by a fluid."[5,7] Clearly, traditional gel chemistry is dominated by polymers. Recently much interest is developed on gel systems in which the fibrous aggregates are made of small molecules that are interlinked in a non-covalent fashion. It is these "supramolecular" gels, composed of low molecular weight gelators.

12.1.2 TYPES OF GELS

Broadly, gels are classified into two categories, namely, physical gels and chemical gels. Self-assembly of low-molecular-weight organic molecules in appropriate solvents through weak intermolecular interactions such as hydrogen bonding, electrostatic interactions, van der Waals interactions, and π–π interactions, results in physical gel.[4] Based on the solvent that is congealed, physical gels can be further classified into hydrogels or aqua gels and organogels.[4,8] Hydrogels encapsulate water molecules within the cavities of the self-assembled network.[9] Organogels immobilize organic solvents within the domains of the well-organized 3D networks. They are categorized as thermally reversible or irreversible, depending on the mechanism of interaction for the formation of their network in the system.[10] Chemical gels are formed by the extensive use of covalent bonds.[9]

The intermolecular forces among the low-molecular-weight gelators are so delicate that it responds to any stimulus from external including heat,[11] light,[11,12] ultrasound,[11] electric or magnetic fields,[11] pH,[13] and oxidative/reductive reactions.[11,14] "Supramolecular" gels are network structures with assorted properties and applications which are indeed due to low molecular weight molecules. The prime propulsion behind the formation of these gels is non-covalent bonds like solvophobic interactions, π–π stacking, hydrogen bonding, etc. These weak interaction forces among the gelators make supramolecular gel peculiar.[6] The most common strategy to develop a rigid gel structure is to use multiple weak interaction forces.[15,16]

12.1.3 GEL STRUCTURING PROCESS

Based on the number of components involved in the gel formation, gels are classified as:

(i) Single component gel system
(ii) Multicomponent gel system.

12.1.3.1 SINGLE-COMPONENT GEL SYSTEM

The gel network formation primarily involves the dissolution or dispersion of the gelator molecules, followed by their assembly utilizing the non-covalent interactions. The above phenomenon does not require any external energy. Self assembly of suitable functional moieties leads to hierarchical assembly which can suitably entrap solvent molecules.[17]

Figure 12.1 represents the formation of self-assembled fibrillar networks from gelators. Step A shows the nucleation of the gelators, followed by the formation of fibrous structure (step B). The formed fibers entangle to form 3D networks with the aid of non-covalent interactions. The fibers can also take off into any suitable structural direction, landing to the bundling of fibers by lateral overlap[17] or even compartmentalization.[18] The individual motifs may also form hollow tubular nanostructures,[19,20] fibers,[20,17] spheres, coils, and sheets[20] making use of the intermolecular forces which is the deciding factor of the mechanical property.[17] The properties of the supramolecular gels largely depend on several parameters, which could be optimized. Apart from the solvent molecules entrapped and the gelator concentration, it is the functional tag of the gelator and additives incorporated during gelation impart differences in the gel structure,[19] strength, electrical, and biological properties.[6]

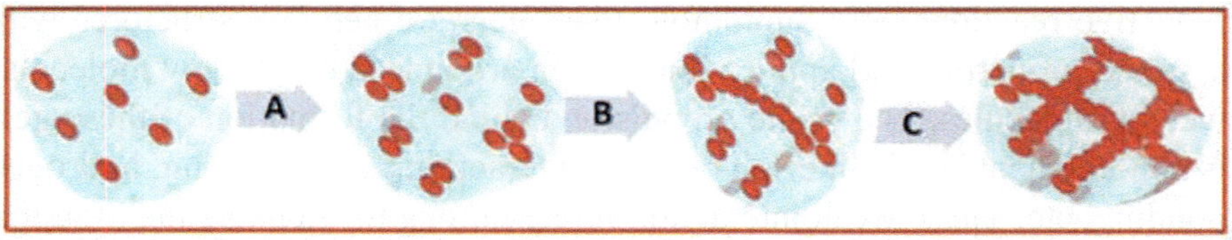

FIGURE 12.1 The phenomenon of gel formation A = nucleation of the gelators; B = fiber formation; C = gel formation.

12.1.3.2 MULTICOMPONENT GEL SYSTEM

Multicomponent gel systems are built by involving more than one type of gelator.[21] The three classes of multicomponent systems are: (1) two gelators involved and both cannot form gel separately, only in combined phase they self-sort to form gel; (2) a two-gelator system and both are capable of forming gels independently; and (3) a gelator and a non-gelling additive.[22] When more than one gelator is used in the process of self-assembly, the property of the gel is affected by the process of gelation, that is, mainly on the rate of gelation and rate of mixing.[17] A deep study of the gelator components at the molecular level can reveal the possible non-covalent interactions among them. Hence, the properties and structure of the gel system could be well speculated and tuned.[23] Hence, it is difficult yet a challenging task to foresee the self-sorting process of a multicomponent gelator system. Also, the structure of the gel matrix formed depends on various factors, including the 3D projection of the gelators.[23] Hierarchical compartmentalization by multilevel self-sorting was also reported. The in situ formed two gelators from a non-assembling building block, self-sort themselves into two separate fibers. These fibers were self-sorted into separated microdomains, which then formed compartmentalized gel frameworks.[18]

12.2 GEL

A lot of gel systems have been explored using small molecules and low-molecular-weight gelators. Owing to the availability of library of functional motifs and the possibility of architecting new compounds, the probability of designing gels is very high. But they could be classified depending upon the type of gelator (Fig. 12.2) used and the non-covalent interactions involved in the gelation process. The different types of gels are given in Figure 12.3.

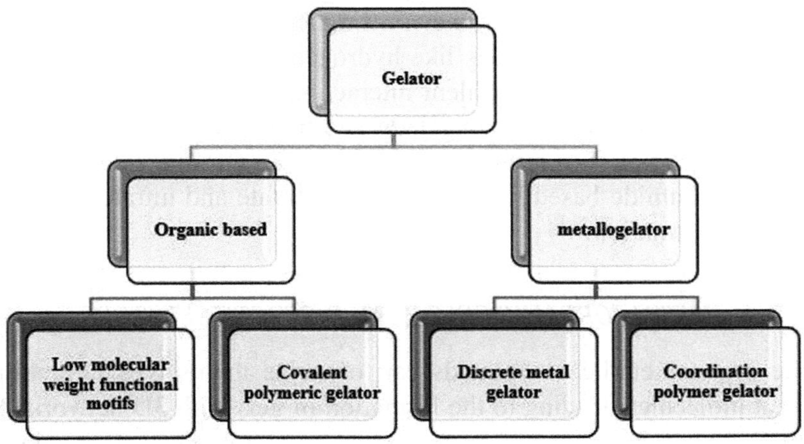

FIGURE 12.2 Classification of gelators.

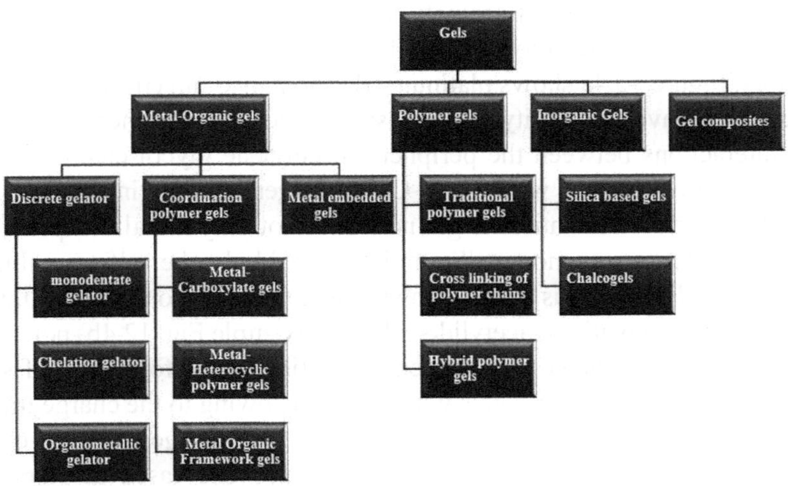

FIGURE 12.3 Different types of gel systems.

12.2.1 METAL–ORGANIC GELS

Metal-organic gels (MOGs) are self-constructed 3D networks utilizing mainly the metal ion and the organic moiety. These gels are known for their highly porous nature and large surface area, with the scope for application in the field of adsorption and sensing. The interaction of the metal ion with the organic moiety in MOGs is broadly of two types. Type 1 involves the self-assembly of metal-organic compounds as gelators using non-covalent interactions.[24] Type II metal-organic gels include the incorporation of metal nanoparticles in the matrix of type I gels. Metal-organic gel also includes 3D association of metal-organic frameworks directed by coordination bonds and weak non-covalent interactions like hydrogen bonding and π–π stacking.[25] Apart from the other non-covalent interactions, it is also the nature of the metal coordination with the nongelator and metal salt involved are the key factors for gelation. For example, chlorides of Zn (II) and Cd (II) forms gel with pyridyl amide-based ligands, whereas acetate and nitrate salts did not induce gel formation.[26]

12.2.1.1 METAL–ORGANIC GELS WITH DISCRETE GELATORS

Some organometallic compounds can organize themselves immobilizing solvent molecules, leading to the formation of gels with 3D network. Apart from hydrogen bonding and π–π interaction, metal–metal interaction also could provide additional stability to metal-organic gels in certain gels.[5]

Ligands customized with H-bonding motifs, van der Waals interaction and appropriately oriented aromatic ring could act as better gelator when coordinated to the metal ion. Figure 12.4 shows the series of metal-organic gelators. Figure 12.4a shows platinum (II) and palladium (II) dipyridyl gelators, which have the ability to form supramolecular polymeric structures. The interactions between the peripheral triethylene glycol chain, aromatic π–π interaction, and the weak C–H–C–l hydrogen bonding interactions play a vital role in the formation of gel network. Not only metal complexes are potential gelators, organometallic moieties could also be self-assembled to gel systems. Ferrocene is a well-known organometallic compound. Ethynylpyrene-modified platinumacetylide gelators (example Fig. 12.4b) potentially entrap mixture of alkane and benzene, resulting in the formation of stable gels. They also form hybrid gels with graphene, owing to the charge transfer interaction among them. A lot of ferrocene–peptide gelators have been synthesized (Figs. 12.4c–4d). Most of the ferrocene-based gelators have three main compartments: (1) the ferrocenyl component which supports

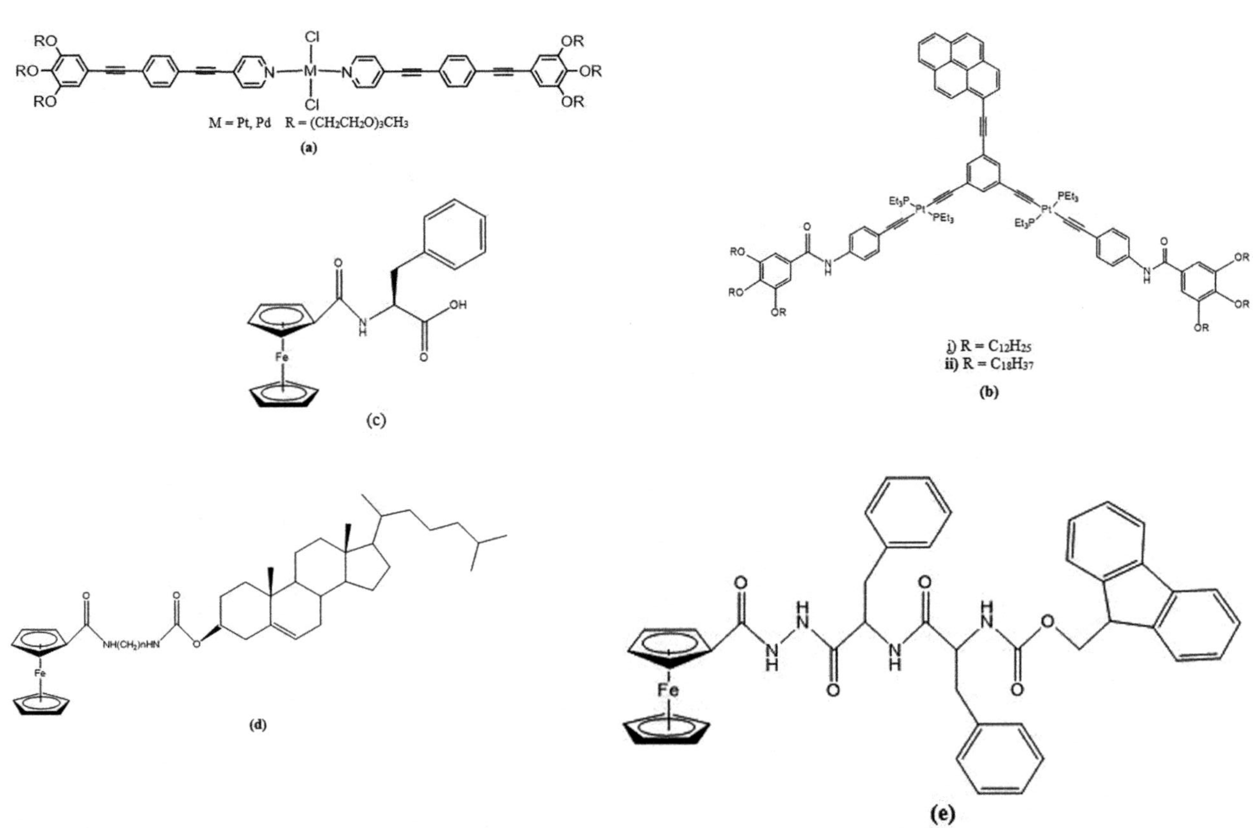

FIGURE 12.4 Structures of metal-organic gelators (a) oligophenyleneethynylene (OPE)-based amphiphilic *trans*-Pt^{2+}, (b) ethynyl-pyrene-modified platinumacetylide organometallic gelators, (c) ferrocenyl phenylalanine, (d) ferrocenyl diamino with cholesteryl moiety, (e) ferrocene-peptide gelators protected by fluorenyl-9-methoxycarbonyl.

the gel by its π–π interaction, (2) peptide linkage, the source for hydrogen bonding, and (3) the hydrophobic or the aromatic system.[27]

12.2.1.2 COORDINATION POLYMER GELS

Coordination polymer gels are a class of metal-organic gels, which are formed from coordination polymer gelators. Construction of coordination polymer gels requires a metal ion and a minimum of two coordination centers. Hence, the structure of the polymeric framework formed depends on the geometry of the metal ion and the structure of the gelators.

The dynamic nature of coordinate bond involved in metal–ligand is explored for:

(i) Building of controlled polymeric gels,
(ii) Producing stimuli-responsive coordination polymer gels,
(iii) Applications as sensors, drug delivery systems, and optoelectronics.[28]

Studies on the gelation of coordination polymer prepared from amino-pyridine conjugates and in situ preparation of silver nanoparticles have been conducted. The active site in the ligand is the presence of tertiary nitrogen, which could serve as a donor for the incoming metal ion and has the potential to form metal complex. From the study, it was observed that Ag^{+} has the potential to drive the process of gelation. They orient the coordination polymerization, acts as a good cross-linker and a starting material for the formation of silver nanoparticles. The process of gelation was driven by tetracoordinated silver. Overall, the gelation process depends on the nature of (1) cation, (2) anion, (3) solvent, and (4) ratio of ligand:metal.[29]

12.2.1.2.1 Metal-Organic Gels with Bridging Ligands

There are a group of metal-organic gels formed using a variety of bridging ligands. Availability of the metal ion species and the wide scope of tuning the structure of bridging ligands offers a platform to design a range of coordination polymer gels. The most commonly encountered class of bridging ligands are (1) carboxylate ligands, (2) heterocyclic ligands, and (3) ligands with hybrid donors.

Bridging carboxylate ligands have the potential to coordinate with the metal ions forming coordination polymer. The most commonly studied metal ions which possess the ability to coordinate with dicarboxylates are Fe^{3+},

Al^{3+}, and Cr^{3+}. The nature of the coordination polymer is governed by several parameters like the metal-to-ligand ratio, nature of solvent, and nature of the counter ion. The coordination polymers formed could be amorphous or crystalline in nature. Faster reactions lead to the formation of an amorphous coordination polymer, whereas the slow process leads to the assembly of the organic ligands and the metal ions, resulting in a crystalline framework. The formed polymers could suitably entrap the solvent, resulting in the formation of gels.[27] Aluminum (III) formed gel with 12.5a and 12.5c exhibiting luminescence. The formed gels were used for sensing of picric acid with less than 1 ppm.

Ligands with heterocyclic rings are potential donors that coordinate with the metal ions. The ligand 12.5d forms metal-organic gels with cobalt bromide. The gel was found to be a potential sensor for selectively detecting toxic vapors like hydrochloric acid, thionyl chloride, oxalyl chloride, and phosgene. Sonicating a mixture of ligand 12.5e with zinc triflate in 2:1 ratio leads to the conversion of coordination polymer microparticles to nanofibers with the potential to entrap organic solvents and form gels. Tripodal terpyridine moieties were also utilized in the preparation of metal-organic gels. When a cyclam group was incorporated between two terpyridine groups 12.5f, it shows good coordinating affinity toward Ni^{2+} ion, resulting in gel formation.

Apart from the pure carboxylate and heterocyclic bridging ligands, hybrid donors are also equally explored for their ability to coordinate with metal ion leading to gels. The ligands like 12.5g with two donor groups were used in the synthesis of metal-organic gels with heterometallic ions, where the Pd^{2+} coordinates with the pyridine moiety and the other metal ion with the carboxylate moiety. Similarly, Zn^{2+} forms metal-organic gels with ligands containing hybrid donors 12.5h in the presence of triethyl amine. The gel framework is stabilized by the coordination between the metal ion and the hybrid donors, π–π stacking interaction due to aromatic moiety and metal–metal interaction. These gels are good visual sensors for the detection of anions, as they could easily interact with Zn^{2+} ion.[27]

12.2.2 COVALENT POLYMER GELS

Polymer gels are formed through the cross-linking of long polymer chains. Silicone-based gels and polymer hydrogels fall under this type. The polymeric chains consist of covalent backbone and the cross-linking could be hydrogen bonding, coordination bonding, and dynamic covalent bonding.

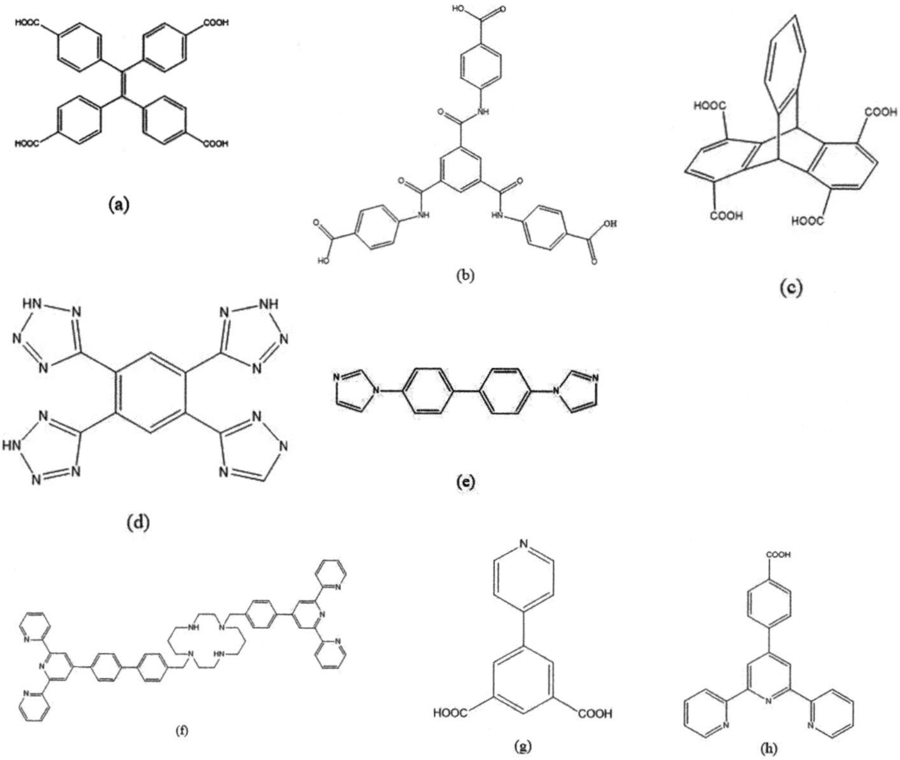

FIGURE 12.5 Structure of different types of bridging ligands (a) tetrakis(4-carboxyphenyl) ethylene, (b)tris-amide-tris-carboxylate ligand, (c) 1,4,5,8-triptycenetetracarboxylic acid, (d) hybrid ligand 5- (pyridin-4-yl)isophthalic acid, (e) cyclam bis-terpyridine, (g) tetrazole-appended ligand, (h) 4,4 –bisimidazolylbiphenyl.

A lot of work has gone into the exploration of possible dynamic covalent bonding. Figure 12.6 represents the various dynamic covalent bonds utilized for the formation of polymer gels.[27]

12.2.3 INORGANIC GELS

The inorganic gels dating back to 1846 mainly include silica-based gels and chalcogels. Inorganic gels are mainly obtained in the colloidal medium. Silica and metal oxides in their colloidal state involved in the process of gelation. Silica-based gels are obtained by the cross-linking of silica nanoparticles using silicon alkoxide. The pH, proportion of water, and silicon alkoxide could be optimized for complete hydrolysis and condensation. Gel–bacteria composites prepared using silica-based gels were used for the elimination of

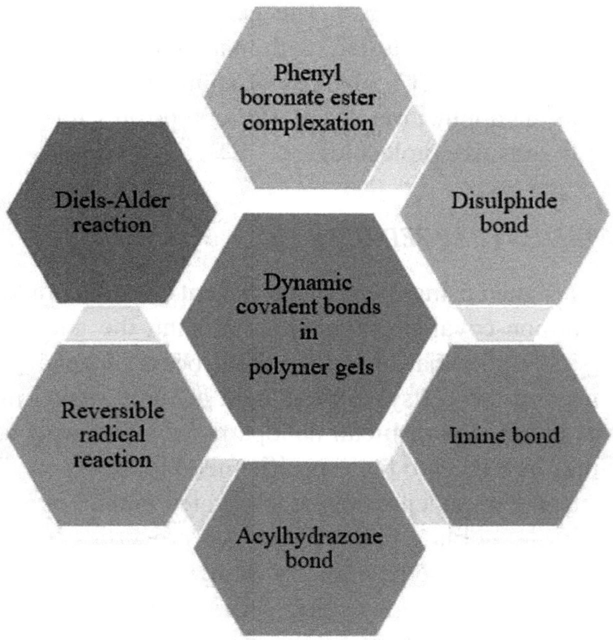

FIGURE 12.6 The dynamic covalent bonds involved in polymer gel formation.

water pollutants. Owing to the less weight and increased mechanical properties, silica aerogels have been much explored for lightweight materials. Fumed silica-based gels have also been prepared and employed as working electrolytes in valve-regulated lead-acid batteries. Sulfide and selenide clusters were also studied for their gelation behavior. The advantages of possessing a hydrophobic surface and stability in humid conditions have been explored. Chalcogels suits to be used for biological applications like nitrogen fixation. Non-platinum chalcogels are suitable candidate materials for energy applications.[27]

12.3 GEL NANOCOMPOSITES

The structure and properties of the gel matrix are distinctive.[20] The gelator molecules self-assemble to form various nanostructures, which also results in the development of microdomains. These microdomains with suitable pockets entrap the solvent molecules in their nearby environment.[20,30] Such pockets in the entangled gel system also can trap with them nanomaterials in a host–guest fashion forming gel nanocomposites. A variety of nanomaterials could be incorporated as guests.[20] The gel nanocomposites would hence serve as better candidate materials with synergistic effects.[31–34] These gel nanocomposites are used for a plethora of applications. In this chapter, the focus is only on the application of gel nanocomposites for the detection of environmentally sensitive molecules.

12.3.1 METAL-ENTRAPPED GELS

Gels could be designed using a single component or multicomponent gelator. It is mainly the non-covalent interactions holding the gel network. Metal species possess the potential to initiate the process of gelation. Additional stabilizing forces in the gel system could be achieved using metal components either as metal ions or metal nanoparticles.[5] The process of gelation could sometimes be triggered by a specific metal ion.[35]

Also, the metal component when incorporated into an organic moiety of the gel network, the gel is gifted with add-on properties like conductivity, optical activity, and magnetic property. It can also sometimes lead to a change in the morphological structure.[5] Instigation of the gel formation is also brought about by the metal nanoparticles.[26] There are also reports on the weakening of the gel upon incorporation of metal species. Hence, functionalization of the gel is possible using metal components. This in turn offers the metal-incorporated gel with wide scope for applications.[5] The metal component in the gel system can be of the following types:

1. Metal nanoparticles
2. Metal ion as cross-linker
3. Metal–ligand coordinates bond as cross-linker

The unique properties associated with the size and shape of the metal nanoparticles have rooted deeply in various applications of science. Hence, the amalgamation of metal nanoparticles with the self-assembled gel network may lead to the development of novel composite materials with complementing properties and potential applications. The gel due to its unique morphology and porosity has the potential to reduce and stabilize the metal nanoparticles. Therefore, the metal nanoparticles could be formed during the gelation process. The surface morphology and the porosity of the gels could be architected by the proper choice of the gelator and the gelation protocol. Hence, it is viable to obtain metal nanoparticles of different sizes and shapes using the supramolecular gel as templates.[30] The supramolecular gels play a dual role of acting both as templates and trapping sites for the metal nanoparticles.[30,36,37] Also, the 3D network with the pool of functional groups can facilitate the nucleation of metal nanoparticles.[36,37]

Amino acids are the most commonly employed gelator which has the potential to generate metal nanoparticles. Employing cysteine-based compounds for the synthesis of metal gel framework serves for two purposes: stabilization of the metal nanoparticles, and exerts a residual interaction with the gelators in the gels system. These "residual" interactions also add to the mechanical strength of the gel system. When the size and concentration of nanoparticles increase, there occurs a reduction in the interactions among the gelator molecules. Hence, the strength of the gel is lowered.[38]

Also, amino acid with pyrrole moiety acts as a good gelator. The pyrrole group takes up two roles: it acts as a reducing agent to reduce the metal ion to metal nanoparticles and oxidizes to form polypyrrole conductive networks.[39] Silver nanoparticles have been prepared using curcumin-based organogelators.[36] Hydrogel incorporated with metal nanoparticles (NPs) has kindled researchers in the field of tissue engineering.[40]

Literature provides examples of silver nanoparticles incorporated in the MOGs, where the nanoparticles were grown without disturbing the matrix of metal-organic gel. One of the strategies that could be employed for the synthesis of metal nanoparticles/MOGs hybrid is the choice of the ligand. The ligand chosen should be capable of coordinating with the metal ion as well as a potential reducing agent to generate metal nanoparticles.[24] *In situ* synthesis of gold nanoparticles MOG (Fe) hybrids in the presence of tannic acid was successfully achieved. The synthesized hybrid gel was found to exhibit peroxidase-like activity and incredible chemiluminescence enhancement ascribed to the presence of gold nanoparticles. Apart from enzyme mimicking systems, the application of the above-mentioned hybrid gel was extended to quantitatively analyze the organophosphorous pesticides in real water samples.[41]

Ongaratto et al.,[42] synthesized fatty N-acylamino hydrazide derived from fatty acids and amino acids as gelator and template for the in situ generation of gold nanoparticles. The gelator with serine formed opaque gel and gelator derived from alanine, valine, and phenylalanine resulted in a gel with translucid properties. Microwave-assisted synthesis of gold nanoparticles using auric chloride and organogelator was carried out in the absence of any external reducing reagents. The in situ generation and stabilization of the gold nanoparticles are due to the presence of hydrazine moiety in the gelator. The gelator system with suitable hydrazine coordinates to the gold ion forming complexes, followed by the reduction of gold ions using microwave irradiation. The presence of long-chain fatty acids also contributes to the narrowing down of the size of metal nanoparticles.[42] Hence, metal-embedded gel could be prepared using a suitable gelator with potential reducing and stabilizing groups.[42] Therefore, compared to the pristine gels, the MOG hybrids possess a plethora of applications due to the synergistic enhancement in their properties.[24]

12.3.2 GEL NANOCOMPOSITES WITH CARBON-BASED MATERIAL

With the development of different gel systems, suitable nanocomposites were prepared using 0D and 1D nanostructures. The gel matrix as hosts and the added nanostructures as guests, there exists a non-covalent host–guest interaction.[20] Hence, the suitable choice of the gelator and the guest compound are important to attain a uniformly distributed composite material leading to robust gel composites. The versatile application of nanocomposite material in various fields, stems from the property of synergistic effect. It was found from literature that composite gels exhibit more remarkable properties than the native gel.[24] As discussed above, the nanoparticle-based gel composites were found to be more suitable for optoelectronic or catalytic applications by acting as nanofillers in the composite matrix. Incorporation of 1D and 2D nanocarbon into gel matrix would lead to robust gel material as these nanocarbons can strengthen the gel matrix. For example, guests with scope for π–π stacking could be accommodated by aromatic gelators. Also, these composites are more focused on energy applications and electronic devices.[20]

12.4 DETECTION OF ENVIRONMENTALLY SENSITIVE MOLECULES

The weak non-covalent interactions are mainly responsible for the network structure in the gel. There are a lot of corroboration for applications of gel in many areas like environmental remediation, catalysis, tissue engineering,[11,43] etc. One important area which requires lot of attention is sensing of environmentally sensitive molecules. There is a great demand for the architecture of "smart gels" which can sense the external stimuli. The gel reacts to the stimulus by the following:

(i) Conversion to sol or gel
(ii) Turn on-off fluorescence
(iii) Change in color of the fluorescence
(iv) Change in visual color of the gel either with retention or destruction of the gel structure.[44]

Figure 12.7 depicts the response of the gel to the external stimulus.

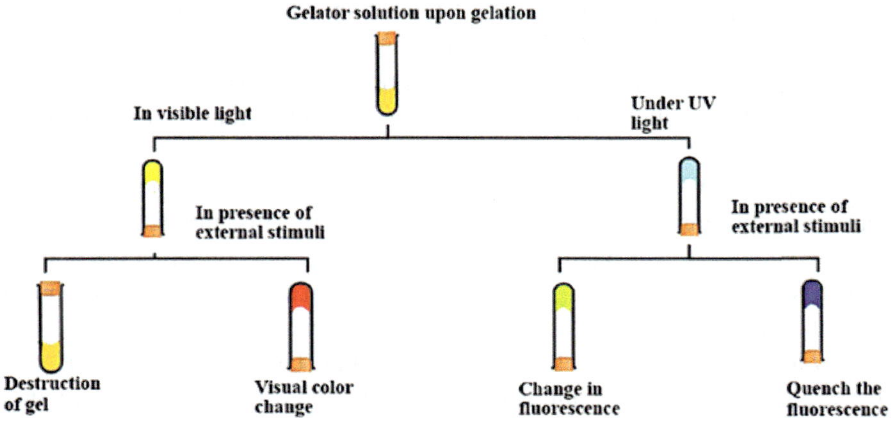

FIGURE 12.7 The various stimuli response behavior of gels.

These kinds of visual responsiveness for the external stimuli can broaden its horizon in the field of sensing. The gel material could serve for versatile analyte sensing. The advantages of employing gels as sensors are given in Figure 12.8.

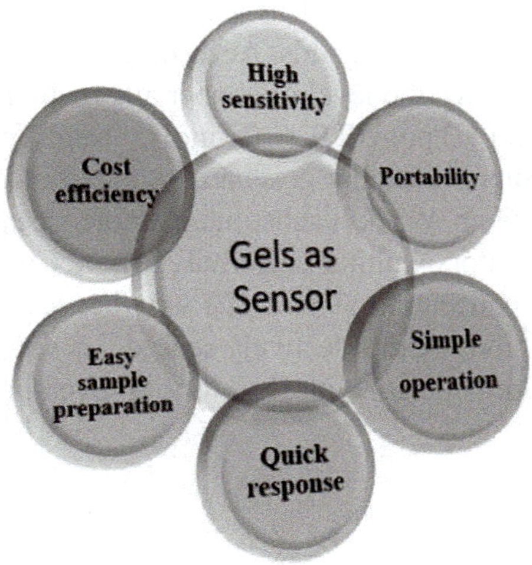

FIGURE 12.8 The advantages of gels as sensors.

12.4.1 HEAVY METAL SENSING

The presence of toxic heavy metal ions in biological and environmental samples is still a problem.[45] Many advanced analytical techniques like atomic absorption spectrometry (AAS), mass spectrometry (MS), inductively coupled plasma MS (ICP–MS), atomic emission spectrometry (AES), and X-ray fluorescence are available to determine the levels of heavy metals contaminants in water, which are expensive and unavailable at all points of requirements. So, there is a desideratum for an in situ sensor to assess the presence of heavy metals.[46] It is a well-known chemistry that cationic species can reversibly interact with the π-electron system. This interaction could be very well exploited to plan for the design of smart gels which could act as a selective sensor for cations.[47] The gelators for such application will be chosen in such a way they have aromatic moiety which serves dual purpose. They act as π–π interaction site as well as π-electron donor. The cations Hg^{2+}, Cd^{2+}, and Ag^{+} are soft acids, so they interact efficiently with the π electrons. Hence, they can act as a highly selective sensor for sensing these cations.[48,47] Moreover, the binding of mercury ions does not lead to the disrupt of the gel structure. Copper, despite serving in many physiological processes, excess of it displays toxic effects including neurodegenerative diseases. Mercury, even in very low concentration, when entered into the human body, creates serious health issues. These metal ions are proved to be highly threatening to human and to the environment. Sensing of copper ions using supramolecular gel, often leads to the disruption of the gel matrix. This may be due to the complexing ability of Cu^{2+} ions with the gelator ligands. According to the U.S. Environmental Protection Agency (EPA), the level of copper and mercury in drinking water cannot exceed beyond 1.3 ppm (2.0×10^{-5} mol L^{-1}) and 2 ppb (9.97×10^{-9} mol L^{-1}), respectively. Supramolecular gels based on indolin-2-one and quinoline moieties with long chain *N*- alkyl substituted amide group were synthesized, among which one gelator exhibited selectivity toward Cu^{2+} and Hg^{2+} ions. The minimum gelation concentration was 1% w/v, hence called supergelators. Only Cu^{2+} and Hg^{2+} ions were selectively sensed.[11] A series of aminothiazole ligand forms metal-organic gel with Hg^{2+} ions as given in Figure 12.9.[27]

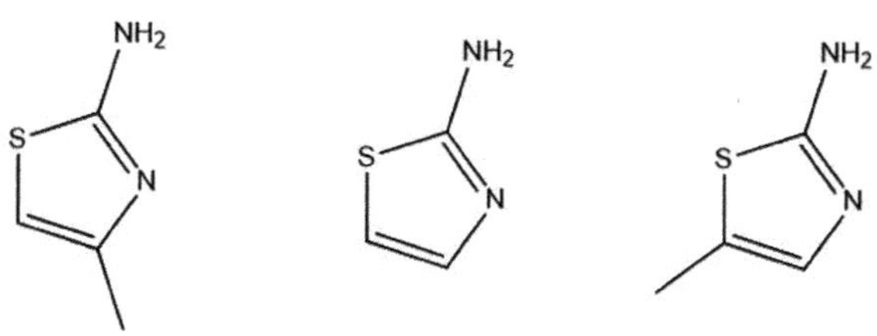

9a 9b 9c

FIGURE 12.9 The structure of the aminothiazole gelators.

Yadav et al.,[27] studied the gelation abilities of the aminothiazoles with various metal acetate. They found that mercuric acetate formed coordination polymers with the aminothiazole ligands 12.9(a–c). The polymers formed by the ligands 12.9a and 12.9b exhibited gelation properties at room temperature without any external stimuli. Among them, it was 12.9a, which formed a strong gel with aminothiazoles. From NMR studies, it was found that the suitably positioned methyl group in 12.9a is involved in the process of gelation. The addition of mercuric acetate to the aminothiazoles 12.9a, caused a significant shift in the signal of the methyl proton, indicating its participation in hydrogen bonding. The choice of acetate salts over the halides may be due to the ability to coordinate with the metal ion through the oxygen atom. Hence, the synthesized coordination polymer-based metallogel was reported for the detection and quantitative removal of Hg^{2+} ions.[27]

12.4.2 CYANIDE SENSING

Among the various ionic analytes, cyanide is considered to be highly hazardous to human[49–51] and to the environment. It is the pollutant expelled from textile, paper, and plastic industries. The maximum acceptable level of cyanide in drinking water is 0.2 ppm.[43] This has driven the scientific community to develop a sensitive and selective sensor to detect its presence. Despite the development of various analytical and spectroscopic tools for the detection of cations and anions, there exists a need for a simple and low-cost method.[43,52–55] Gel-based cyanide sensors are in the infancy stage with scanty work published. The addition of anion affects the gel nature, leading to the collapse of the metallogel.[26] Cadmium (II) metallogels are good candidates for sensing of cyanide due to the strong coordination of cyanide compared to other halide anions.[26] Development of visual sensors for ionic analytes is possible, as they could alter the chemical and optical properties of the gel. The external ionic analyte interacts with the network structure of the gel. This can happen in two ways. First, the common trait observed is that the added ionic analyte breaks the gel network structure. Secondly, the ionic analyte has the potential to alter the network, resulting in a gel-to-gel transition.[43] The anions interact with the binding site of the gelator through electrostatic interaction or via hydrogen bonding.[43] The gelators with suitable anion binding sites, respond to the anions either through electrostatic interaction or via hydrogen bonding.[43]

During cyanide sensing, the cyanide rummages the metal ion present in the gel resulting in anyone of the following process depending on the nature of the gelator:

(i) Destruction of the assembled structures
(ii) Gelators reassemble to form gel without the metal ion.

Gelators bind with cyanide ion using the following binding sites:

1. Metal ion: The gelators containing the metal ion as the binding site are formed by the coordination of the metal ion to an appropriate ligand. Here, the metal ion binds with the incoming cyanide ion through electrostatic interaction.
2. Hydrogen bond donor: The gelator moieties with hydrogen bond donor can efficiently form intermolecular hydrogen bonding with the cyanide ion. This leads to deprotonation from the hydrogen bond donor.

There is a dearth of understanding of the gelation behavior, hence design-based molecules for visual sensing are much less explored.[43] Recently, metal-free ionogel-based sensors exhibit visual detection of cyanide ions with the retention of the gel matrix. The ionogels are imparted with multiple self-assembly driving forces. Hence, these types of ionogel-based sensors could be developed as sensor kits to monitor the presence of cyanide ions in water samples.[15]

The supramolecular polymer with long-alkyl chained acylhydrazone derivative (Fig. 12.10) possesses aggregation-induced emission property. The presence of acylhydrazone moiety serves as a site for the coordination of a suitable metal ion. The long alkyl chain imparted the gel with strong van der Waals interaction and the aromatic rings responsible for the π–π stacking. They utilized the polymeric material for the detection of Hg^{2+} and Fe^{3+} in water. The Hg^{2+} and Fe^{3+} ions are coordinatively combined with the polymeric metal to form metallogels with the quenching of fluorescence. Upon addition of various anions to the mercury-based metallogel, it was cyanide anion that turned the fluorescence on.[16]

12.4.3 EXPLOSIVE SENSING

A very commonly known explosive is picric acid which is more powerful than that of trinitrotoluene. It is a highly soluble toxic compound imposing serious threats against humans and environment. The use of these compounds is inexorable in rocket fuel manufacturing and pharmaceuticals. The gelator used for sensing of picric acid bears multiple sites available for non-covalent interactions to form a self-assembled system.[56]

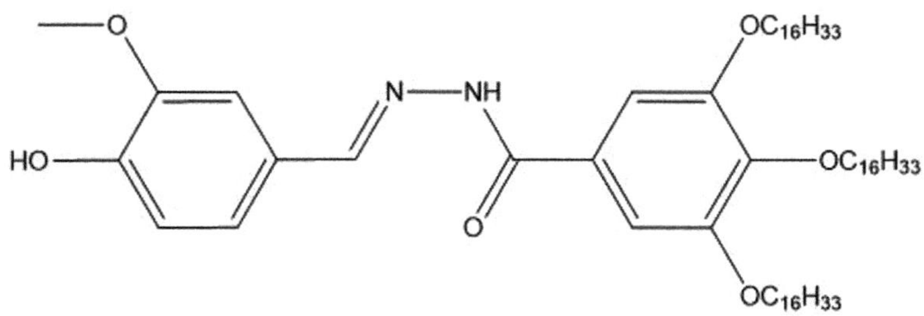

FIGURE 12.10 The structure of long-alkyl chained acylhydrazone.

Cholesterol-linked heterocyclic motifs have also been designed and synthesized for sensing picric acid. For example, benzothiazole[57] and imidazole-based[58] systems have been developed for visual sensing of picric acid. The compounds 12.11a and 12.11b exhibit gelation in various solvents. The cholesteryl moiety, through its hydrophobic interactions, organizes the molecular aggregation, in various solvents.[57,58] The absence of such cholesteryl motifs (Fig. 12.11c) does not result in gel formation, which explains the importance of the cholesteryl system in the process of aggregation.[58] The π–π stacking among the benzothiazole group also facilitates the self-aggregation process in the formation of gel using compound 12.11a.[57] Similarly, the imidazole moiety severs as a hydrogen bonding site and the anthraquinone establishes the π stacking leading to self-aggregation of the gelator 12.11b.[58] The nitrobenzene gel of compound 12.11a and 12.11b acts as a selective sensor for sensing picric acid over the other nitroaromatics. The sensing process involves the gel-to-sol conversion with an identifiable color change. The color change is from pale yellow to deep yellow with the nitrobenzene gel of 12.11a,[57] and reddish brown to yellow with 12.11b.[58] In gelator 12.11d, the heterocyclic aromatic ring is responsible for π–π stacking and dipole–dipole interaction, the tetrazole ring involves in hydrogen bonding, and the long alkyl chains offers hydrophobic interaction. The compound forms gel in dimethyl sulfoxide and is a "supergelator" as it has a low critical gelation concentration. Upon addition of picric acid, the gel was destructed leading to a gel-to-sol conversion. The destabilization was due to the disruption of the hydrogen bonding in the gel structure. Amidst various nitro aromatics, the gel was able to sense picric acid selectively.[56]

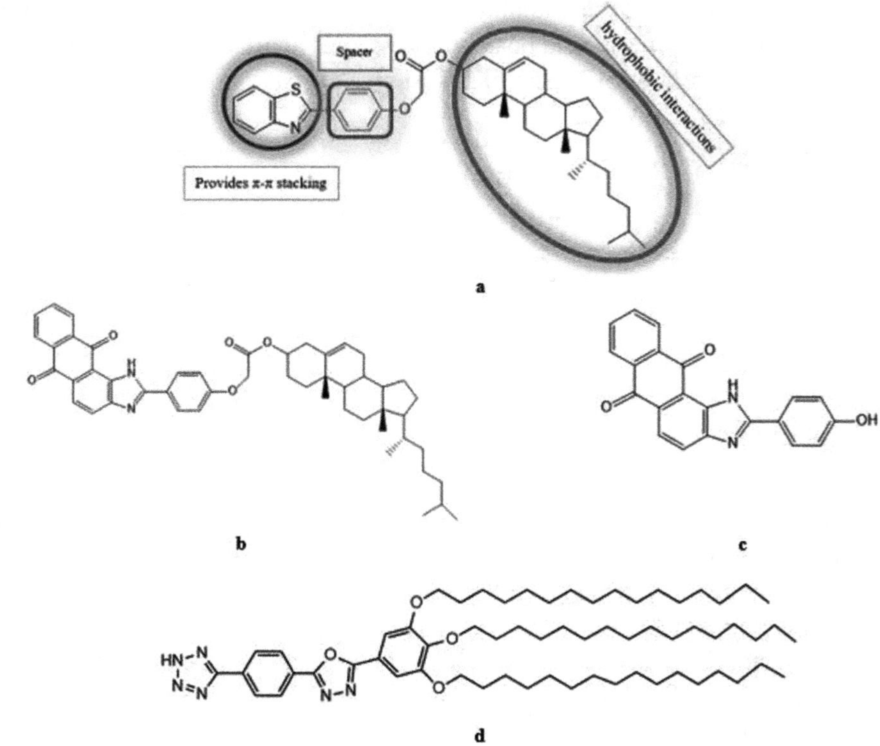

FIGURE 12.11 Structure of the gelators. (a) Cholesterol-appended benzothiazole. (b) Anthraquinone derived cholesterol linked imidazole. (c) Structure of (b) without the cholesteryl motif (d) low molecular organogelator with 3,5-substituted-1,3,4- oxadiazole and tetrazole units with long flexible alkyl chain units.

12.4.4 NITRITE SENSING

Human beings can never circumvent the exposure to nitrite as it is used in the food industry as preservative. The excess of this nitrite leads to the destruction of red blood cells and may be carcinogenic when reacted with other substances in certain cases. Simple spectroscopic methods are available to easily detect the presence of these nitrites, but they suffer from interferences[59] and difficulty in sample preparation.[60] Hence, they lack sensitivity. Electrochemical sensors are well known for their sensitivity and stability in their operation. It is well known that the bare glassy carbon electrodes manifest less sensitivity for the detection of analytes. Many graphene-based materials and nanocomposites have been designed to vanquish the sensitivity issue. The higher surface area offering a suitable binding site for the analytes to be sensed makes these materials viable for sensing applications. The disadvantage lies in the fact that they are developed under high conditions of temperature and requires lot of treatment after modification. To overcome this difficulty, metal-organic gels were designed by the self-assembly of ligands and metal ions utilizing the non-covalent interactions. They meet the required criteria for electrochemical sensing by possessing high surface area and accessibility. Even then, the application of MOG for electrochemical sensing is much less explored due to its viscoelastic solid-like property. The copper (II) metal-organic gel using 2,6-bis(2-benzimidazolyl) pyridine was synthesized. The gel responded to the addition of ammonia by converting into a blue-colored sol, which was due to the greater coordinating ability of ammonia compared to the ligand. The glassy carbon electrode was modified with metal-organic xerogel for the sensing of nitrite. After polishing the glassy carbon electrode, the metal-organic xerogel suspension was cast onto the surface of the electrode. The greater active sites in the metal-organic gel compared to the glassy carbon electrode were indicated by the enhanced current response during nitrite sensing. Hence, further modification of the metal-organic gels with suitable additives involving nanoparticles and carbon-based materials may lead to novel, robust composite materials with conducting properties.[59] Supramolecular gels are also known for their visual sensing capacity. Luminescent metal-organic gels are also available to selectively sense nitrite. For example, the metal-organic gel, Tb-containing 4-[2'2':6"2"-terpyridine]-4'-ylbenzoic acid was developed under milder conditions. It exhibits good stability toward heat, acid, and alkali. It also possesses the property of luminescence due to the antenna effect. The luminescence was quenched in the presence of nitrite rapidly due to the transfer of energy from Tb^{3+} to NO_2^-. Hence, this could be made available for practical sensing applications in the form of test strips.[61]

12.4.5 CHEMICAL VAPOR SENSING

Organic amines play a vital role in day-to-day life since it is the chief raw material for various consumer products like food additives, cosmetic industries, and serves as biomarkers for uremia and lung cancer. Not only the synthetic world, but organic amines also are released from the biological world during cell growth and decomposition.[62] Hence, many chemical vapors including organic amines, organic and inorganic volatile acids,[63] and hydrazines[64] impose serious threats to the environment and to the health of humans.[63] Therefore, there exists a great demand for sensing such chemical vapors.

Especially, sensing of amine vapors is extremely important as it is involved in many applications like diagnosis of diseases, food safety, etc.[64] The bearable limit of amine vapors like isopropylamine and butylamine is 5 ppm.[64] Lot of sensors for the detection of amine compounds from various samples are available. But there is a lack in progress for sensing in the vapor phase.[64] It is known that self-assembled materials respond to chemical vapors and gases by exhibiting property variation.[65] Table 12.1 shows the binding site/interaction of some chemical vapors which could be well utilized to design sensors.

TABLE 12.1 The vapors/gases toxic to the environment and its binding site/interaction in the gel matrix.[63]

Vapors /gases	Binding site/interaction
CO_2	Nitrogen rich surfactant
Formaldehyde, ammonia, and acids	Luminophores
Nitro compound vapors	Charge transfer interaction

Source: Adapted from Ref. [63]

Regardless of such understanding in the interaction of the chemical vapors and the expected outcome, the progress in sensing of these vapors by supramolecular gels is very limited.[65] Gel materials with lot of cavities would be ideal for capturing the organic vapors.[64] Xerogels and aerogels with suitable porous morphology with rich channels and binding sites make them captivating to observe and sense organic vapours.[62] Also, fluorescent nature of gel could be the property capitalized for sensing of vapors as it offers simplicity and less time-consuming. But the efficiency and speed of sensing would depend on the concentration and the volatile nature of the organic compounds.

TABLE 12.2 A Review on the Applications of Gel-Based Materials for Sensing Environmentally Sensitive Molecules.

Material	Environmental pollutant	Detection method	Effect of the external stimuli	Reference
Pyridine appended poly(alkyl ether) based ionogels	CN^-	Visual color change with retention of the gel matrix	Gel to gel	[15]

TABLE 12.2 *(Continued)*

Material	Environmental pollutant	Detection method	Effect of the external stimuli	Reference
Long-alkyl-chained acylhydrazone-based coumarin supramolecular polymer	CN^-	Fluorescent detection	Gel to gel	[16]
Benzimidazole and acylhydrazone naphthol based gel	Multi-analyte sensor (CN^-)	Fluorescent detection	Gel-to-gel	[69]
Amide derivatives of trimesic acid-based gels	Multi-analyte sensor (CN^-)	Change in gel morphology	Gel to gel	[70]
Bis-acylhydrazone supramolecular gel	CN^-	Fluorescent detection(quenching)	Retention of gel (gel-to- gel)	[71]
Pyridylazo derivatives with dicyanovinyl appendage (phenol ether)	CN^-	Visual change	Gel to sol	[72]
Bis-cyanostilbene-based organogels	CO_2	Fluorescence sensing	Gel to gel	[66]
Organic gelator derived from butyl naphthalimide having a naphthyl group attached at the 3-position with the amido linkage	CO_2	Visual change (Anion-induced strategy)	Sol to gel	[67]
Cholesteryl naphthalimide-based gelators	CO_2	Visual change (Anion-induced strategy)	Anion-induced strategy	[68]
Quinoline-indolin-2-one based gel	Cu^{2+}, Hg^{2+}, and CN^-	Visual color change (Hg^{2+} and CN^-) Gel to sol (Cu^{2+})	Retention of gel (gel-to- gel)	[11]

TABLE 12.2 *(Continued)*

Material	Environmental pollutant	Detection method	Effect of the external stimuli	Reference
4-aminophenyl functionalized naphthalimide derivative and tri-(pyridine-4-yl)-functionalized trimesic amide-based gel	Hg^{2+}	Fluorescent detection	Retention of gel (gel-to-gel)	[47]
Naphthalimide functionalized-pillar[5]arene based-gels	Hg^{2+}	Fluorescent detection	Retention of gel (gel-to-gel)	[48]
N-(pyridinium-4-yl)-naphthalimide and n-pentanoic acid based gels	Hg^{2+}	Fluorescent detection	Retention of gel (gel-to-gel)	[73]
Organogelator naphthoformyl hydrazone derivative	Hg^{2+}, Cu^{2+}, and Fe^{3+}	Fluorescent detection	Retention of gel (gel-to-gel)	[74]
Quinine derivative of Dextrin/PVA based hybrid gel	$Cu^{2+}$and Pb^{2+}	Fluorescent detection	—	[75]
spiropyran moiety based polymeric gel	Cu^{2+}	Calorimetric detection	—	[76]
ZnS-supramolecular organogel hybrid films	Organic monoamines and diamines	Fluorescence sensing	—	[64]
Fluorenone-based organogel	Amine vapors	Fluorescence sensing	—	[77]
Metal-organic gels of bis(benzimidazole)-based ligands with copper(II)	Nitrite	Electrochemical sensing	—	[60]
Branched-polyethyleneimine (BPEI)-derived, inherently luminescent carbon dots crosslinked with pentaacrylate	Nitrite	Fluorescence sensing		[78]

TABLE 12.2 *(Continued)*

Material	Environmental pollutant	Detection method	Effect of the external stimuli	Reference
Tb(III)-based metal-organic gel(MOG)	Nitroaromatics	Fluorescence sensing	—	[54]
Polyacrylamide gel	TNT vapor	Electrochemical gas sensing	—	[42]

12.4.6 CARBON DIOXIDE SENSING

With the growth in industries and automobiles, there is a drastic increase in the well-known greenhouse gas, carbon dioxide. It poses several threats to the environment, contributing to global warming. Being an odorless and colorless gas, it goes undetected by human senses. It also creates irreversible damage to the health of human beings at higher concentration. Hence, a suitable sensor detects and quantitatively determines the concentration of carbon dioxide.[66] Special portable sensing devices with visual detection are of utmost importance in closed places of work like industries, mines, and wells.[67] Carbon dioxide interprets with the sensor or alters the sensor medium, thereby leading to a change in the absorption and emission property of the sensor.[66] The gelator compounds (Fig. 12.12a–12.12c) were synthesized and their gelation ability was studied in different solvents using heating and cooling methods. The gels were utilized for the sensing of CO_2 gas. Anion-induced strategy was employed to detect the presence of CO_2. The anions like fluoride play an important role of activating the gelator toward the capture of CO_2. The gelators 12.12b, 12.12c, and 12.12(d–f) were found to selectively respond to the fluoride ion over chloride, bromide, and iodide, respectively. The addition of fluoride ion to the gelator 12.12b and 12.12c induces an intramolecular charge transfer effect resulting from the deprotonation of the amide NH proton. This was very well reflected in the decrease in fluorescence intensity corresponding to the gelator and an increase in a new emission peak. Upon bubbling with CO_2, the new emission peak was quenched, restoring the emission peak at 458 nm. Also, a visual change was observed when carbon dioxide was passed through the solution of 12.12b containing 20 equivalents of fluoride ion. The visual change was qualitatively explained by the decrease of the charge transfer band around 492 nm and an increase in the absorption band at around 372 nm. Hence, the fluoride ion disrupts the gel systems formed by both the gelator compounds, followed by the sol-to-gel conversion upon exposure to CO_2.[67,68]

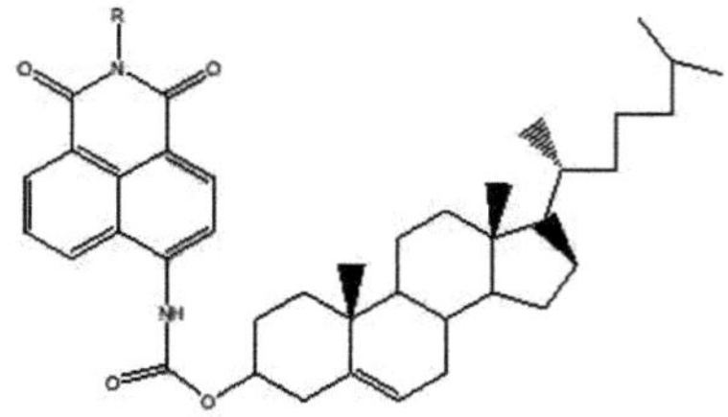

12a: R = -$(CH_2)_{17}CH_3$
12b: R = -$(CH_2)_2O(CH_2)_2OH$
12c: R = -$(CH_2)_2O(CH_2)_2O(C{=}O)CH_3$

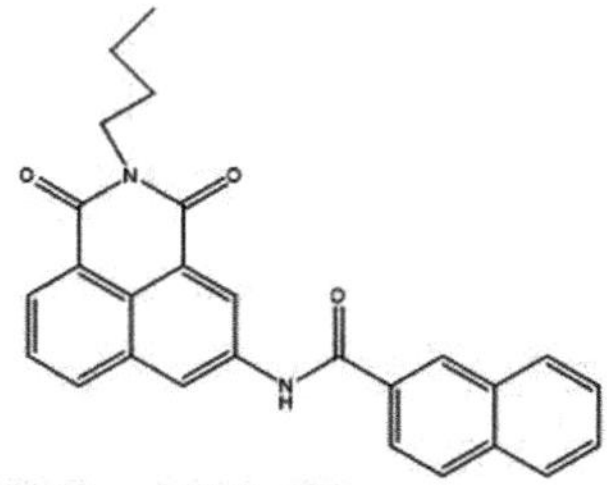

12d: R = -$(CH_2)_{17}CH_3$
12e: R = -$(CH_2)_2O(CH_2)_2OH$
12f: R = -$(CH_2)_2O(CH_2)_2O(C{=}O)CH_3$

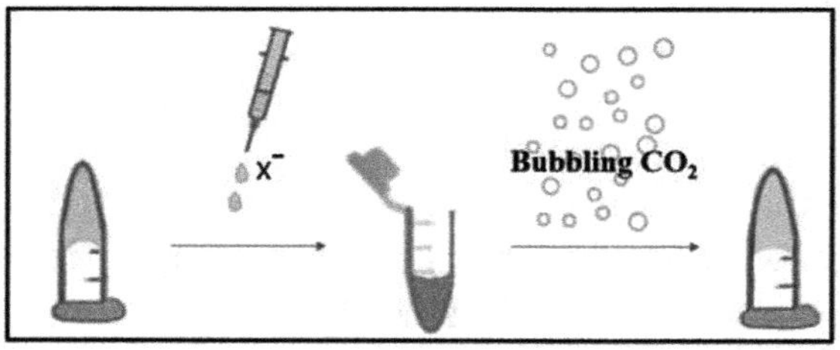

FIGURE 12.12 Schematic representation of anion triggered sensing of CO_2.

12.5 CHALLENGES AND FUTURISTIC OUTLOOK

The non-covalent chemistry involved in gel materials is yet to be explored, and the process of gelation remains abstruse. Despite an in-depth knowledge of possible interactions, forecasting the phenomena of gelation for a gelator is a challenge to the scientific community. Augment to this, many gels have come up as a result of serendipitous discoveries. Still, the contribution to the field of "gels" is on an increasing note which would help to a more feasible approach of its engineering.[5]

Reproducing these non-covalent interactions could be made possible only through proper documentation of the synthetic and sample preparation protocols. Hence, the non-covalent chemistry "once not accepted" could now go beyond the proof-of-concept. A lot of input is required in the development of cheap, portable, and highly sensitive sensing strips based on the supramolecular gels for sensing environmentally sensitive molecules. The gel nanocomposites comprising carbon-based networks could be explored in the field of electrochemical sensors for sensing environmentally sensitive molecules.

There is significant untapped potential on the recycling of the gel after the recognition of the environmentally sensitive molecule and it is still in the stage of infancy with only very few reports on it.

12.6 SUMMARY

Gel materials with desired properties could be architected by rationally designing the gelator and the gelation process. The use of multiple gelators would impart the gel with assorted properties. The gel-based sensor has a lot of advantages than solution-based sensors. The gel-based sensors could be designed as a portable sensor kit requiring no sophisticated instruments for the visual detection of environmental pollutants. The notable features of the supramolecular gels are the following (1) easy method of synthesis, (2) high response to external stimuli, (3) tailor-made, and (4) practical applicability. The application of supramolecular gels is not only confined to the sensing of environmentally sensitive molecules, but also reports on the usage of these supramolecular gels for the removal of environmental pollutants are on an increasing verge.[47] For example, few xerogels were employed for the adsorption and removal of environmental pollutants. The supramolecular gel has opened up new horizons for the design of novel materials, reaching out to practical applications.

ACKNOWLEDGMENT

The authors greatly acknowledge the Principal and Management of Lady Doak College, Madurai, for their constant support and encouragement. Researchers contributed toward the sensing of environmentally sensitive molecules using supramolecular gel-based materials are highly acknowledged.

KEYWORDS

- **supramolecular gels**
- **sensors**
- **environmentally sensitive**
- **cyanide**
- **heavy metal**
- **hydrogel**

REFERENCES

1. Taylor, M. J.; Tomlins, P.; Sahota, T. S. Thermoresponsive Gels. *Gels* **2017,** *3* (1), 4.
2. Panja, S.; Adams, D. J. Gel to Gel Transitions by Dynamic Self-Assembly. *Chem. Commun.* **2019,** *55* (68), 10154–10157.
3. Okesola, B. O.; Smith, D. K. Applying Low-Molecular Weight Supramolecular Gelators in an Environmental Setting–Self-Assembled Gels as Smart Materials for Pollutant Removal. *Chem. Soc. Rev.* **2016,** *45* (15), 4226–4251.
4. Prathap, A.; Sureshan, K. M. Sugar-Based Organogelators for Various Applications. *Langmuir* **2019,** *35* (18), 6005–6014.
5. Wu, H.; Zheng, J.; Kjøniksen, A. L.; Wang, W.; Zhang, Y.; Ma, J. Metallogels: Availability, Applicability, and Advanceability. *Adv. Mater.* **2019,** *31* (12), 1806204.
6. Christoff-Tempesta, T.; Lew, A. J.; Ortony, J. H. Beyond Covalent Crosslinks: Applications of Supramolecular Gels. *Gels* **2018,** *4* (2), 40.
7. Hou, J.; Sapnik, A. F.; Bennett, T. D. Metal–Organic Framework Gels and Monoliths. *Chem. Sci.* **2020,** *11* (2), 310–323.
8. Li, Y. F.; Li, Z.; Lin, Q.; Yang, Y. W. Functional Supramolecular Gels Based on Pillar [n] Arene Macrocycles. *Nanoscale* **2020,** *12* (4), 2180–2200.
9. Karoyo, A. H., Wilson, L. D. Physicochemical Properties and the Gelation Process of Supramolecular Hydrogels: A Review. *Gels* **2017,** *3* (1), 1.
10. Kiliona, K. P. S.; Zhou, M.; Zhu, Y.; Lan, P.; Lin, N. Preparation and Surface Modification of Crab Nanochitin for Organogels Based on Thiol-ene Click Cross-Linking. *Int. J. Biol. Macromol.* **2020,** *150*, 756–764.
11. Mandegani, F.; Zali-Boeini, H.; Khayat, Z.; Scopelliti, R. A Smart Low Molecular Weight Gelator for the Triple Detection of Copper (II), Mercury (II), and Cyanide Ions in Water Resources. *Talanta* **2020,** *219*, 121237.
12. Draper, E. R.; Adams, D. J. Photoresponsive Gelators. *Chem. Commun.* **2016,** *52* (53), 8196–8206.
13. Externbrink, M.; Riebe, S.; Schmuck, C.; Voskuhl, J. A Dual pH-Responsive Supramolecular Gelator with Aggregation-Induced Emission Properties. *Soft Matter.* **2018,** *14* (30), 6166–6170.
14. Falcone, N.; Kraatz, H. B. Ferrocene Peptide-based Supramolecular Gels: Current Trends and Applications. *Advances in Bioorganometallic Chemistry*, 2019, 57–74.
15. Sarkar, B.; Prabakaran, P.; Prasad, E.; Gardas, R. L. Pyridine Appended Poly (Alkyl Ether) Based Ionogels for Naked Eye Detection of Cyanide Ions: A Metal-Free Approach. *ACS Sustain. Chem. Eng.* **2020,** *8* (22), 8327–8337.
16. . Hu, J. H.; Yin, Z. Y.; Gui, K.; Fu, Q. Q.; Yao, Y.; Fu, X. M.; Liu, H. X. A Novel Supramolecular Polymer Gel-Based Long-Alkyl-Chain-Functionalized Coumarin Acylhydrazone for the Sequential Detection and Separation of Toxic Ions. *Soft Matter.* **2020,** *16* (4), 1029–1033.
17. Draper, E. R.; Adams, D. J. Low-Molecular-Weight Gels: The State of the Art. *Chem* **2017,** *3* (3), 390–410.
18. Wang, Y.; Lovrak, M.; Liu, Q.; Maity, C.; le Sage, V. A.; Guo, X.; van Esch, J. H. Hierarchically Compartmentalized Supramolecular Gels Through Multilevel Self-Sorting. *J. Am. Chem. Soc.* **2018,** *141* (7), 2847–2851.
19. Conte, M. P.; Singh, N.; Sasselli, I. R.; Escuder, B.; Ulijn, R. V. Metastable Hydrogels from Aromatic Dipeptides. *Chem. Commun.* **2016,** *52* (96), 13889–13892.

20. Bhattacharya, S.; Samanta, S. K. Soft-Nanocomposites of Nanoparticles and Nanocarbons with Supramolecular and Polymer Gels and their Applications. *Chem. Rev.* **2016,** *116* (19), 11967–12028.
21. Draper, E. R.; Adams, D. J. How Should Multicomponent Supramolecular Gels be Characterised? *Chem. Soc. Rev.* **2018,** *47* (10), 3395–3405.
22. Buerkle, L. E.; Rowan, S. J. Supramolecular Gels Formed from Multi-Component Low Molecular Weight Species. *Chem. Soc. Rev.* **2012,** *41* (18), 6089–6102.
23. Ghosh, D.; Farahani, A. D.; Martin, A. D.; Thordarson, P.; Damodaran, K. K. Unraveling the Self-Assembly Modes in Multicomponent Supramolecular Gels Using Single-Crystal X-ray Diffraction. *Chem. Mater.* **2020,** *32* (8), 3517–3527.
24. Li, Y.; Guo, M. X.; He, L.; Huang, C. Z.; Li, Y. F. Green One-Pot Synthesis of Silver Nanoparticles/Metal–Organic Gels Hybrid and its Promising SERS Application. *ACS Sustain. Chem. Eng.* **2019,** *7* (5), 5292–5299.
25. Zheng, X.; Zhang, H.; Rehman, S.; Zhang, P. Energy-Efficient Capture of Volatile Organic Compounds from Humid Air by Granular Metal Organic Gel. *J. Hazard. Mater.* **2021,** *411*, 125057.
26. Ghosh, D.; Deepa; Damodaran, K. K. Metal Complexation Induced Supramolecular Gels for the Detection of Cyanide in Water. *Supramol. Chem.* **2020,** *32* (4), 276–286.
27. Zhang, Jianyong, Ya Hu, and Yongguang Li. "Gel Chemistry." *Lecture Notes in Chemistry.* **2018,** 96, 61–118.
28. Sutar, P.; Maji, T. K. Coordination Polymer Gels: Soft Metal–Organic Supramolecular Materials and Versatile Applications. *Chem. Commun.* **2016,** *52*, 8055.
29. Tatikonda, R.; Bulatov, E.; Özdemir, Z.; Haukka, M. Infinite Coordination Polymer Networks: Metallogelation of Aminopyridine Conjugates and *In Situ* Silver Nanoparticle Formation. *Soft Matter.* **2019,** *15* (3), 442–451.
30. Divya, K. P.; Miroshnikov, M.; Dutta, D.; Vemula, P. K.; Ajayan, P. M.; John, G. *In Situ* Synthesis of Metal Nanoparticle Embedded Hybrid Soft Nanomaterials. *Accounts of chemical research* **2016,** *49* (9), 1671–1680.
31. Choudhury, P.; Dinda, S.; Das, P. K. Fabrication of Soft-Nanocomposites from Functional Molecules with Diversified Applications. *Soft Matter.* **2020,** *16* (1), 27–53.
32. Sanka, R. V.; Krishnakumar, B.; Leterrier, Y.; Pandey, S.; Rana, S.; Michaud, V. Soft Self-Healing Nanocomposites. *Front. Mater.* **2019,** *6*, 137.
33. Wang, Y.; Yu, Q.; Bai, Y.; Zhang, L.; Zhou, F.; Liu, W.; Cai, M. Self-Constraint Gel Lubricants with High Phase Transition Temperature. *ACS Sustain. Chem. Eng.* **2018,** *6* (11), 15801–15810.
34. Wang, H.; Chen, Q.; Zhou, S. Carbon-Based Hybrid Nanogels: A Synergistic Nanoplatform for Combined Biosensing, Bioimaging, and Responsive Drug Delivery. *Chem. Soc. Rev.* **2018,** *47* (11), 4198–4232.
35. Qin, Z. S.; Dong, W. W.; Zhao, J.; Wu, Y. P.; Zhang, Q.; Li, D. S. A Water-Stable Tb (III)-Based Metal–Organic Gel (MOG) for Detection of Antibiotics and Explosives. *Inorg. Chem. Front.* **2018,** *5* (1), 120–126.
36. Rathinam, B.; Huang, Z. Y.; Liu, B. T. Curcumin-Derived One-and Two-Component Organogelators and their Performance as Template for the Synthesis of Silver Nanoparticles. *Arabian J. Chem.* **2020,** *13* (6), 5679–5690.
37. Zhu, J.; Wang, R.; Geng, R.; Zhang, X.; Wang, F.; Jiao, T.; Peng, Q. A Facile Preparation Method for New Two-Component Supramolecular Hydrogels and their Performances in Adsorption, Catalysis, and Stimuli-Response. *RSC Adv.* **2019,** *9* (39), 22551–22558.

38. Cametti, M.; Džolić, Z. New Frontiers in Hybrid Materials: Noble Metal Nanoparticles–Supramolecular Gel Systems. *Chem. Commun.* **2014,** *50* (61), 8273–8286.
39. Wang, P.; He, G.; Ji, J.; Li, J.; Zhou, K.; Tian, L.; Li, G. A Reductive Supramolecular Hydrogel: A Platform for Facile Fabrication of Diverse Metal-Nanoparticle-Decorated Conductive Networks with Spatiotemporal Control. *ChemPlusChem* **2020,** *85* (8), 1704–1709.
40. Tan, H. L.; Teow, S. Y.; Pushpamalar, J. Application of Metal Nanoparticle–Hydrogel Composites in Tissue Regeneration. *Bioengineering* **2019,** *6* (1), 17.
41. He, L.; Jiang, Z. W.; Li, W.; Li, C. M.; Huang, C. Z.; Li, Y. F. *In Situ* Synthesis of Gold Nanoparticles/Metal–Organic Gels Hybrids with Excellent Peroxidase-Like Activity for Sensitive Chemiluminescence Detection of Organophosphorus Pesticides. *ACS Appl. Mater. Interfaces* **2018,** *10* (34), 28868–28876.
42. Ongaratto, R.; Conte, N.; D'Oca, C. R. M.; Brinkerhoff, R. C.; Ruas, C. P.; Gelesky, M. A.; D'Oca, M. G. M. *In Situ* Formation of AuNPs Using Fatty N-acylamino Hydrazide Organogelators as Templates. *New J. Chem.* **2019,** *43* (1), 295–303.
43. Panja, S.; Panja, A.; Ghosh, K. Supramolecular Gels in Cyanide Sensing: A Review. *Mater. Chem. Front.* **2021,** *5* (2), 584–602.
44. Feng, Y.; Jiang, N.; Zhu, D.; Su, Z.; Bryce, M. R. Supramolecular Oligourethane Gel as a Highly Selective Fluorescent "on–off–on" Sensor for Ions. *J. Mater. Chem. C* **2020,** *8* (33), 11540–11545.
45. Mako, T. L.; Racicot, J. M.; Levine, M. Supramolecular Luminescent Sensors. *Chem. Rev.* **2018,** *119* (1), 322–477.
46. Ferrari, A. G. M.; Carrington, P.; Rowley-Neale, S. J.; Banks, C. E. Recent Advances in Portable Heavy Metal Electrochemical Sensing Platforms. *Environ. Sci. Water Res. Technol.* **2020,** *6* (10), 2676–2690.
47. Yang, H. L.; Sun, X. W.; Zhang, Y. M.; Wang, Z. H.; Zhu, W.; Fan, Y. Q.; Lin, Q. A Bi-component Supramolecular Gel for Selective Fluorescence Detection and Removal of Hg^{2+} in Water. *Soft Matter.* **2019,** *15* (46), 9547–9552.
48. Lin, Q.; Mao, P. P.; Fan, Y. Q.; Liu, L.; Liu, J.; Zhang, Y. M.; Wei, T. B. A Novel Supramolecular Polymer Gel Based on Naphthalimide Functionalized-Pillar [5] Arene for the Fluorescence Detection of Hg^{2+} and I^- and Recyclable Removal of Hg^{2+} via Cation–π Interactions. *Soft Matter.* **2017,** *13* (39), 7085–7089.
49. Kaushik, R.; Ghosh, A.; Singh, A.; Gupta, P.; Mittal, A.; Jose, D. A. Selective Detection of Cyanide in Water and Biological Samples by an Off-the-Shelf Compound. *ACS Sensors* **2016,** *1* (10), 1265–1271.
50. Jackson, R.; Logue, B. A. A Review of Rapid and Field-Portable Analytical Techniques for the Diagnosis of Cyanide Exposure. *Anal. Chim. Acta* **2017,** *960*, 18–39.
51. Udhayakumari, D. Chromogenic and Fluorogenic Chemosensors for Lethal Cyanide Ion. A Comprehensive Review of the Year 2016. *Sens. Actuators B Chem.* **2018,** *259*, 1022–1057.
52. Wang, P.; Liang, B.; Xia, D. A Linear AIE Supramolecular Polymer Based on a Salicylaldehyde Azine-Containing Pillararene and its Reversible Cross-Linking by Cu^{II} and Cyanide. *Inorg. Chem.* **2019,** *58* (4), 2252–2256.
53. Liu, T.; Huo, F.; Li, J.; Cheng, F.; Yin, C. Two Novel CN Turn on Fluorescent Imaging Materials with Twin Binding Groups. *Sens. Actuators B Chem.* **2017,** *239*, 526–535.

54. Chao, J.; Xu, M.; Liu, Y.; Zhang, Y.; Huo, F.; Yin, C.; Wang, X. A Pyrene-Based Turn-On Fluorescence Probe for CN^- Detection and Its Bioimaging Applications. *ChemistrySelect* **2019,** *4* (11), 3071–3075.
55. Kang, J.; Huo, F.; Zhang, Y.; Chao, J.; Glass, T. E.; Yin, C. A Novel Near-Infrared Ratiometric Fluorescent Probe for Cyanide and its Bioimaging Applications. *Spectrochimica Acta Part A: Molecular and Biomolecular Spectroscopy* **2019,** *209*, 95–99.
56. Nath, S.; Pathak, S. K.; Pradhan, B.; Gupta, R. K.; Reddy, K. A.; Krishnamoorthy, G.; Achalkumar, A. S. A Sensitive and Selective Sensor for Picric Acid Detection with a Fluorescence Switching Response. *New J. Chem.* **2018,** *42* (7), 5382–5394.
57. Mondal, S.; Raza, R.; Ghosh, K. Cholesterol Linked Benzothiazole: A Versatile Gelator for Detection of Picric Acid and Metal Ions such as Ag^+, Hg^{2+}, Fe^{3+} and Al^{3+} under Different Conditions. *New J. Chem.* **2019,** *43* (26), 10509–10516.
58. Mondal, S.; Ghosh, K. Anthraquinone Derived Cholesterol Linked Imidazole Gelator in Visual Sensing of Picric Acid. *ChemistrySelect* **2017,** *2* (17), 4800–4806.
59. Luo, H.; Lin, X.; Peng, Z.; Zhou, Y.; Xu, S.; Song, M.; Zheng, X. A Fast and Highly Selective Nitrite Sensor Based on Interdigital Electrodes Modified with Nanogold Film and Chrome-Black T. *Front. Chem.* **2020,** *8*, 366.
60. Britschgi, L.; Villez, K.; Schrems, P.; Udert, K. M. Electrochemical Nitrite Sensing for Urine Nitrification. *Water research X.* **2020,** *9*, 100055.
61. Yuan, D.; Zhang, Y. D.; Jiang, Z. W.; Peng, Z. W.; Huang, C. Z.; Li, Y. F. Tb-Containing Metal-Organic Gel with High Stability for Visual Sensing of Nitrite. *Mater. Lett.* **2018,** *211*, 157–160.
62. Fan, J.; Chang, X.; He, M.; Shang, C.; Wang, G.; Yin, S.; Fang, Y. Functionality-Oriented Derivatization of Naphthalene Diimide: A Molecular Gel Strategy-Based Fluorescent Film for Aniline Vapor Detection. *ACS Appl. Mater. Interfaces* **2016,** *8* (28), 18584–18592.
63. Xue, P.; Ding, J.; Shen, Y.; Gao, H.; Zhao, J.; Sun, J.; Lu, R. Aggregation-Induced Emission Nanofiber as a Dual Sensor for Aromatic Amine and Acid Vapor. *Journal of Materials Chemistry C.* **2017,** *5* (44), 11532–11541.
64. Ll Xia, H.; Liu, G.; Zhao, C.; Meng, X.; Li, F.; Wang, F.; Chen, H. Fluorescence Sensing of Amine Vapours Based on ZnS-supramolecular Organogel Hybrid Films. *RSC Adv.* **2017,** *7* (28), 17264–17270.
65. Cheng, Q.; Wang, Z.; Hao, A.; Xing, P.; Zhao, Y. Aromatic Vapor Responsive Molecular Packing Rearrangement in Supramolecular Gels. *Mater. Chem. Front.* **2020,** *4* (8), 2452–2461.
66. Ma, Y.; Cametti, M.; Džolić, Z.; Jiang, S. AIE-Active Bis-cyanostilbene-Based Organogels for Quantitative Fluorescence Sensing of CO_2 Based on Molecular Recognition Principles. *Journal of Materials Chemistry C.* **2018,** *6* (34), 9232–9237.
67. Zhang, X.; Song, Y.; Liu, M.; Li, H.; Sun, H.; Sun, M.; Yu, H. Visual Sensing of CO_2 in Air with a 3-Position Modified Naphthalimide-Derived Organogelator Based on a Fluoride Ion-Induced Strategy. *Dyes and Pigments* **2019,** *160*, 799–805.
68. Zhang, X.; Li, H.; Mu, H.; Liu, Y.; Guan, Y.; Yoon, J.; Yu, H. Cholesteryl Naphthalimide-Based Gelators: Their Applications in the Multiply Visual Sensing of CO_2 Based on an Anion-Induced Strategy. *Dyes and Pigments* **2017,** *147*, 40–49.
69. Yao, H.; Wang, J.; Song, S. S.; Fan, Y. Q.; Guan, X. W.; Zhou, Q.; Zhang, Y. M. A Novel Supramolecular AIE Gel Acts as a Multi-Analyte Sensor Array. *New J. Chem.* **2018,** *42* (22), 18059–18065.

70. Ghosh, A.; Das, P.; Kaushik, R.; Damodaran, K. K.; Jose, D. A. Anion Responsive and Morphology Tunable Tripodal Gelators. *RSC Adv.* **2016,** *6* (86), 83303–83311.
71. Qu, W. J.; Yang, H. H.; Hu, J. P.; Qin, P.; Zhao, X. X.; Lin, Q.; Wei, T. B. A Novel Bis-Acylhydrazone Supramolecular Gel and its Application in Ultrasensitive Detection of CN^-. *Dyes Pigm.* **2021,** *186*, 108949.
72. Panja, A.; Ghosh, K. Pyridylazo Derivatives with Dicyanovinyl Appendage in Selective Sensing of CN in Sol-Gel Medium. *ChemistrySelect* **2018,** *3* (6), 1809–1814.
73. Lin, Q.; Mao, P. P.; Fan, Y. Q.; Jia, P. P.; Liu, J.; Zhang, Y. M.; Wei, T. B. Novel Multi-Analyte Responsive Ionic Supramolecular Gels Based on Pyridinium Functionalized-Naphthalimide. *Soft Matter.* **2017,** *13* (40), 7360–7364.
74. Chen, Y. Y.; Gong, G. F.; Fan, Y. Q.; Zhou, Q.; Zhang, Q. P.; Yao, H.; Lin, Q. A Novel AIE-Based Supramolecular Polymer Gel Serves as an Ultrasensitive Detection and Efficient Separation Material for Multiple Heavy Metal Ions. *Soft Matter.* **2019,** *15* (34), 6878–6884.
75. Sharma, A. K.; Kaith, B. S.; Singh, A.; Chandel, K. Enzymatic Construction of Quinine Derivative of Dextrin/PVA Based Hybrid Gel Film for the Simultaneous Detection and Removal of Copper and Lead Ions in Real Water Samples. *Chem. Eng. J.* **2020,** *382*, 122891.
76. Xiao, X.; Zhang, C.; Chen, L.; Liao, L. Re-usable Colorimetric Polymeric Gel for Visual and Facile Detection of Multiple Metal Ions. *React. Funct. Polym.* **2021,** *160*, 104824.
77. Su, H.; Liu, R.; Shu, M.; Tang, M.; Wang, J.; Zhu, H. Fluorenone-Based Organogel and Self-Assembled Fibrous Film: Synthesis, Optical Properties and Reversible Detection of Aniline Vapor. *Dyes and Pigments.* **2019,** *162*, 52–58.
78. Baruah, U.; Manna, U. The Synthesis of a Chemically Reactive and Polymeric Luminescent Gel. *Chem. Sci.* **2021,** *12*, 2097–2107.

CHAPTER 13

Dielectric and AC Conductivity Studies of Ag: CdZnTe–PVA Films

KIRAN JOHN U. AND SIBY MATHEW

Department of Physics, Sacred Heart College, Thevara, Kochi, Kerala, India

ABSTRACT

The hybrid freestanding polyvinyl alcohol (PVA) films embedded with ternary alloy CdZnTe and Ag:CdZnTe were synthesized by in situ chemical method. The modifications in the dielectric properties of PVA films with CdZnTe alloy quantum dot dopants and Ag co-doping were investigated. The dielectric properties and AC conductivity were analyzed over the frequency range from 100 Hz to 4 MHz at room temperature. The frequency dependence of the real and imaginary parts of the dielectric constant, loss tangent, and AC conductivity was examined. Both the real and imaginary parts of the dielectric constant shows frequency dispersions and decreasing tendency toward higher frequencies. The conductivity curves obey Jonscher's power-law and power-law parameters were obtained. The conductivity values fall in the range 10^{-6}–10^{-7} S/cm for the doped samples. The dielectric and power-law parameters were improved for Ag: CdZnTe-PVA freestanding films. The complex impedance plots for CdZnTe-PVA and Ag: CdZnTe-PVA films were analyzed and for both samples, semi-circles were obtained in the higher frequency region. The semi-circles obtained in the complex plots suggest single relaxation process and the possible modeling by an equivalent parallel RC circuit. The Cole–Cole parameters were calculated and higher values were exhibited by Ag:CdZnTe-PVA films when compared to CdZnTe-PVA

Foundation and Growth of Macromolecular Science: Advances in Research for Sustainable Development. Meegle S. Mathew, PhD, Józef T. Haponiuk, PhD & Sabu Thomas, PhD, DSc (Eds.)

films. The improvements in the dielectric properties indicate that the hybrid freestanding films can be considered as potential candidates as dielectric materials.

13.1 INTRODUCTION

In recent years, researchers were working on eco-friendly, low-cost, and high-dielectric materials with tunable properties. Polymer nanocomposites (PNCs) with two or more distinct nanofillers can act as potential materials in this regard. Tailoring of structural, optical, and dielectric properties was possible with the variation of filler concentration and composition in the polymer matrix.[1,2] The polymer matrix embedded with nanofillers like metal-semiconductor, metal oxides, and alloy semiconductor composites can be explored for improved and tunable properties. These hybrid PNCs are essential in the field of integrated, organic, and flexible electronics. The semiconductor quantum dots (QDs) together with metal nanoparticles in polymer matrices can be considered as potential materials for memory devices, dielectrics, solar cells, OLED, LED, display applications, bio-labeling, and various optoelectronic devices. The quantum confinement effect of nanofillers plays a key role in modifying the electrical and optical characteristics of polymer composites. The PNC with various nanofillers like CdS, ZnS, ZnO, CdSe, ZnSe, PbSe, and Ag_2S in polymers like PVA, PVP, PMMA, PVDF, PSS, etc., have been investigated by many research communities and reported for their improved optical and electrical properties.[3,6]

The alloyed semiconductor QDs add the advantage of tailoring of physiochemical properties not only with the size but also with the composition ternary alloyed cadmium zinc telluride (CZT) QD is one among the significant alloyed semiconductor with high quantum yield and stability. There are reports that the toxicity of the binary CdTe QDs can be effectively reduced with Zn alloying.[7,8] The polymer polyvinyl alcohol (PVA) is ecofriendly and acts as a good host for the dispersion of fillers. PVA is water soluble, biocompatible, high durability, flexible, and cost-effective product. The electrical and dielectrics performance of PVA is very poor.[5,9,10] The metal-doped QD exhibits enhanced optical and electrical properties. The conducting filler enhances the dielectric properties due to interfacial polarization called Maxwell–Wagner–Miller polarization. The conductive fillers like silver (Ag), aluminum (Al), copper (Cu), and manganese (Mn) are usually dispersed in polymers to produce composites. The dielectric and optical properties of the

PNCs can be effectively tailored with the concentration and composition of the nanofillers.[2,3]

In the present work, the frequency-depended AC conductivity and dielectric responses of CdZnTe-PVA PNCs were investigated over the frequency range 100 Hz–4 MHz at room temperature. The response of CdZnTe-PVA PNCs with silver doping was also analyzed. The power law parameters, complex impedance plots, and Cole–Cole parameters were determined for both samples.

13.1.1 OVERVIEW OF DIELECTRIC PROPERTIES

Dielectrics are materials with very few electrons to take part in the electric conductivity. In an external electric field, dielectric material polarizes in order to form dipoles. A dipole is an entity with equal positive and negative charges separated by small distance. In general, we can classify the dipoles mainly into three which are instant dipole, permanent dipole, and induced dipole. The magnitude of the dipoles is dipole moment and is calculated by the product of one of the charges (q) and separation between the charges (l). The alignment of dipoles to the direction of the external electric field is called the polarization and defined as the total dipole moment per unit volume. Polarization can cause piezoelectric effect and can be used in energy storage applications. Polarization in materials categorized into four types: electronic, ionic, orientation, and interfacial polarization. Depending on the materials' property and external conditions, dielectric material exhibits at least one of the polarizations.[11]

The shift of electron clouds under the external electric filed is the electronic polarization. This type of polarization happens in molecules or atoms, compared to other polarizations and it is relatively small. Ionic polarization is observed in material like NaCl which have ionic bonds. When an electric field is applied to an ionic material the cations and anions shifts in opposite directions and in effect produces non-zero dipole moment. Dipolar or orientational polarization occurs in materials with permanent dipoles. Under an external electric field, these randomly oriented dipoles align in the direction of electric filed giving a net electric dipole moment. Interfacial polarization or Maxwell–Wagner polarization occurs in material with multiple phases like in PNCs. The accumulation of charge carriers at the interface of the polymer and nanomaterial can cause interfacial polarization. The interfacial polarization or MaxwellWagner polarization can influence the dielectric permittivity of the material depending on the nanomaterial. The accumulation of large number of space charge carriers at the interface increases the dielectric permittivity. Due to the different relaxation times at the interface of nanomaterials in the polymer can cause an increase in dielectric loss.

The accumulation of charge carriers can be controlled by the concentration of the nanomaterials in the polymer matrix. The dielectric properties of the nanocomposites can be modified with the surface functionalization of nanomaterials and size of the nanomaterials. The nanomaterials in polymer matrices can bring significant changes to the physiochemical and dielectric properties of the PNC system.[3,11,12]

Dielectric properties in materials are influenced by either one of the four polarization mechanisms described. Since the four polarizations have different relaxation frequencies the dielectric parameters of the material show a strong frequency-depended behavior. In order to investigate the frequency-depended dielectric parameters, dielectric spectroscopic technique is employed. This will give the dielectric parameters at different frequencies and temperatures which will be useful to design dielectric appliances. The dielectric material sandwiched between two metal plates can store a certain amount of energy due to polarization and can function like a parallel plate capacitor. The dielectric permittivity is directly proportional to the capacitance of the dielectric material and can be found using the capacitance value. As the polarization of the material varies, a part of energy stored is dissipated due to thermal or conduction effects. The dielectric loss tangent (tanδ) measurement is useful to analyze the dissipated energy loss. The frequency variation of loss tangent shows peaks depending on the structure and composition of the material. Frequency-dependant conductivity is another important parameter and it is better known by universal dynamic response.[3,4,13,14]

13.1.2 POLYMER NANOCOMPOSITES

Polymers are considered to be good electrical insulators with high electrical resistance and breakdown strength. The polymer matrices can act as suitable capacitors and dielectric materials with the advantages like light weight, low absorption, low dielectric loss, etc. The polymers like polypropylene, polyethylene terephthalate, polyethylene sulfide, etc., are widely used commercial dielectric polymers. But one of the main drawbacks of polymer dielectric is its low dielectric constant value. The typical dielectric constant of polymers falls under 10. Thus, in order to resolve this problem, PNCs with improved dielectrics were put forward.[15–18]

The PNCs are developed with highly active nanomaterial as the fillers in the regular polymer matrix. This hybrid material possesses the merits of the polymer matrix and the quantum confinement effect of the nanomaterial. Depending on the nanomaterial as the choice of filler, the PNC exhibits enhanced structural optical and dielectric properties. Different types of nanofiller can be used inside the polymer in order to modify the physical, chemical, and electronic properties. The properties of the nanocomposites vary depending on the shape, size, and nature of the nanomaterial. Other important factors are uniform distribution and dispersion of nanomaterials in the PNC.[2,19,20] These two quantities play a vital role to enhance the optical, mechanical, dielectric, and conductivity parameters of the material. Therefore, the uniform dispersion and distribution are essential to produce high functional dielectric PNCs. But achieving uniform dispersion and distribution in PNC are challenging and need greater effort. The nanomaterial has high surface energy and to minimize the energy nanomaterials aggregate to form agglomerates in the polymer and it became difficult to obtain uniform dispersion.[11,21–23]

The method of preparation has significant effects on the nature and specificity of the nanocomposite polymer matrices. The method should compromise between the uniform dispersion and distribution of the nanomaterial filler in the host matrices. There exist three general methods to synthesis the nanocomposite polymer matrices. The first method is the in situ polymerization method in which the nanofillers are incorporated in the monomer and the resultant mixture gets polymerized.[24] Homogenous dispersion of nanofillers takes place in this method. The second one is the sol–gel method which is the bottom-up approach in which the condensation and hydrolysis of metal alkoxides take place. The sol–gel reactions of silica–polymer composites can be summarized as

$$n\ Si(OR)_4 + 4n\ H_2O \rightarrow n\ Si(OH)_4 + 4n\ ROH$$

$$n\ Si(OH)_4 \rightarrow 3n\ SiO_2 + 2n\ H_2O$$

The method is suitable for polymers with hydrogen bond acceptors. This approach can be used for nanoparticles with alcohol group functionals and water-soluble polymers.[3] The third approach is the direct mixing of the polymers and nanomaterial. Due to the simplicity and control toward the nanofillers, this method is often used to synthesis PNCs. The mixing of presynthesized nanoparticles and polymers can mechanically mix with or without the assistance of solvents.[2,19]

Depending on the choice of nanofiller for the host matrices, the nanocomposites can exhibit enhanced structural, spectroscopic, and electronic functions. The nanomaterials can be classified mainly into three: quantum well, quantum wires, and QDs. Quantum well structure emerges when two dimensions of the material are in the nano regime. If only one dimension in the nano regime it is called nano wires or quantum wires. The QDs are the structures with all three dimensions were confined to the nano regime. With the emergence of nanotechnology, there exists a wide variety of nanostructures like nanofibers, nanoneedles, nanoflower, fullerene, etc.[19,20] The materials in the nano regime behaves differently from their bulk size properties. In the nano regime quantum confinement effects were become significant. The conductive fillers like Ag, Au nanoparticles are usually used in host polymer matrices to achieve high permittivity.[25,26] The nanostructures of carbon like carbon QDs, carbon quantum rods, graphene, etc., are potential nanofillers. The semiconductor QDs are another category of nanofillers used to make high dielectric constant nanocomposites. The commonly used nanoparticles are ZnO, ZnS, ZnSe, CdSe, etc. Ceramic nanomaterials like $BaTiO_3$-based nanocomposites were much effective to have large dielectric constant and low loss values. SiO_2- and TiO_2-based nanofillers were also used to develop hybrid gate dielectric materials.[3–6,27–30]

13.1.3 POLYMER NANOCOMPOSITE DIELECTRICS-REVIEW

Polymer nanocomposite materials have highly encouraged due to their enhanced optical and electronic properties. There has been intense research carried out to produce various PNCs for improved functionalities. The metal nanoparticle-based nanocomposite was explored as excellent fillers in polymers for engineered applications. The dielectric properties of this combination can be changed by controlling particle size, concentration, and ensuring uniform dispersion. Metal nanoparticles like silver, gold, etc., are usually chosen in the host matrix for applications like capacitors, conductive pastes, dielectric films, etc.[19,25,26]

Toor et al., synthesized nanocomposite dielectric by embedding polyvinylpyrrolidone (PVP) encapsulated Au nanoparticles in the polyvinylidene fluoride (PVDF) polymer matrix and investigated the frequency-depended dielectric permittivity and loss tangent. They observed increase in dielectric values and decrease in dielectric loss with surface functionalization of Au nanoparticles with PVP. This solid-state dielectric material is suggested for energy storage applications.[31] Sahu et al., investigated the dielectric relaxation behavior of silver nanoparticle and graphene oxide (GO) embedded polyvinyl alcohol nanocomposite films. The nanocomposite films were fabricated using solvent casting technique. They achieved high dielectric constant (ε′) value of 3×10^4 for Ag- and GO-embedded PVA films due to the increased dipoles. A remarkable increase in the A.C. conductivity value was obtained for the nanocomposite hybrid films due to the increase in interconnected conduction networks and hoping mechanism. In the nanocomposite films, dielectric values were highly temperature depended and exhibited negative temperature coefficient. The fabricated Ag- and GO-embedded PVA hybrid films were recommended for energy storage applications.[32] Quantum dot (QD) semiconducting materials are another class of materials incorporated in polymer matrices to form PNCs. Tunable optical and electronic properties are possible with varying composition, size, and shape of the QDs. Barandiaran et al., studied the photo-responsive and dielectric properties of cadmium selenide (CdSe)-PVA films by changing the capping agent. The CdSe QDs, synthesized by changing the capping agents and incorporated in PVA can form hybrid films. The nanocomposite films exhibited excellent photoluminescence properties. The dielectric parameters showed frequency dispersions and the properties can be tunable by changing in capping agents.[2] Reddy et al., used green synthesized Zinc Sulfide (ZnS) QDs as nanofiller in PVA to develop PNC films using solution casting method. The morphology of the PNC was agglomerated for high concentration of ZnS nanofiller. The dielectric constants and loss values were enhanced in the ZnS-PVA nanocomposite films and this is explained by the interfacial polarization. The nanocomposite films were suggested for energy storage applications.[20]

Zinc oxide (ZnO) is very effective filler for polymer hosts to improve structural, electrical, and optical properties. ZnO nanorods incorporated PVA matrices were fabricated by M.M Abutalib and observed increase in dielectric values with the increase in ZnO.[10] Polyaniline-based Fe_3O_4 and graphene PNC, polypyrrole, and polythiophene polymer composites, etc., were explored for the electromagnetic interference (EMI) shielding and microwave absorption applications.[24] Choudhary et al., investigated the structural, optical, and dielectric properties of ZnO nanoparticles incorporated PEO-PVP blends to form PNC. Significant changes in the dielectric parameters observed in the polymer blend composites with increase in ZnO nanofillers. The polymer blend nanocomposites were suggested as potential candidates for microelectronics.[6] Afsharimani et al., reports PVA/SiO_2 PNC as hybrid gate dielectric materials. The prepared PVA/SiO_2 films exhibited larger capacitance, less hydrophilicity, and considerable leakage currents compared to pure PVA polymer.[30]

Mary et al., reported freestanding Ag_2S/CuS PVA films with improved dielectric properties for organic electronic applications. The sol-gel method was adopted to synthesis Ag_2S/CuS alloy PVA films by varying the composition of the constituents. The frequency variation of the dielectric constant showed Maxwell—Wagner—Sillars (MWS) polarization. The conductivity was found enhanced for considerable amount of constituents and obeys Jonscher's power-law. The complex impedance plots showed semicircular nature which is modeled by an equivalent parallel RC circuit. The nanocomposite freestanding films showed greater dielectric properties and are suitable for flexible and organic electronics.[3] Enhanced dielectric properties were reported for metal-semiconductor combination in polymer matrices. Zhang et al., reports dielectric enhancement in Zn–ZnO metal semiconductor core-shell structure incorporated in polyvinylidene fluoride (PVDF). The enhancement is attributed to the duplex interfacial polarizations of metal-semiconductor interface and semiconductor-polymer interface. The dielectric loss for the composite films was very low and the materials are well-suitable for applications in embedded passive electronic components. [33]

13.2 EXPERIMENTAL PART

Hybrid freestanding polyvinyl alcohol (PVA) films embedded with ternary alloy CdZnTe and Ag: CdZnTe were synthesized by in situ chemical method. CdZnTe QDs were synthesized as per the reported method with few modifications.[7,8] Required amount of cadmium acetate, zinc acetate, and 3-mercaptopropionic acid (3-MPA) was weighed and mixed with 100 mL double-distilled water under constant stirring. The pH of the mixture was maintained at 8.2 adding 1 M NaOH solution under stirring. Then three weight percentage PVA and required quantities of sodium tellurite and sodium borohydride were added to the mixture under vigorous stirring. The stirring was maintained for 4 h with temperature at 90°C under open-air conditions. The typical molar ratio of Cd, Zn, Te, MPA, and $NaBH_4$ was 1:1:0.25:2.7:2.5. The quantity of reducing agent sodium borohydride was 10 times than sodium tellurite's molar concentration. To prepare the 0.1 wt% Ag-doped CdZnTe-PVA nanocomposite 50 μL of 0.1 M $AgNO_3$ solution added to the CdZnTe-PVA solution and stirred for another 3 h at 60°C. The solutions prepared were cast into containers and heated at 65°C in a hot air oven for 8 h and the PNC freestanding films were peeled off from the container. Schematic representation of the synthesis of Ag: CdZnTe-PVA composite shown in Scheme 13.1.

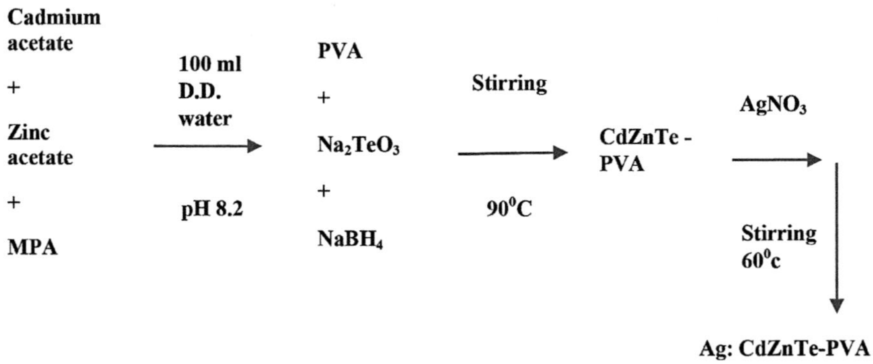

SCHEME 13.1 The preparation scheme of Ag:CdZnTe-PVA polymer nanocomposites.

The structural investigations of the polymer composites were carried out using PANalytical AERIS powder X-ray diffractometer in the 2θ range 10–90° with Cu Kα radiation having a wavelength 1.5406 Å. The absorbance of the samples was recorded in Shimadzu UV-1800 UV-vis spectrometer. Transmission electron microscopy images were taken in an FEI Tecnai G2 30-S TWIN electron microscope with an accelerating potential 300 kV. The dielectric and A.C. conductivity measurements were done over a frequency range of 100 Hz–4 MHz at room temperature using LCR Meter HIOKI 3532-50.

13.3 RESULTS AND DISCUSSION

The diffraction patterns obtained for the CdZnTe and Ag: CdZnTe polymer composites are shown in Figure 13.1. The peak at 2θ = 19.4° is the characteristic polycrystalline nature of polyvinyl alcohol (PVA).[1] The small peaks were observed at 2θ = 22.5° and 41 for both PNC indicate the formation of CdZnTe clusters. The presence of such agglomerated clusters can influence the PVA structure and can produce less intense and broadened peaks. The respective position of the peaks indicates the (111) and (220) planes correspond to cubic CdZnTe.[7,23]

The peaks corresponding to CdZnTe were broadened, due to the quantum confinement of the particles and amorphous nature of the PVA matrix.[7,34] The CdZnTe QDs have large surface energy and high surface-to-volume ratio. To minimize surface energy and to achieve stability, the QDs form agglomerates in polymer matrix. The appearance of small peaks of CdZnTe also indicates the non-uniform distribution and dispersion.[5,11,20,23] The crystal planes of silver were not observed in Ag-doped CdZnTe polymer composites because the weight percentage of silver was very low.[5,20]

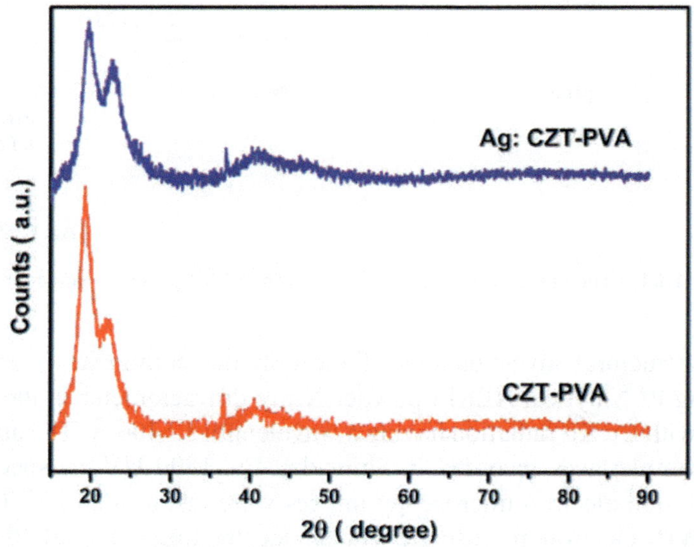

FIGURE 13.1 XRD patterns of the polymer nanocomposites.

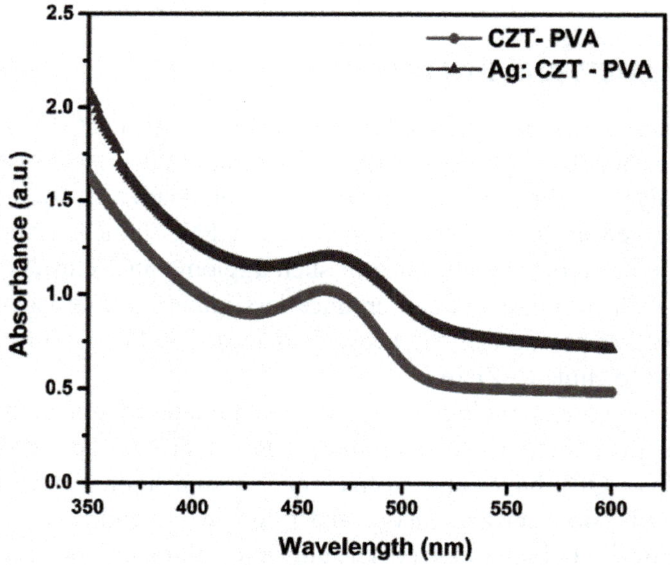

FIGURE 13.2A Absorption spectra of polymer nanocomposites.

The absorption spectrum of the polymer composites was presented in Figure 13.2A. The sliver-doped CdZnTe-PVA composites showed an enhanced absorbance with redshifted absorbance edge compared to CdZnTe-PVA composites. The absorption coefficient is calculated from the absorbance of the respective nanocomposites. The optical band gap of the samples evaluated using the absorption spectra. Tauc plot method was employed to examine the effect of Ag doping on the band gap of the CdZnTe-PVA sample as shown in Figure 13.2B. The band gap energy, incident photon energy, and absorption coefficient are related by the expression.[4]

$$\alpha h\upsilon = A(h\upsilon - E_g)^n \tag{13.1}$$

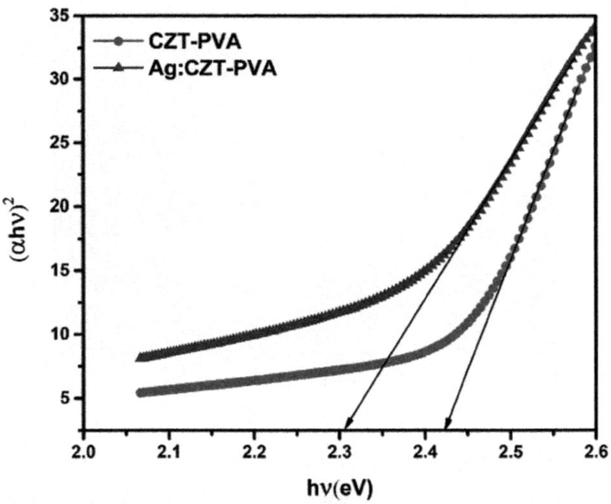

FIGURE 13.2B Tauc plot for the polymer nanocomposites.

where, α is the absorption coefficient, $h\upsilon$ is the incident photon energy, E_g is the band gap energy of the material, A is a constant and the exponent *n* depends on the type of transition and taken as n = ½ considering direct transition. The band gap energy is calculated by extrapolating the linear portion of the $(\alpha h\upsilon)^2$ vs $h\upsilon$ graph on the $h\upsilon$ axis to $\alpha = 0$ as shown in Figure 13.2B. The band gap value of undoped CdZnTe-PVA was found to be 2.42 eV and for Ag-doped CdZnTe-PVA, it was reduced to 2.3 eV.

The TEM images acquired for the PNC are shown in Figure 13.3. The formation of large-sized clusters was seen in the TEM images and one of the reasons for this is the agglomeration. The smaller size and high surface energy of the nanomaterials lead to the agglomeration. The agglomeration depends on concentration and synthesis conditions of the nanomaterials.[1,5,20] The cluster size gets larger when silver was introduced as the co-filler due to increase in interfacial interaction. It was observed that the silver doping promoting the agglomeration process. The nanoclusters of CdZnTe-PVA and Ag: CdZnTe were found in a spherical shape and the dopant particles were well encapsulated by PVA.

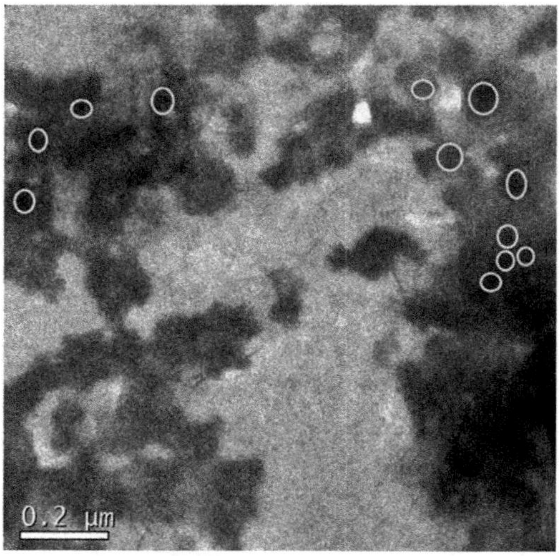

FIGURE 13.3A TEM images of CdZnTe-PVA.

The average size of the clusters in CdZnTe-PVA polymer composites is found to be 55 nm and for the Ag-doped CdZnTe-PVA sample, it was around 85 nm.[5,14,23] The presence of agglomerated clusters was predicted in XRD measurements which is evident from the TEM images. The agglomerated morphology can make significant changes in the dielectric and conductivity parameters.

In A.C. conductivity studies, sinusoidal voltages of different frequencies are applied across the sample. The real (ε′) and imaginary (ε″) part of the complex dielectric constant (ε*) given by Ref. [1,3].

$$\varepsilon^* = \varepsilon' + i\varepsilon'' \quad (13.2)$$

$$\varepsilon' = C_p d/\varepsilon_0 A \quad (13.3)$$

$$\varepsilon'' = \varepsilon' \tan\delta \tag{13.4}$$

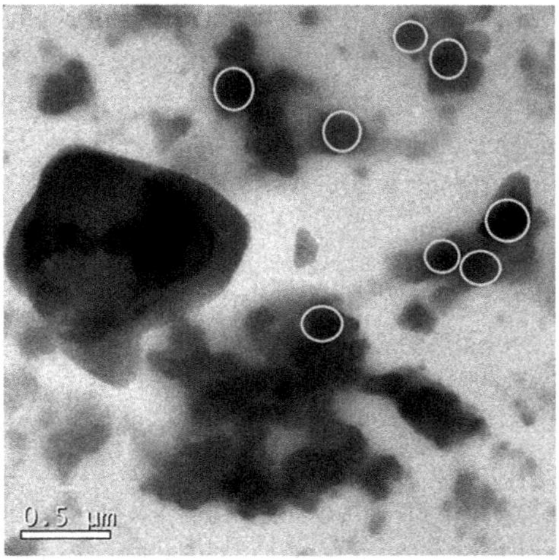

FIGURE 13.3B TEM images of Ag:CdZnTe-PVA.

AC conductivity is given by:

$$\sigma_{AC} = 2\pi f \varepsilon_0 \varepsilon' \tan\delta \tag{13.5}$$

where, "d" is the sample thickness, "A" is the area, "f" is the frequency of the applied signal, "Cp" is the measured capacitance, "ε_0" is the permittivity of free space and "tanδ" is the loss tangent given by $\tan\delta = |1/\tan\theta|$.

Jonscher's power law explains the relation between the frequency and conductivity given by Ref. [3,14]

$$\sigma_T = \sigma_0 + A\,\omega^s \tag{13.6}$$

where, σ_T - total conductivity; σ_0 - DC conductivity; "A" and "s" are material- and temperature-depended parameters, respectively. The DC conductivity term also known as frequency-independent conductivity, which appear as a plateau region in the typical log–log plot of conductivity versus frequency graphs. The second term in the RHS of Equation 13.6, "$A\,\omega^s$" known to be the dispersive term and appears in the high frequency region. The dispersive term can be labeled as AC conductivity σ_{AC}. The frequency-dependent and frequency-independent behavior of conductivity were explored by Jonscher with the universal power law equation. The fitting of Jonscher's equation to respective conductivity curve gives the values of parameters σ_0, A and s known as power law parameters. The power-law exponent "s" is a measure of correlation between frequency and AC conductivity and in general "s" has value between zero and one. But there have been cases reported to have S having values greater than unity. Also, "s" is temperature-dependent parameter and correlates the hopping mechanism of charge carriers.[14]

The complex impedance is given by

$$z^* = z' + z'' \tag{13.7}$$

where, $z' = z_{real} = |z|\cos\theta$ Real part of the impedance, $z'' = z_{imaginary} = |z|\sin\theta$ imaginary part of the impedance, $\theta = \tan^{-1}(z''/z')$ is the Phase angle. The complex plots of z′ versus z″ are effective tools to analyze the relaxation process and possibility of equivalent of circuit modeling. The semicircle fitting in the complex plots or Nyquist plots can be considered to be a relaxation process. The fitting parameters, bulk resistances (R_b), bulk capacitance (C_b), and relaxation time (τ) known as Cole–Cole parameters were found using the relations given below

$$2\pi f_p R_b C_b = 1 \tag{13.8}$$

$$\tau = 2\pi R_b C_b \tag{13.9}$$

The variation in real (ε′) and imaginary (ε″) part of dielectric constant over frequency 100 Hz–4 MHz (logarithmic scale) for the samples at room temperature is given in Figure 13.4A and 13.4B. The dielectric values of undoped PVA are very less compared to doped composites as shown in Figure 13.4A and 13.4B. The samples CdZnTe-PVA and Ag:CdZnTe-PVA have a similar variation over frequency except the silver-doped polymer composites show higher dielectric values. The silver doping improves the network formation and as seen in the TEM images. Ag: CdZnTe-PVA has large agglomerate morphology compared to CdZnTe-PVA. These improvements resulted in higher dielectric values. The decreasing trend of the dielectric constants toward the increase in frequency can be explained using the interfacial polarization or Maxwell–Wagner–Silllars (MWS) polarization. The accumulation of charge carriers at the interface between the host matrix and dopants in external electric field causes interfacial polarization. Both CdZnTe and Ag:CdZnTe polymer composites show a strong interfacial polarization.[3]

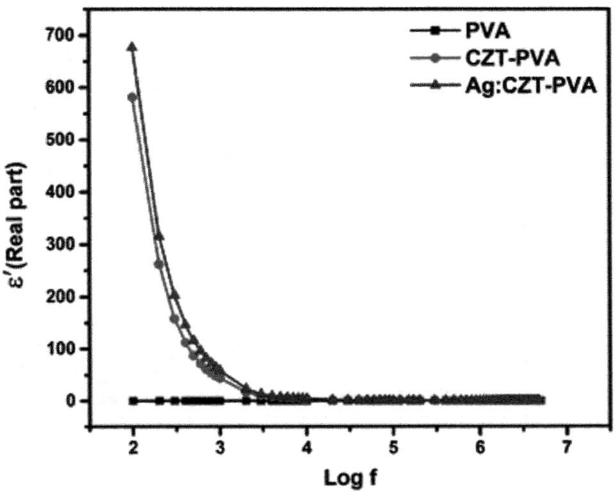

FIGURE 13.4A The variation in real (ε′) part of dielectric constant with log f.

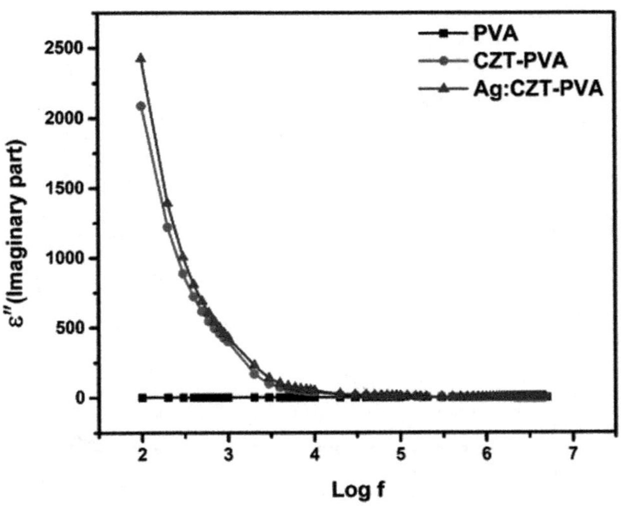

FIGURE 13.4B The variation in imaginary (ε″) part of dielectric constant with log f.

The loss tangent (tanδ) is found to be varying with the frequency as shown in Figure 13.5A. The loss tangent of both polymer composites is peaking around intermediate frequency region but it is less for Ag-doped CdZnTe-PVA composites. This kind of nature in loss tangent indicates the relaxation process in the polymer composites.

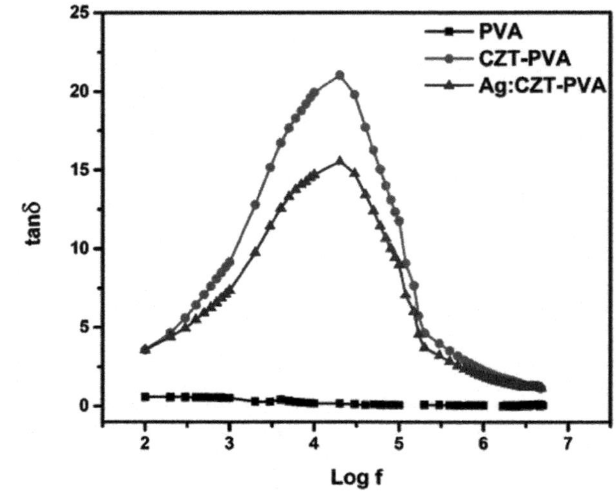

FIGURE 13.5A Frequency variation of tan δ.

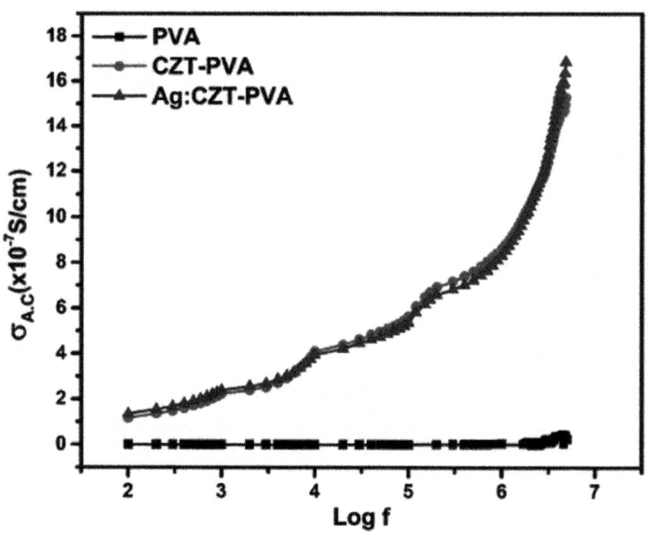

FIGURE 13.5B Frequency variation of conductivity.

The A.C. conductivity is found to have a strong dependence on frequency Figure 13.5B. As observed from the conductivity graph, the conductivity of pure PVA is very less and falls in a range of 10^{-8}–10^{-10} S/cm. The doping with semiconductor QDs and metal–semiconductor combination in PVA improves the conductivity range to a higher order of 10^{-6}–10^{-7} S/cm. σ_{AC} for both PNC shows a similar nature including the frequency-independent plateau region and frequency-dependent dispersive region except for the improved values of Ag:CdZnTe-PVA polymer composites.[3,14] The conductivity curves of both polymer composites obey Jonscher's power-law and from the power-law fitting, the power-law parameters were extracted for Ag-doped and undoped samples shown in Table 13.1. The power-law parameters, DC conductivity, constant A, and exponent s exhibits larger values for Ag-doped sample. The increase in the value of exponent "s" for Ag:CdZnTe-PVA indicates that the samples have higher correlation with frequency compared to CdZnTe-PVA composites.

TABLE 13.1 Power-Law Parameters.

Sample	σ 0	S	A
CdZnTe - PVA	1.26×10^{-7}	0.40	1.31×10^{-9}
Ag:CdZnTe - PVA	2.24×10^{-7}	0.53	1.48×10^{-10}

The complex impedance plots drawn for the polymer composites are given in Figure 13.6. The higher frequency region of both the PNC has a semicircular shape. Fitting of semicircles to the complex curves of PNC can deliver the Cole–Cole parameters which are tabulated in Table 13.2. From the Cole–Cole parameters, it is observed that Silver–CdZnTe combination have much enhanced values than the single CdZnTe in polymer matrix.

TABLE 13.2 Cole–Cole Parameters.

Sample	Rb (KΩ)	Cb (pF)	τ (μ Sec)
CdZnTe - PVA	11.35	8.7	0.62
Ag:CdZnTe - PVA	16	9	0.9

The semicircular region in the complex plots gives the indication of relaxation process in the PNC. Since only one semicircular region appears for each polymer composites, the possible reason for this is the single relaxation process. The semicircular nature in complex plot can be effectively modeled to a parallel RC circuit.[3] The product RC is the relaxation time τ which is the product of bulk resistance R_b and bulk capacitance C_b.

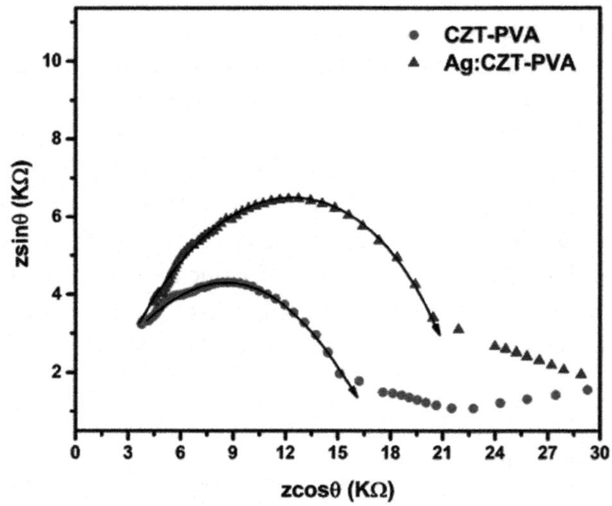

FIGURE 13.6 Complex impedance plot.

13.4 CONCLUSIONS

Freestanding PNC of polyvinyl alcohol embedded with ternary CdZnTe QDs and a combination of Ag-CdZnTe have been synthesized using chemical in situ preparation method. The structural investigations were done using XRD which confirms the polycrystalline nature of PVA and cubic crystal structure of CdZnTe QDs. Redshifted and enhanced absorbance were observed for Ag:CdZnTe PNCs compared to CdZnTe-PVA sample. The band gap energy is found to decrease with silver doping. The TEM images reveal the agglomerated spherical morphology for both PNC with good encapsulation.

The real and imaginary part of the dielectric constant shows frequency dispersions and the values are found to be decrease with increase in frequency. Ag-CdZnTe sample shows a strong MWS polarization, resulting in high dielectric values. Silver doping minimizes the loss tangent value and it is peaking around intermediate frequency region giving an indication of relaxation process. The conductivity curves of PNC obey Jonscher's power-law and the power-law parameters σ_0, s, and A were extracted from the fitting. The Cole–Cole parameters were found out from the complex impedance plot by fitting semicircles in the higher frequency region of the polymer composites which indicates the single relaxation process. The semicircle appearing in the complex plots can effectively be modeled by an equivalent parallel

RC circuit with the characteristic relaxation time τ. Silver doping has greatly improved dielectric and conductivity values of the hybrid films which find potential applications in dielectric and allied fields.

ACKNOWLEDGMENT

Kiran John U. acknowledge KSCSTE, Trivandrum, Kerala, for the research fellowship.

KEYWORDS

- **polymer nanocomposite**
- **CdZnTe alloy semiconductor quantum dots**
- **dielectric constant**
- **AC conductivity**
- **complex impedance plots**

REFERENCES

1. Nangia. R.; Shukla, N. K.; Sharma, A. Dielectric Relaxation and AC Conductivity Behaviour of $Se_{80}Te_{15}Bi_5$/PVA Nanocomposite Film. *Polym. Test* **2019,** *79*, 106088.
2. Barandiaran, I.; Gutierrez, J.; Etxeberria, H.; Tercjak, A.; Kortaberria, G.; Tuning Photoresponsive and Dielectric Properties of PVA/CdSe Films by Capping Agent Change. *Compos. A Appl. Sci. Manuf.* **2019,** *118*, 194–201.
3. Aippunny, A. M. K.; Shamsudeen, S. M.; Valparambil, P.; Mathew, S.; Vishwambharan, U. N. Freestanding Ag_2S/CuS PVA Films with Improved Dielectric Properties for Organic Electronics. *J. Appl. Polym. Sci.* **2016,** *133* (25), 43568.
4. Nayak, D.; Choudhary, R. B. Augmented Optical and Electrical Properties of PMMA-ZnS Nanocomposites as Emissive Layer for OLED Applications. *Opt. Mater.* **2019,** *91*, 470–481.
5. Ambrosio, R.; Carrillo, A.; Mota, M. L.; De la Torre, K.; Torrealba, R.; Moreno, M.; Vazquez, H.; Flores, J.; Vivaldo, I. Polymeric Nanocomposites Membranes with High Permittivity Based on PVA-ZnO Nanoparticles for Potential Applications in Flexible Electronics. *Polymers* **2018,** *10* (12), 1370.
6. Choudhary, S. Structural, Optical, Dielectric and Electrical Properties of (PEO–PVP)–ZnO Nanocomposites. *J. Phys. Chem. Solids* **2018,** *121*, 196–209.
7. Balakrishnan, J.; Preethi, L. K.; Sreeshma, D.; Jagtap, A.; Madapu, K. K.; Dhara, S.; Rao, K. K. Temperature-and Size-Dependent Photoluminescence in Colloidal CdTe and $Cd_xZn_{1-x}Te$ Quantum Dots. *J. Phys. D Appl. Phys.* **2021,** *54* (14), 145103.

8. Al-Rasheedi, A.; Wageh, S.; Al-Zhrani, E.; Al-Ghamdi, A. Structural and Optical Properties of CdZnTe Quantum Dots Capped with a Bifunctional Molecule. *J. Mater. Sci. Mater. Electron.* **2017,** *28* (12), 9114–9125.
9. Brza, M. A.; Aziz, S. B.; Anuar, H.; Ali, F.; Dannoun, E. M.; Mohammed, S. J.; Abdulwahid, R. T.; Al-Zangana, S. Tea from the Drinking to the Synthesis of Metal Complexes and Fabrication of PVA Based Polymer Composites with Controlled Optical Band Gap. *Sci. Rep.* **2020,** *10* (1), 1–17.
10. Abutalib, M. M. Effect of Zinc Oxide Nanorods on the Structural, Thermal, Dielectric and Electrical Properties of Polyvinyl Alcohol/Carboxymethyle Cellulose Composites. *Physica B Condens. Matter* **2019,** *557*, 108–116.
11. Li, B.; Zhong, W. H. Theoretical Analysis of Dielectric Relaxation in Polymer Nanocomposites. In *Polymer Nanocomposites for Dielectrics*; Zhong, W. H., Li, B., Eds.; Pan Stanford Publishing: Singapore, 2017; pp 17–24.
12. Singh, V. P.; Ramani, R.; Singh, A. S.; Mishra, P.; Pal, V.; Saraiya, A. Dielectric and Conducting Behavior of Pyrene Functionalized PANI/P(VDF-co-HFP) Blend. *J. Appl. Polym. Sci.* **2016,** *133* (41).
13. Hussain, S.; Cao, C.; Nabi, G.; Khan, W. S.; Tahir, M.; Tanveer, M.; Aslam, I. Optical and Electrical Characterization of ZnO/CuO Heterojunction Solar Cells. Optik **2017,** *130*, 372–377.
14. Greenhoe, B. M.; Hassan, M. K.; Wiggins, J. S.; Mauritz, K. A. Universal Power Law Behaviour of the AC Conductivity Versus Frequency of Agglomerate Morphologies in Conductive Carbon Nanotube-Reinforced Epoxy Networks. *J. Polym. Sci. B Polym. Phys.* **2016,** *54* (19), 1918–1923.
15. Hassan, Y. A.; Hu, H. Current Status of Polymer Nanocomposite Dielectrics for High-Temperature Applications. *Compos. A Appl. Sci. Manuf.* **2020,** 106064.
16. Naskar, A. K.; Keum, J. K.; Boeman, R. G. Polymer Matrix Nanocomposites for Automotive Structural Components. *Nat. Nanotechnol.* **2016,** *11* (12), 1026–1030.
17. Ho, J. S.; Greenbaum, S. G. Polymer Capacitor Dielectrics for High Temperature Applications. *ACS Appl. Mater. Interfaces* **2018,** *10* (35), 29189–29218.
18. Gnonhoue, O. G.; Velazquez-Salazar, A.; David, É.; Preda, I. Review of Technologies and Materials Used in High-Voltage Film Capacitors. *Polymers* **2021,** *13* (5), 766.
19. Fu, S.; Sun, Z.; Huang, P.; Li, Y.; Hu, N. Some Basic Aspects of Polymer Nanocomposites: A Critical Review. *Nano Mater. Sci.* **2019,** *1* (1), 2–30.
20. Reddy, P. L.; Deshmukh, K.; Chidambaram, K.; Ali, M. M. N.; Sadasivuni, K. K.; Kumar, Y. R.; Lakshmipathy, R.; Pasha, S. K. Dielectric Properties of Polyvinyl Alcohol (PVA) Nanocomposites Filled with Green Synthesized Zinc Sulphide (ZnS) Nanoparticles. *J. Mater. Sci. Mater. Electron.* **2019,** *30* (5), 4676–4687.
21. Mansour, D. E. A.; Abdel-Gawad, N. M.; El Dein, A. Z.; Ahmed, H. M.; Darwish, M. M.; Lehtonen, M. Recent Advances in Polymer Nanocomposites Based on Polyethylene and Polyvinylchloride for Power Cables. *Materials* **2021,** *14* (1), 66.
22. Hore, M. J. Polymers on Nanoparticles: Structure & Dynamics. *Soft Matter*. **2019,** *15* (6), 1120–1134.
23. Soliman, T. S.; Vshivkov, S. A. Effect of Fe Nanoparticles on the Structure and Optical Properties of Polyvinyl Alcohol Nanocomposite Films. *J. Non-Cryst. Solids* **2019,** *519*, 119452.

24. Ganguly, S.; Bhawal, P.; Ravindren, R.; Das, N. C. Polymer Nanocomposites for Electromagnetic Interference Shielding: A Review. *J. Nanosci. Nanotechnol.* **2018,** *18* (11), 7641–7669.
25. Morsi, M. A.; Oraby, A. H.; Elshahawy, A. G.; Abd El-Hady, R. M. Preparation, Structural Analysis, Morphological Investigation and Electrical Properties of Gold Nanoparticles Filled Polyvinyl Alcohol/Carboxymethyl Cellulose Blend. *J. Mater. Res. Technol.* **2019,** *8* (6), 5996–6010.
26. Elashmawi, I. S.; Menazea, A. A. Different Time's Nd: YAG Laser-Irradiated PVA/Ag Nanocomposites: Structural, Optical, and Electrical Characterization. *J. Mater. Res. Technol.* **2019,** *8* (2), 1944–1951.
27. Hariharan, P. S.; Subhashini, N.; Vasanthalakshmi, J.; Anthony, S. P. A Facile Method for the Synthesis Fluorescent Zinc Chalcogenide (ZnO, ZnS and ZnSe) Nanoparticles in PS and PMMA Polymer Matrix. *J. Fluoresc.* **2016,** *26* (2), 703–707.
28. Li, W.; Song, Z.: Qian, J.; Tan, Z.; Chu, H.; Wu, X.; Nei, W.; Ran, X. Enhancing Conjugation Degree and Interfacial Interactions to Enhance Dielectric Properties of Noncovalent Functionalized Graphene/poly (Vinylidene Fluoride) Composites. *Carbon* **2019,** *141*, 728–738.
29. Zheng, M. S.; Zheng, Y. T.; Zha, J. W.; Yang, Y.; Han, P.; Wen, Y. Q.; Dang, Z. M. Improved Dielectric, Tensile and Energy Storage Properties of Surface Rubberized BaTiO3/Polypropylene Nanocomposites. *Nano Energy* **2018,** *48*, 144–151.
30. Afsharimani, N.; Nysten, B. Hybrid Gate Dielectrics: A Comparative Study Between Polyvinyl Alcohol/SiO_2 Nanocomposite and Pure Polyvinyl Alcohol Thin-Film Transistors. *Bull. Mater. Sci.* **2019,** *42* (1), 26.
31. Toor, A.; So, H.; Pisano, A. P. Improved Dielectric Properties of Polyvinylidene Fluoride Nanocomposite Embedded with Poly (vinylpyrrolidone)-Coated Gold Nanoparticles. *ACS Appl. Mater. Interfaces* **2017,** *9* (7), 6369–6375.
32. Sahu, G.; Das, M.; Yadav, M.; Sahoo, B. P.; Tripathy, J. Dielectric Relaxation Behavior of Silver Nanoparticles and Graphene Oxide Embedded Poly (vinyl alcohol) Nanocomposite Film: An Effect of Ionic Liquid and Temperature. *Polymers* **2020,** *12* (2), 374.
33. Zhang, Y.; Wang, Y.; Deng, Y.; Li, M.; Bai, J. Enhanced Dielectric Properties of Ferroelectric Polymer Composites Induced by Metal-Semiconductor Zn-ZnO Core–Shell Structure. *ACS Appl. Mater. Interfaces* **2012,** *4* (1), 65–68.
34. Qian, J.; Lu, X.; Wang, C.; Cui, H.; An, K.; Long, L.; Hao, N.; Wang, K. Controlling Over the Terminal Functionalities of Thiol-Capped CdZnTe QDs to Develop Fluorescence Nanosensor for Selective Discrimination and Determination of Fe (II) Ions. *Sens. Actuators B Chem.* **2020,** *322*, 128636.

CHAPTER 14

Electric Circuit Modeling of Impedance Spectroscopic Characteristics of GFRP Nanocomposites with Hybrid Carbon Nanofillers

B. M. MADHU[1*], RASHMI[1], R. R. N. SAILAJA[2], and RAJAN J. SUNDARA[1]

[1]*V.T.U. Research Centre, Department of Electrical and Electronics Engineering, Siddaganga Institute of Technology, Tumakuru, Karnataka, India*

[2]*The Energy and Resources Institute, Southern Regional Centre, Bangalore, Karnataka, India*

ABSTRACT

Nanocomposites with hybrid carbon nanofillers have emerged as an alternative to traditional composites displaying superior properties for use in a wide range of industrial applications. The incorporation of two nanofillers creates a better-interconnected network of fillers that leads to an increase in electrical conductivity, thermal conductivity, and mechanical strength. The network of fillers helps to enhance the adhesion strength and binding properties of the polymer matrix. Though many experimental investigations and theoretical estimations are presented in the literature, there is little information on the use of macroscale equivalent circuit models to account for the phase transformation that occurs in the nanocomposites due to the nanofillers or their hybrid combinations. In this chapter, the use of circuit modeling is enumerated for understanding the electrical properties of the glass fiber reinforced epoxy

Foundation and Growth of Macromolecular Science: Advances in Research for Sustainable Development. Meegle S. Mathew, PhD, Józef T. Haponiuk, PhD & Sabu Thomas, PhD, DSc (Eds.)

composites. The impedance and phase angle measurements in the frequency range from 10 Hz to 8 MHz form the basis of the circuit modeling process. The effect of temperature on the resulting circuit models is also considered over a temperature range of 25–120°C. The electrical circuit components are developed by a comparative multistep procedure, and by varying the circuit elements a good match of the results of the impedance spectra by simulation experimental is achieved. The equivalent circuits help to understand the physical mechanisms and phase transitions. These mechanisms occur due to the presence of conductive nanofillers of different weight percentages and temperature changes. The simulation of nanocomposites based on epoxy resin and nanofillers is presented.

14.1 INTRODUCTION

In 2004, S. Iijima demonstrated the possibility of growing carbon nanotubes (CNT) and the rediscovery of graphene using Scotch tape uncovered the immense possibility of numerous applications of graphene.[1–4] The need for alternatives became imminent since the cost of conventional polymers started increasing and it was also compounded by the problems of limited availability of certain raw materials. Fiber-reinforced materials were established as viable alternatives as conventional polymers in various applications, due to their higher strength and lower cost.[5–8] With the development of glass fiber reinforced polymer (GFRP) composites with carbon-based nanofillers, mainly targeting the aerospace and automobile industry, these high-performance multifunctional composites find a variety of industrial applications today.[9–11]

Further advancements in GFRP with multiscale reinforcement lead to the emergence of many new applications.[12–14] The incorporation of carbon fillers like MWCNT into the glass fiber reinforced polymer matrix is a well-established method to increase the fiber/matrix interactions that would improve the mechanical strength by interlocking, improved epoxy wettability.[15] Many innovations in the nanocomposites have shown outstanding and substantial changes in the properties of the composites and this has enabled a wider application of the composites. Different nanofillers have been investigated and discussed in the literature to reinforce the polymer matrix and these include the nanoclay, alumina, silica, titanium dioxide, and carbonaceous fillers. Multiwalled carbon nanotubes (MWCNTs) and graphene nanoplatelets (GNPs) reinforced composites are being increasingly used for their versatility and advantages in terms of the performance and functionality of the composites. Incorporating nanofillers as reinforcements leads to modifications in the processability of the polymer as well as the appearance and physical properties. The major advantages that accrue are the enhancement in mechanical, thermal, and electrical properties with additional advantages like improved chemical resistance, flame retardance, and others.[16–19] Fillers like MWCNTs and GNPs possess excellent intrinsic properties that are mainly responsible for improvements in the performance of the nanocomposites. It is also backed by the advantage of achieving flexibility to control the properties to suit diverse application requirements in fields like aeronautics, power transmission, and distribution to biomimetics, sensors, and wearables.[20] Smart materials could be designed using these nanocomposites by controlling the dispersion of nanofillers that are the prerequisites for stable conductive networks. The electrically conducting composites have wide applications in power transmission and structural applications.[21]

Glass-reinforced composites are extensively used by the industry for their excellent mechanical, electrical, thermal, and tribological properties. Many research studies have elucidated the influence of different forms of MWCNTs as well as GNPs separately on the performance of epoxy resins. However, there are very few studies dealing with the optimal mixing ratios of the two nanofillers.[22,23] Glass-reinforced composites are fabricated with different loadings of GNPs/MWCNTs fillers. The mixing of the two fillers is assisted by mechanical mixing and ultrasonication for proper dispersion in the polymer matrix. Improvement in electrical conductivity is studied using impedance spectroscopy (IS) since it is a proven tool to characterize materials in diverse fields. The frequency-domain electrical impedance data would require a postprocessing stage for the extraction of the most critical information, which will be hidden in the recorded spectra under different conditions. The impedance spectra can be modeled with the help of the equivalent electrical circuit for a better insight into the nature of the dielectric behavior, when different filler combinations are used.[24,25]

In this chapter, the use of impedance spectroscopy is highlighted with the typical example of glass fiber reinforced epoxy nanocomposite containing a hybrid combination of MWCNTs and GNPs nanofillers. The effect and influence of the filler alignment and distribution within the polymer matrix and its impact on the dielectric properties are discussed in this chapter. The importance is stressed on the resistive component ascribed to the bulk resistance of the composite and the resulting capacitance, which are the direct consequences of the measured impedance data. The usefulness of electrical equivalent circuits and the magnitudes of the circuit elements under different filler combinations are examined to stress the role of this technique in modeling the present-day nanocomposites.

14.2 TYPICAL EXPERIMENTAL PARAMETERS

To highlight the importance of the fabrication methods in the resulting dielectric parameters, a typical example of fiber-reinforced epoxy is discussed. Electrical–chemical resistant glass fiber along with the epoxy resin (Araldite-MY740) and hardener (Aradur-HY918) of 85pbw (parts per hundred resin) are used as the host polymer matrix. The nanofillers used are MWCNTs, GNPs, along with the micro filler aluminum trihydrate (ATH). The ATH filler is intended for improving the fire-retardant property of the hybrid nanocomposites. The GNPs filler used is processed by chemical vapor deposition. For the reinforcement, E-glass fiber roving with a density of 2.72 g/cm^3 was used. The fillers MWCNTs and GNPs of requisite mass are initially transferred to the epoxy resin and are thoroughly mixed. The ATH filler along with the hardener was sequentially incorporated into the epoxy with MWCNTs and GNPs filler. The mixture is constantly subjected to mechanical stirring and agitation to facilitate in-situ exfoliation and also to achieve a homogeneous mixture of the fillers used.

The mixture is then transferred to a temperature-controlled closed resin bath of the pultrusion setup. By controlling the pull time and the bath temperature of the pultrusion equipment, glass fibers were coated with the mixture containing epoxy, hardener, and fillers, using an optimized pulling rate of 6 ± 0.5 m/h. This process enables sustained and uniform wetting of fibers. The glass fibers coated with the epoxy mixture are then pulled through a rectangular die of 40 mm width and 3.7 mm thickness while maintaining a temperature of 145 ± 5°C for curing. The resulting glass-reinforced nanocomposite (GNC) is then cut to the required size and transferred to a chamber for postcuring. The chamber is maintained at a temperature of 60°C for 24 h. For simplicity and ease of identification, the glass-reinforced composites are identified as shown in Table 14.1. The details of the filler compositions used are also indicated in the table.

TABLE 14.1 Parameters are Determined by Fitting Results for GNCs to the Circuit Model.

Sl. No.	Sample	Composition	Epoxy + Fillers				Glass fiber
			MWCNTs (wt.% of Ep)	GNPs (wt.% of Ep)	ATH (wt.% of Ep)	Epoxy wt.%	
1.	EP+GF	Epoxy/Glass fiber/ATH	–	–	20	80	80 wt.% Total weight of composites
2.	H1	Epoxy/Glass fiber/ATH/ GNPs/MWCNTs	2	1	20	77	
3.	H2	Epoxy/Glass fiber/ATH/ GNPs/MWCNTs	2	2	20	76	
4.	H3	Epoxy/Glass fiber/ATH/ GNPs/MWCNTs	2	3	20	75	

For measurements, it is always preferable to machine the pultruded composites carefully into desired dimensions using a diamond-tipped cutter. The specimens were dried in a hot-air-circulating convection-type oven for a period of 24 h at 50°C and then preserved in desiccators.

14.3 IMPEDANCE SPECTROSCOPY

Impedance spectroscopy (IS) is a versatile technique for investigating the electrical properties of polymer composites. In the example considered, highly conducting MWCNTs/GNPs fillers are incorporated into the epoxy matrix, and hence it is natural to expect a decrease in electrical impedance of the system due to the formation of the conductive network of fillers. Impedance spectroscopy helps in the monitoring of electrical impedance response of the dielectric material with a great deal of sensitivity. It can detect the changes in the bulk structure of the polymer composites as well as the changes that occur in the microstructure. For impedance spectroscopy, a small ac electric field perturbation with a variable frequency is sufficient. The applied ac signal can polarize the dielectric material, which can be visualized in the form of phase shift of the output signal with respect to source. From a comparison of the response to the frequency spectra, it is possible to arrive at the possible electrical equivalent circuit. Typical variations in the impedance and AC conductivity of the composites with frequency are shown in Figures

14.1(a) and (b). It is observed that the impedance spectra of the composites are not the same since the fillers tend to change the material properties due to their distribution as well as interactions with the epoxy matrix.

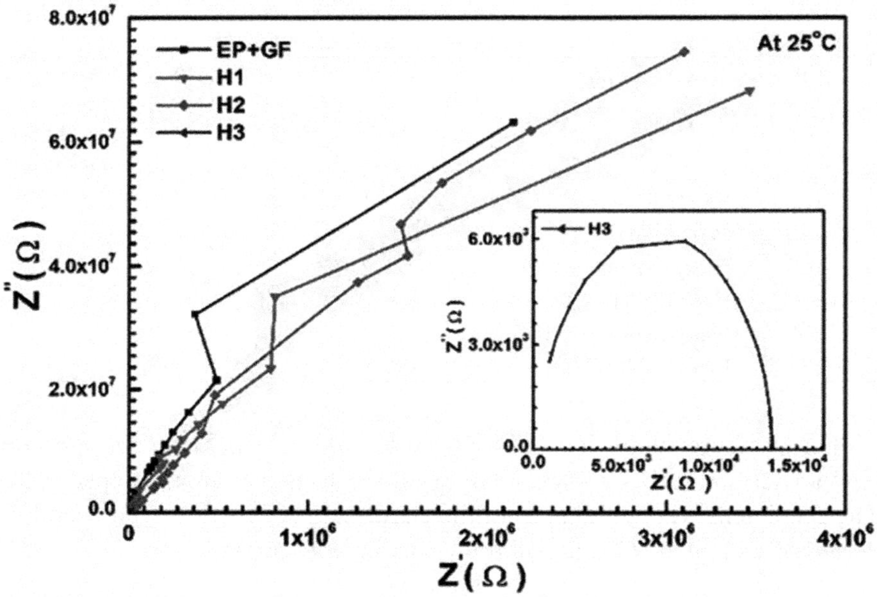

FIGURE 14.1(A) Impedance spectrum of GNCs.

In the percolation theory of electrical conductivity, it is well known that when the wt. % of filler is below a certain critical value, which is called the percolation threshold, the possibilities of the formation of conductive paths are limited. However, when the wt. % is above the critical value, the probability of the formation of conductive networks increases significantly. From the impedance spectroscopy, it is possible to identify the percolation threshold of the nanocomposites.

With the increase in the wt.% of the nanofillers, the impedance spectrum of the composites shows characteristic changes that are evident from Figure 14.1(a). The impedance characteristics in Figure 14.1(b) show the effect of temperature of the nanocomposites on the impedance. Formation of the undistorted semicircle of the impedance spectra of composite H3 is attributed to the combined effect of electronic polarization of the nanocomposites which is ascribed to the presence of conductive nanofillers at or above the percolation threshold and formation of conducting network by the two fillers namely MWCNTs and GNPs. A substantial change in AC electrical conductivity can be observed from Figure 14.1(b), which is due to the electron tunneling mechanism. It is well known that electric charges try to tunnel through the potential barriers and hence contribute to the electrical conductivity. There are possibilities of changes in the tunneling distance or resistance when different wt. % of fillers are used. Similarly, the interfiller particle distances can get reduced to an increase in the wt% of the fillers. It is also possible that there is a continuous formation of long conductive networks due to the homogeneous dispersion of nanofillers used. The fillers individually can also be capable of achieving conducting networks when their wt. % are higher.

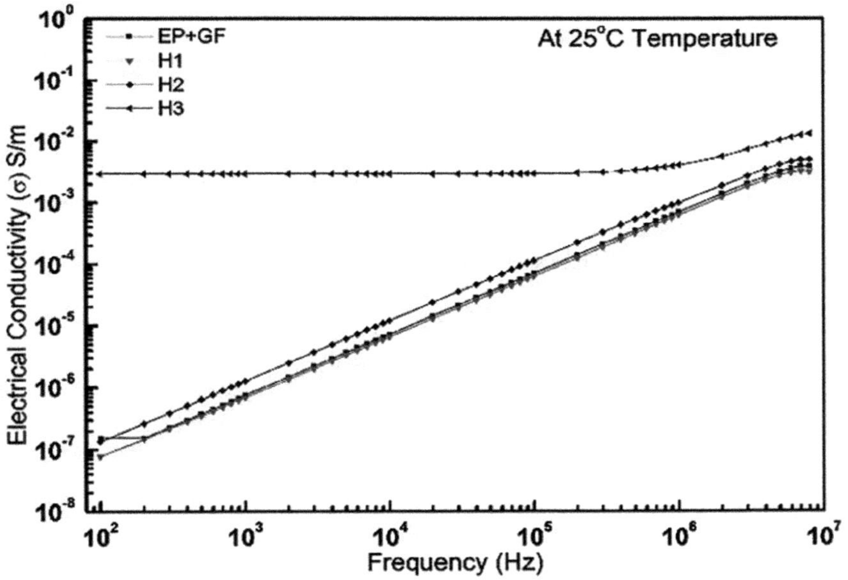

FIGURE 14.1(B) Conductivity versus frequency graphs GNCs.

The presence of two conducting fillers leading to the formation of conductive networks is an indication of the presence of a strong capacitive element in the equivalent circuit model of the composite. The real part of the impedance spectra is reduced by the formation of the conducting network. The network of conductive fillers also helps to enhance the conductivity of the polymer composites. The imaginary part of the impedance shows a decreasing trend due to the effects of temperature and energy losses. These changes can be analyzed with the help of the impedance plots to construct a reliable equivalent circuit model that is a true representation of the polymer composite. The impedance can be measured at different temperatures, and this would act as an input to understand the changes brought about by the temperature in the circuit elements or the equivalent circuit.

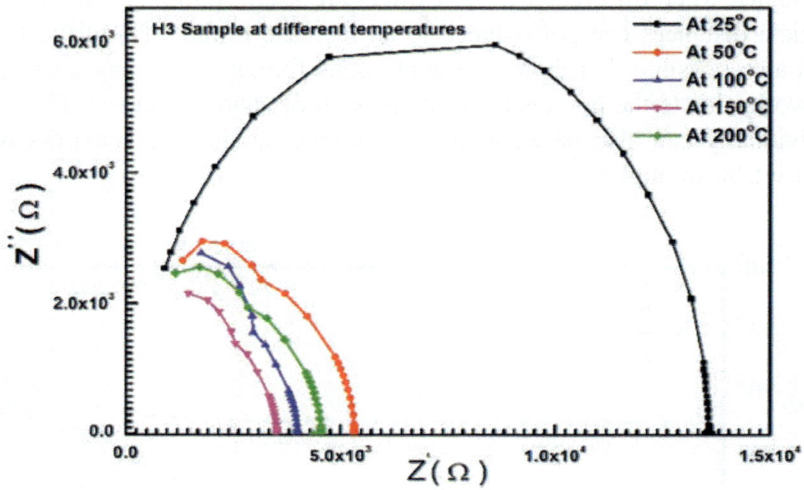

FIGURE 14.2(A) Impedance spectrum of H3 GNC at different temperatures.

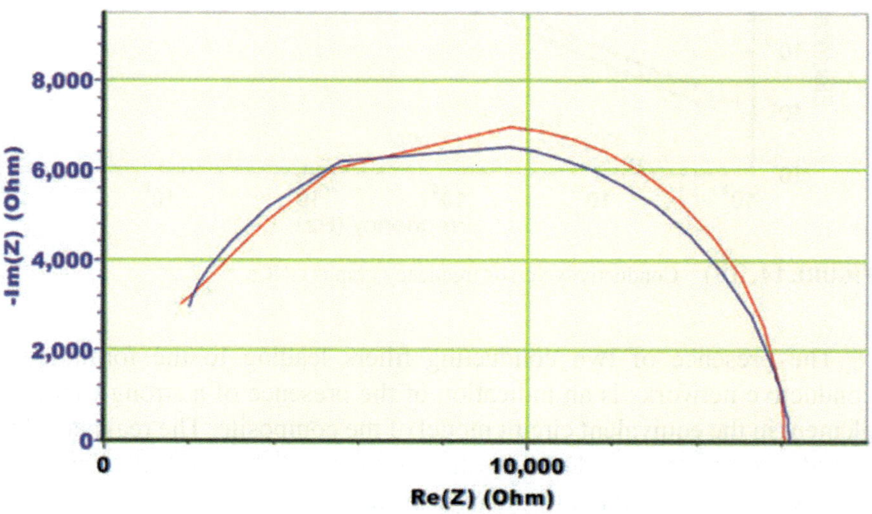

FIGURE 14.2(B) Curve fitting of sample F impedance spectrum.

The ac impedance spectrum of the composite H3 as a function of frequency at different temperatures is shown in Figure 14.2(a). In nanocomposite H3, the formation of a continuous conductive network appears to be evident due to the incorporation of 2 wt. % MWCNTs and 3wt. % GNPs. It is fact that with an increase in filler wt. % and uniform dispersion of the fillers, the interparticle distance would get reduced. Thus, a considerable decrease in the complex impedance (|Z|) is observed. Beyond critical frequency (which is 100 kHz), the electrons get sufficient energy to hop from one conducting cluster to another because of the smaller interparticle distances. This leads to a decrease in |Z| and an increase in the electrical conductivity of the nanocomposite. With this typical scenario regarding the effects of the wt. % of the two fillers and their consequent influence on the impedance and AC conductivity, the process of building an equivalent circuit model of the conducting composites is enumerated in the next section.

14.4 EQUIVALENT ELECTRICAL CIRCUIT MODELING

The determination of the lumped parameter circuital model of the composite is critical to the accuracy of the electrical equivalent model development. A high-precision LCR meter with a suitable electrode system can be used for this purpose. The frequency variations of impedance and the phase angle are the inputs for the determination of the equivalent circuit models. The results of impedance spectra are modeled as a combination of resistors and capacitors connected in series and parallel or complex series–parallel combinations, depending upon the structure of the composite. The experimental results of the impedance spectra undergo postprocessing in a specially developed software tool that facilitates the extraction of useful information. This is the most important phase of the development of the equivalent electrical circuit model. For example, from the frequency variations of the complex impedance values of the epoxy composites, the real and imaginary parts of the complex impedance are computed, and this forms the input to the software, like EC-lab or other software tools. As an additional precaution, the results can also be verified with the Z-View simulator. By iterative processing, the most appropriate equivalent circuit model that fits well into the data is developed.

A typical electrical equivalent circuit developed is shown in Figure 14.3. This circuit is developed by the curve fitting of the impedance data as shown in Figure 14.2(b). There is provision for fitting a large number of more complex electrical circuits for highly complex nanocomposite structures. When different wt% of the nanofillers are used in the nanocomposites, the impedance spectra would get altered and therefore no universal equivalent circuit representation is possible. However, the use of simple electrical circuits would suffice for the majority of the applications for the analysis of nanocomposites with single or binary combinations of fillers. With the incorporation of conducting fillers, the equivalent circuit of the epoxy composites can well be represented by a combination of resistances and capacitances. However, the nature of the circuits would depend on the fillers, their dispersion, and the interactions of the fillers with the glass fiber reinforced epoxy matrix. The electrical equivalent circuit developed is shown in Figure 14.3.

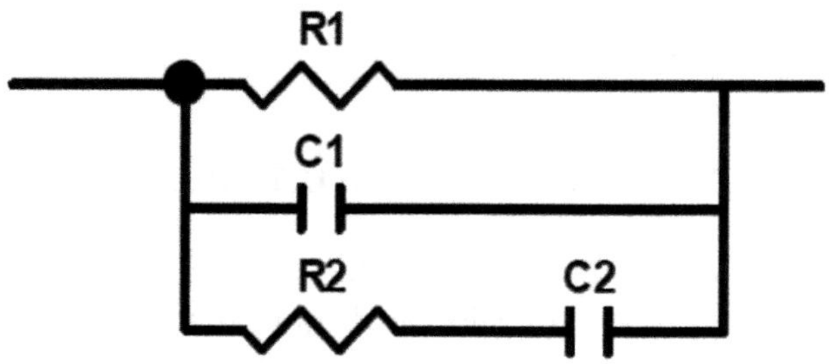

FIGURE 14.3 Equivalent circuit fitting based in the impedance spectra of the glass-reinforced epoxy nanocomposites with hybrid nanofillers.

The impedance spectra of the nanocomposites are generally indicative of two important attributes namely the capacitive and resistive nature of the composites. Due to the existence of a coating of the epoxy around the MWCNTs and GNPs fillers, there are numerous layers of polymers between the filler particles. In the epoxy composites, the fillers MWCNTs and GNPs are randomly distributed leading to the formation of a complex RC network.[26] In the parallel resistor-capacitor R1C1 part of the equivalent circuit shown in Figure 14.3, R1 is the contact resistance that hinders the passage of electrons from one particle to another. The associated capacitance between filler particles with sandwiched epoxy forms the capacitance that is represented by C1. The additional series-connected R2C2 elements are attributed to the possible agglomeration of the nanofillers causing some resistance, and the capacitance is due to the epoxy-glass fiber barriers formed between such agglomerated clusters of the fillers. Based on the circuit analysis, the impedance of the nanocomposite can be computed by equations (2) and (3), and it is possible to estimate the values of the real, imaginary values of the complex impedance. The variations of imaginary impedance as a function of the real impedance would give a semicircular plot about the real axis.

$$Z_{eq} = \frac{Z_1 Z_2}{Z_1 + Z_2} \tag{14.1}$$

$$Z_1 = R_1 + \frac{1}{j\omega C_1} \tag{14.2}$$

$$Z_2 = \frac{R_2 X_{C_2}^2}{R_2 + X_{C_2}^2} - j\frac{R_2 X_{C_2}^2}{R_2 + X_{C_2}^2} \tag{14.3}$$

For all the glass-reinforced epoxy composites discussed, Z1 comprises a parallel combination of R1 and C1, which could be associated with the nature of the homogenization of the glass fiber reinforcement in the epoxy matrix. The impedance Z1 is in parallel with Z2 and it consists of the series elements R2 and C2 that consider the conduction processes due to the network formed by the MWCNTs and GNPs fillers through the barriers formed by the fiber-reinforced epoxy. Resistance characterizes the dissipation of heat and capacitance represents the energy storage and change in the phase aspects of the polymer composites. The AC conductivity spectra of the sample H3 indicate that the addition of nanofillers is much above the percolation threshold.

The conductivity characteristics of the nanocomposites can be investigated for understanding the relaxation behavior of fiber-reinforced epoxy composites with conducting fillers. It can be observed that the conductivity of the composite H3 is almost invariant with a frequency, and this fact is an indication of a strong conducting network path is created inside the nanocomposite. With the addition of 5 wt% combined nanofillers (3 wt% MWCNTs + 2 wt% GNPs), the percolation behavior is confirmed. The values of the circuit parameters in the equivalent circuit are summarized in Table 14.2.

TABLE 14.2 Parameters are Determined by Fitting Results for GNCs to the Circuit Model.

Sample/Circuit Elements	C_1	R_1	C_2	R_2
EP + GF	1.978 pF	7.196 GΩ	0.464 pF	126.323 kΩ
H1	1.74 pF	4.125 GΩ	0.4969 pF	137.856 kΩ
H2	2.654 pF	3.518 GΩ	1.406 pF	111.869 kΩ
H3	4.006 pF	16.012 kΩ	4.116 pF	9.451 kΩ

The simple equivalent circuit representation consisting of resistance (R) and capacitance (C) is shown in Figure 14.3 to explain and understand the nature of the nanocomposites with a hybrid combination of conducting fillers. This circuit model shows satisfactory agreement and fits well into the experimental data. The possible error, which is natural in such computations, can be estimated to understand the efficacy of the models developed. The goodness of fit between the experimental and simulated data, based on the "best fit" approach, was tested by using the chi-square (χ) and average square error (β) methods. The comparison can be performed based on two basic parameters of convergence, namely the normalized average square error β and the normalized standard deviation square error χ that are applied separately to the parameters namely the real and imaginary parts of the impedance. The first parameter β indicates the variance between the experimental and computer-generated spectra, whereas χ indicates the goodness of the simulation. If this value is larger, then the number of frequency points is greater where the local discrepancy would be greater than the average error. The algorithms can be further optimized to achieve better convergence criteria; that is the two parameters should be lower than some predetermined values for both functions to be optimized. In particular, $\chi \leq 10^{-2}$ and $\beta \leq 10^{-1}$ have been assumed in this work.[27–29]

14.5 SUMMARY

Impedance spectroscopy is a powerful technique to understand the complex characteristics of polymer composites. An equivalent circuit model can be developed to simulate the nature of different glass-reinforced polymer composites with a hybrid combination of conducting fillers. The variations in the temperature of the polymer composites and the extension of the frequency range of impedance measurements provide additional flexibility to the analysis. The use of impedance spectroscopy facilitates the characterization of polymer composites through equivalent electrical circuit modeling, by a simple and intuitive approach. IT can also be used for the optimization of fillers to develop polymer composites with predetermined characteristics.

The impedance spectrum-based electrical equivalent circuit model can be useful for the interpretation of the mechanisms responsible for electrical characteristics of polymer composites. There are many more advantages of the equivalent circuit model which will help exploit the usefulness and utility of polymer composites for a wide range of industrial applications.

KEYWORDS

- **nanocomposites**
- **electric circuits modeling**
- **impedance spectroscopy**
- **multiwalled carbon nanotubes**
- **graphene**
- **nanoplatelets**
- **hybrid carbon nanofiller**

REFERENCES

1. Kumar, A.; Sharma, K.; Dixit, A. R. Carbon Nanotube and Graphene-Reinforced Multiphase Polymeric Composites: Review on their Properties and Applications. *J. Mater. Sci.* **2020,** *55* (7), 2682–2724, doi: 10.1007/s10853-019-04196-y.
2. Han, Y. H.; et al. Spark Plasma Sintered Bioceramics–from Transparent Hydroxyapatite to Graphene Nanocomposites: A Review. *Adv. Appl. Ceram.* **2020,** *119* (2), 57–74, doi: 10.1080/17436753.2019.1691871.
3. Ivanov, E.; et al. PLA/Graphene/MWCNT Composites with Improved Electrical and Thermal Properties Suitable for FDM 3D Printing Applications. *Appl. Sci.* **2019,** *9* (6), doi: 10.3390/app9061209.
4. Soares, B. G. Ionic Liquid: A Smart Approach for Developing Conducting Polymer Composites: A Review. *J. Mol. Liq.* **2018,** *262*, 8–18, doi: 10.1016/j.molliq.2018.04.049.
5. Rafiee, M.; Nitzsche, F.; Laliberte, J.; Hind, S.; Robitaille, F.; Labrosse, M. R. Thermal Properties of Doubly Reinforced Fiberglass/Epoxy Composites with Graphene Nanoplatelets, Graphene Oxide and Reduced-Graphene Oxide. *Compos. Part B Eng.* **2018,** *164*, 1–9, 2019, doi: 10.1016/j.compositesb.2018.11.051.
6. Vivek, K. A.; Agrawal, G. D. Organic Solar Cells: Principles, Mechanism and Recent Developments. *Int. Res. J. Eng. Technol.* **2014,** 2319–2322.
7. Lazo, M. A. G.; Teuscher, R.; Tween, R.; Velut, P.; Leterrier, Y.; Månson, J. E. Cost Effective Encapsulation Composites for Building Integrated Photovoltaics, **2014**, 22–26
8. Romanenko, A.; et al. Composites and their Properties. 2012.
9. Aradhana, R.; Mohanty, S.; Nayak, S. K. Novel Electrically Conductive Epoxy/ Reduced Graphite Oxide/Silica Hollow Microspheres Adhesives with Enhanced Lap Shear Strength and Thermal Conductivity. *Compos. Sci. Technol.* **2018,** *169*, 86–94, 2019, doi: 10.1016/j.compscitech.2018.11.008.
10. Cha, J.; Kim, J.; Ryu, S.; Hong, S. H. Comparison to Mechanical Properties of Epoxy Nanocomposites Reinforced by Functionalized Carbon Nanotubes and Graphene Nanoplatelets. *Compos. Part B Eng.* **2018,** *162*, 283–288, doi: 10.1016/j. compositesb.2018.11.011.
11. Zhu, N.; Yuan, L.; Liang, G.; Gu, A.; Mechanism of Greatly Increasing Dielectric Constant at Lower Percolation Thresholds for Epoxy Resin Composites through Building

Three-dimensional Framework from Polyvinylidene Fluoride and Carbon Nanotubes. *Compos. Part B Eng.* **2019,** *171*, 146–153, doi: 10.1016/j.compositesb.2019.04.042.

12. He, D.; et al. Design of Electrically Conductive Structural Composites by Modulating Aligned CVD-grown Carbon Nanotube Length on Glass Fibers. *ACS Appl. Mater. Interfaces* **2017,** *9* (3), 2948–2958, doi: 10.1021/acsami.6b13397.
13. Campo, M.; Redondo, O.; Prolongo, S. G. Barrier Properties of Thermal and Electrical Conductive Hydrophobic Multigraphitic/Epoxy Coatings. *J. Appl. Polym. Sci.* **2020,** *137* (42), 1–8, doi: 10.1002/app.49281.
14. Al-Saleh, M. H. Electrical, EMI Shielding and Tensile Properties of PP/PE Blends Filled with GNP: CNT Hybrid Nanofiller. *Synth. Met.* **2016,** *217*, 322–330, doi: 10.1016/j.synthmet.2016.04.023.
15. Zeng, S.; Shen, M.; Duan, P.; Yang, L.; Lu, F.; Hao, L. Tunable Mechanical Properties of MWCNT–glass Fiber Fabric Reinforced Epoxy Composites by Controlling MWCNTs Dispersing Conditions. *Compos. Interfaces* **2018,** *25* (10), 901–918, doi: 10.1080/09276440.2018.1439642.
16. Cury, C. P. H., Gundappa, S. K.; Fernando, W. Nanocomposites : Synthesis, Structure, Properties and New Application Opportunities. *Mater. Res.* **2009,** *12* (1), 1–39, doi: 10.1590/S1516-14392009000100002.
17. Gu, H.; Ma, C.; Gu, J.; Guo, J.; Yan, X. An Overview of Multifunctional Epoxy Nanocomposites. *J. Mater. Chem. C* **2016,** *4*, 5890–5906, doi: 10.1039/C6TC01210H.
18. Rafique, I.; Kausar, A.; Anwar, Z.; Muhammad, B. Exploration of Epoxy Resins, Hardening Systems, and Epoxy/Carbon Nanotube Composite Designed for High Performance Materials: A Review. Polym. *Plast. Technol. Eng.* **2016,** *55* (3), 312–333, doi: 10.1080/03602559.2015.1070874.
19. Trihotri, M.; Dwivedi, U. K.; Malik, M. M. Study of Low Weight Percentage Filler on Dielectric Properties of MWCNT-epoxy. *Nanocomposites* **2016,** *6* (3), 1–9, doi: 10.1142/S2010135X16500247.
20. Li, K.; Zhao, R.; Xia, J.; Zhao, G. L. Reinforcing Microwave Absorption Multiwalled Carbon Nanotube–Epoxy Composites using Glass Fibers for Multifunctional Applications. *Adv. Eng. Mater.* **2020,** *22* (3), doi: 10.1002/adem.201900780.
21. Kumosa, M.; et al. In *18th International Conference on Composite Materials Polymer Matrix Composites in High Voltage Transmission Line Applications*, International Conference on Composite Materials.
22. Zakaria, M. R.; Abdul Kudus, M. H.; Md, H.; Akil, Mohd Thirmizir, M. Z. Comparative Study of Graphene Nanoparticle and Multiwall Carbon Nanotube Filled Epoxy Nanocomposites Based on Mechanical, Thermal and Dielectric Properties. *Compos. Part B Eng.* **2017,** *119*, 57–66, doi: 10.1016/j.compositesb.2017.03.023.
23. Batakliev, T.; et al. Effects of Graphene Nanoplatelets and Multiwall Carbon Nanotubes on the Structure and Mechanical Properties of Poly (Lactic Acid) Composites: A Comparative Study. *Appl. Sci.* **2019,** *9* (3), doi: 10.3390/app9030469.
24. Monti, M.; et al., Toward the Microstructure-properties Relationship in MWCNT/Epoxy Composites: Percolation Behavior and Dielectric Spectroscopy. *Compos. Sci. Technol.* **2014,** *96*, 38–46, doi: 10.1016/j.compscitech.2014.03.008.
25. Meeuw, H.; Körbelin, J.; von Bernstorff, D.; Augustin, T.; Liebig, W. V.; Fiedler, B. Smart Dispersion: Validation of OCT and Impedance Spectroscopy as Solutions for in-situ Dispersion Analysis of CNP/EP-Composites. *Materialia* **2018,** *1*, 185–197, doi: 10.1016/j.mtla.2018.06.002.

26. Sanli, A.; Müller, C.; Kanoun, O.; Elibol, C. Piezoresistive Characterization of Multi-walled Carbon Nanotube-Epoxy based Flexible Strain Sensitive Films by Impedance Spectroscopy Piezoresistive Characterization of Multi-walled Carbon Nanotube-Epoxy Based flexible Strain Sensitive films by Impedance Spectroscopy. *Compos. Sci. Technol.* **2015,** *122*, 18–26, doi: 10.1016/j.compscitech.2015.11.012.
27. Qistina, O.; Salmiaton, A.; Choong, T. S. Y.; Taufiq-Yap, Y. H.; Izhar, S. Optimization of Carbon Nanotube-coated Monolith by Direct Liquid Injection Chemical Vapor Deposition based on Taguchi Method. *Catal.* **2020,** *10* (1), doi: 10.3390/catal10010067.
28. Tavakoli, M.; Safa, F.; Abedinzadeh, N. Binary nanocomposite of Fe 3 O 4/MWCNTs for Adsorption of Reactive Violet 2: Taguchi Design, Kinetics and Equilibrium Isotherms. *Fuller. Nanotube. Carbon Nanostructures* **2019,** *27* (4), 305–316, doi: 10.1080/1536383X.2018.1563543.
29. Azqhandi, M. H. A.; Foroughi, M.; Yazdankish, E. A Highly Effective, Recyclable, and Novel Host-guest Nanocomposite for Triclosan Removal: A Comprehensive Modeling and Optimization-Based Adsorption Study. *J. Colloid Interface Sci.* **2019,** *551*, 195–207, doi: 10.1016/j.jcis.2019.05.007.

CHAPTER 15

The Role of Synthetic Polymers in the Aquatic Environment and Its Implications in *Danio Rerio* as a Model Organism

RICHA SHREE, NATRAJAN CHANDRASEKARAN, AMITAVA MUKHERJEE, and JOHN THOMAS

Centre for Nanobiotechnology, Vellore Institute of Technology (VIT), Vellore, Tamil Nadu, India

ABSTRACT

Presence of microplastics in the marine environment is considered a global threat to several marine animals. Marine plastic debris occurs in seas and oceans globally. In addition, pollutants such as polycyclic aromatic hydrocarbons (PAH) and polychlorinated biphenyls (PCBs) are capable of adsorbing to the plastic surface. Metal pollution is common within harbors and marinas, and originates from multiple sources such as the usage of metal-based antifouling paints, industrial waste, and fuel combustion. In the present study, we investigated the impact of polystyrene MPs (100 nm), silver (Ag) nanoparticle, and their combination on adult zebra fish (*Danio rerio*). Accumulation, antioxidant defense (ROS, CAT, and Protein), cellular toxicity, and histological analysis were also carried out. The toxic effects of silver and polystyrene combined particle were studied up to 120 hrs. The LC50 value of the combined nanoparticles at 96 h was found to be 48.2 ppm. The behavior of fish was noted at different concentrations (1 ppm, 3 ppm, 9 ppm, and 27 ppm) for every 24 h for 4–5 days. That is

Foundation and Growth of Macromolecular Science: Advances in Research for Sustainable Development. Meegle S. Mathew, PhD, Józef T. Haponiuk, PhD & Sabu Thomas, PhD, DSc (Eds.)

approx. 96–120 h. Mortality was observed with increasing concentration from 1 to 50 ppm for 96-h exposure. Experiments were conducted for 120 h. Gills and intestine were taken from the treated animals for biochemical analysis and cellular analysis. The oxidative damages caused by Ag-PS were associated with a large number of histological changes. The histological changes reveal a clear stress and apoptosis condition in gill and intestine tissues under a microscope since the polystyrene particles have the ability to bind silver nanoparticle causing toxicity to living organisms. Fourier transform infrared spectroscopy (FT-IR) was performed to investigate the presence of functional groups of silver nanoparticles (Ag-NPs). Silver-polystyrene aggregate (Ag-PS) particle induced great toxicity in zebra fish under higher concentrations. It was found to accumulate in high quantity in the gill and intestines of zebra fish.

15.1 INTRODUCTION

Plastics are materials which can be manufactured at low cost. Because of their light weight, they find various applications like food packaging, consumer products, medical devices, and construction in daily life. It is therefore necessary to assess their biosafety.[1,2] As for susceptibility for microplastics (MP), few investigations found to sorb other chemicals and pollutants such as metal from the surrounding environment affecting both the spatial distribution and the biological interaction of adhered pollutants.[3] Ingestion of pollutants by organisms in the aquatic environment, for example, when organisms mistakenly consume MPs with the adhered contaminants, subject to dietary uptake,[4] and the cellular vector effect in which MPs in the micro- or nanosize ranges are taken up into cell,[5] possibly by endocytotic or phagocytotic processes, allows adhered contaminants to enter the cell.

Plastic pellets deployed in the San Diego Bay area for up to one year accumulated different amounts of nine metals not reaching saturation within the given time frame.[6] From few sources no studies have yet investigated the potential for MP to affect the bioavailability and flux of metals at the organism level. The potential for metals to interact with MPs has been largely overlooked, as plastic surfaces are reported in ecotoxicological studies[7] and silver in particular exhibits strong surface binding characteristics, which requires recognition and mitigation in experimental design.[8,9]

The zebra fish model is the most accepted model for performing toxicity since of their size, rapid growth, rapid embryonic development, and large clutch size; it also facilitates a variety of in vivo toxicity tests within a short period of time and is cost effective with great flexibility.[10] The zebra fish model can be used to assess toxic effects because of the similarity to higher vertebrates including humans at the genomic developmental and anatomical level.[11,12] The zebra fish model has been used to demonstrate the toxicity test for many nanoparticles including silver, gold titanium dioxide, and silica nanoparticles.[13,14]

The main objective of this work was focused on characterization and toxicity of combined polystyrene with silver nanoparticle and also on investigating the potential effect of MPs that affect the bioavailability and flux of metals at the organism level where the zebra fish model is used.

15.2 MATERIALS AND METHODS

15.2.1 SYNTHESIS OF SILVER NANOPARTICLE

Silver nanoparticle was purchased from Sigma-Aldrich. All the glassware used in the study were washed thoroughly with acid to ensure an endotoxin-free environment. A total of 100 nm polystyrene was also used in the study. The stock solution (1000 ppm) was prepared by using silver nanoparticle powder purchased from Sigma-Aldrich. Solution was made by suspending silver nanoparticle powder (40 mg) in ultrapure filtered water, that is, Milli Q (40 mL). The suspension was then exposed to a high-intensity ultrasonic processor for uniform distribution. The suspension was sonicated for 15 min at ice-cold temperature for 30–45 min. The power output was 100 watt and the output frequency was 20 kHz. The stock solution of nanoparticle was stored in a dark bottle at 6°C until use.

15.2.2 SYNTHESIS OF SILVER NANOPARTICLE WITH POLYSTYRENE MICROPLASTIC PARTICLE

The stock solution of polystyrene microplastics (PS) 100 nm (25,000 ppm) was used. 0.2 ml was diluted with 4.8 ml of milli Q water (for 1000 ppm) and the solution of PS containing silver nanoparticle was prepared by mixing solution of PS microplastics (1000 ppm stock of 3 mL) in the already prepared silver nanoparticle (1000 ppm stock of 40 mL). The mixture solution of Ag-PS was placed for overnight stirring in a magnetic stirrer.

15.3 CHARACTERIZATION OF SILVER AND POLYSTYRENE PARTICLE

15.3.1 DYNAMIC LIGHT SCATTERING AND ZETA POTENTIAL MEASUREMENTS

The hydrodynamic diameter and zeta potential of the nanoparticles were characterized by dynamic light scattering (DLS) using a HORIBA Nanoparticle Analyzer. The different stock concentrations were characterized and tested. Measurements were taken after 12 and 24 h. The size and zeta potential were also measured.

15.3.2 SCANNING ELECTRON MICROSCOPE

SEM was employed to visualize the size and shape of silver nanoparticles and polystyrene microplastic particle and also the interaction of polystyrene with silver nanoparticles. The solution was dried on a small cover slip for 24–48 h. After the solution on the cover slip dried, it was observed under a scanning electron microscope.

15.3.3 FOURIER TRANSFORM INFRA-RED ANALYSIS

Fourier transform infra red (FT-IR) analysis was employed to identify the chemical constituents in the region of 400–4000 cm^{-1} of the Ag-NPs, PS, and Ag-PS. Single drops of each sample of silver nanoparticles, polystyrene microplastics, and combination of silver and polystyrene were placed in FT-IR separately for analysis.

15.3.4 UV-VISIBLE SPECTROPHOTOMETER

UV-vis spectroscopy was used for quantitative analysis of metal ion and conjugated organic compounds. The stability of particles was determined by measuring the absorbance using a UV-vis spectrophotometer between 300 and 700 nm.

15.4 ZEBRA FISH MAINTENANCE AND EXPOSURE DESIGN

Zebra fish (*Danio rerio*) were purchased from Chennai, Tamil Nadu, India. The fish were maintained in 10,000 L glass tanks with continuous aeration at room temperature. During the toxicity assessment, feed was not provided. Ag-PS was used for studying its toxicity in zebra fish. The behavior of fish was noted at different concentrations (1 ppm, 3 ppm, 9 ppm, 27 ppm) for every 24 h for 4–5 days. That is approximately 96–120 h. The experiments were conducted in triplicates. Five fish were taken in each tank. The tank without the addition of the nanoparticles served as the control. Dead fish were removed and stored for further analysis.

15.5 BIOCHEMICAL ANALYSIS

At the end of 120 h, fish in each group were collected. Gills and intestine were collected to study the biochemical changes in control and infected fishes. Gill and intestine tissues were homogenized in a homogenizer in ice-cold phosphate buffer solution (1x PBS) for 2 min. The homogenates were immediately centrifuged at 10,000 g at 4°C for 10 min.

ROS activity, antioxidant enzyme (CAT), and protein activities were measured. The ROS activity was measured by the method of Wang et al.[15] with slight modification. Caltalse (CAT) was determined following the method of Beutler,[16] while protein was estimated with a bio-spectrophotometer. The total protein content was measured by the Bradford method,[17] with BSA as the standard. The absorbance values were determined and noted. The generation of reactive oxygen species (ROS) such as intracellular reactive oxygen species (ROS) produces hydroxyl radical OH− and superoxide anion, which can be detected using cell-DCFH-DA (permeable dye, 2-7-dichlorofluorescein-diacetate).[15] It also acts as a bioindicator of stress on biological organisms, which is caused by certain external factors (Sevcu et al., 2011). A nonfluorescent cell membrane-permeable dye-DCFH-DA reacts with intra cellularly produced ROS that emits fluorescent light. This converted fluorescence was measured further by a standard protocol defined by Wang and Joseph (1999). After the addition of 2 μL of 100M DCFH-DA to 2 ml of the supernatant, the samples were kept under dark condition at room temperature for 30 min. The excitation and emission wavelength of DCFH-DA was analyzed using Eclipse fluorescence spectrophotometer purchased from Cary, ModelG9800A, Agilent technologies, USA at 530 and 485 nm.

15.6 HISTOLOGICAL ASSESSMENT

Gills and intestine (27 ppm) were dissected from infected and control fishes. The samples were placed in 10% formaldehyde for the fixation of the sample. Hematoxylin and eosin were used for staining the sections. The sections were then examined under a light microscope.

15.7 RESULTS AND DISCUSSION

15.7.1 CHARACTERIZATION OF SILVER NANOPARTICLE AND POLYSTYRENE MICROPLASTICS

Dynamic light scattering (DLS): In the present study, zeta potential of aggregated Ag-PS particle was determined for 1000 ppm (Z avg. – 116 nm; PI - 0.414), 100 ppm (Z avg. - 317; PI - 0.455) at 12 h (Fig. 15.1(a) and (b)) and later 1000 ppm (Z avg. - 151; PI - 0.376) and 100 ppm (Z avg. - 312.3; PI - 0.518) at 24 h (Fig. 15.1(c) and (d)), 26°C by HORIBA nanoparticle analyzer. Basically, DLS is used for physicochemical characterization of prepared nanoparticle, using the radiation scattering technique for the analysis of biological activities. DLS can probe the size distribution of small particles ranging from submicron down to one nanometre in solution or suspension and is used to measure particles between the range of 2 and 500 nm.[18] It was confirmed that the particle was stabilized at 12 and 24 h of incubation before interacting with the aquatic environment and hence there was no need to add any other agent as a stabilizer.

Scanning electron microscopy (SEM) analysis: A representative SEM image of silver nanoparticles (Ag-NPs), polystyrene microplastic (PS), and silver aggregate polystyrene particles (Ag-PS) is shown in Figure 15.2. The shapes of silver nanoparticle were found to be predominantly large crystalline, whereas elliptical or spherical shapes were found for polystyrene particle. The aggregate silver and polystyrene particle were found as large crystals. Among various microscopic techniques, SEM is a surface imaging method fully capable of resolving different particle sizes, size distribution, nanomaterial shapes, and the surface morphology of the synthesized particles at micro- and nanoscales.[18]

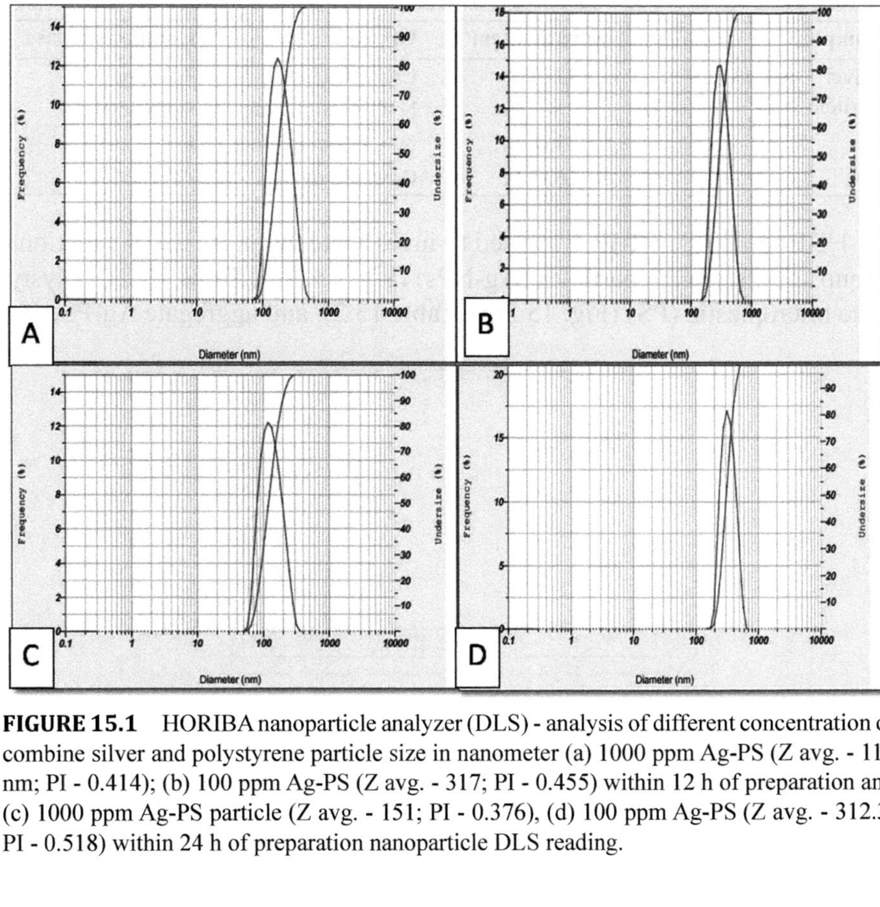

FIGURE 15.1 HORIBA nanoparticle analyzer (DLS) - analysis of different concentration of combine silver and polystyrene particle size in nanometer (a) 1000 ppm Ag-PS (Z avg. - 116 nm; PI - 0.414); (b) 100 ppm Ag-PS (Z avg. - 317; PI - 0.455) within 12 h of preparation and (c) 1000 ppm Ag-PS particle (Z avg. - 151; PI - 0.376), (d) 100 ppm Ag-PS (Z avg. - 312.3; PI - 0.518) within 24 h of preparation nanoparticle DLS reading.

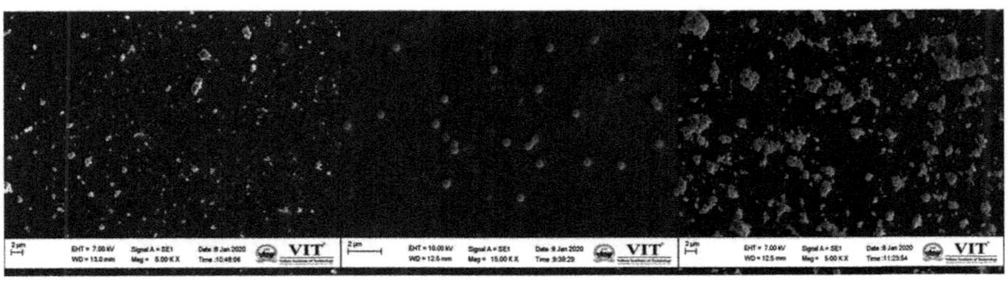

FIGURE 15.2 SEM images: (a) crystalline shape-silver particle image, (b) elliptical or spherical shape – polystyrene image, (c) large crystalline-silver-polystyrene aggregated particle image.

TABLE 15.1 FT-IR Analysis–Aggregated Ag-PS Particle.

Sample	Absorption (cm^{-1})	Group	Compound class
Silver + polystyrene particle	3270.68	C-H stretching	Alkyyne
	2116.49	N=C=S stretching	isothiocynate
	2361.44	N=C=O stretching	Isocynate
	1633.41	C=C stretching	Alkene

FT-IR analysis: FT-IR was used to investigate the presence of functional groups of silver nanoparticles (Ag-NPs) (Fig. 15.3(a), Table 15.2), polystyrene microplastic (PS) (Fig. 15.3(b), Table 15.3), and aggregate Ag-PS.

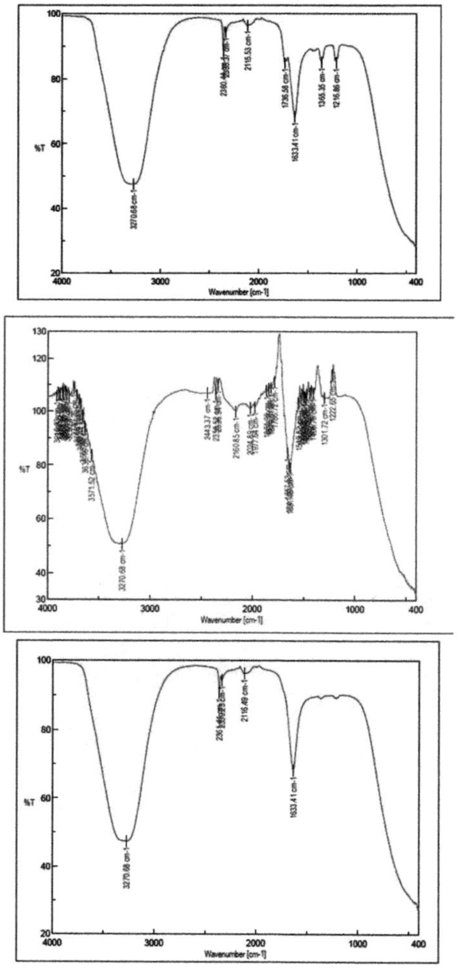

FIGURE 15.3 FT-IR spectra of (a) silver nanoparticle, (b) polystyrene microplastic, and (c) aggregate Ag-PS particle.

TABLE 15.2 FT-IR Analysis–Silver Nanoparticle.

Sample	Absorption (cm^{-1})	Group	Compound class
Silver Nanoparticle	3270.68	C-H stretching	Alkyyne
	2443.37	C=N stretching	Nitrile
	2160.85	N=N=N stretching	Azide
	1301.72	C-O stretching	Aromatic ester
	1222.65	S=O stretching	Sulfate

TABLE 15.3 FT-IR Analysis–Polystyrene Microplastic.

Sample	Absorption (cm^{-1})	Group	Compound class
Polystyrene micro particle	3270.68	C-H stretching	Alkyyne
	2395.37	O=C=O stretching	Carbondioxide
	2115.53	N=C=S stretching	Isothicynate
	1633.41	C=C stretching	Alkene
	1216.86	S=O stretching	Sulphoamide

UV analysis: In the present study, stability of particles was observed by UV-vis spectrophotometer between ranges of 300–700 nm. The results are shown in Figures 15.4(a), (b), and (c).

%Mortality Calculation = (no. of animal live/no. of animal) × 100

Toxicity of aggregated Ag-Ps particles: Mortality was observed with increasing concentration from 1 to 50 ppm for 96-h exposure. Experiments were conducted for 120 h. The LC50 value of combined nanoparticles at 96 h was found to be 48.2 ppm. The behavior of fish was noted at different concentrations (1 ppm, 3 ppm, 9 ppm, 27 ppm) for every 24 h for 4–5 days. That is approximately 96–120 h. Mortality was observed with increasing concentration from 1 to 50 ppm for 96-h exposure.

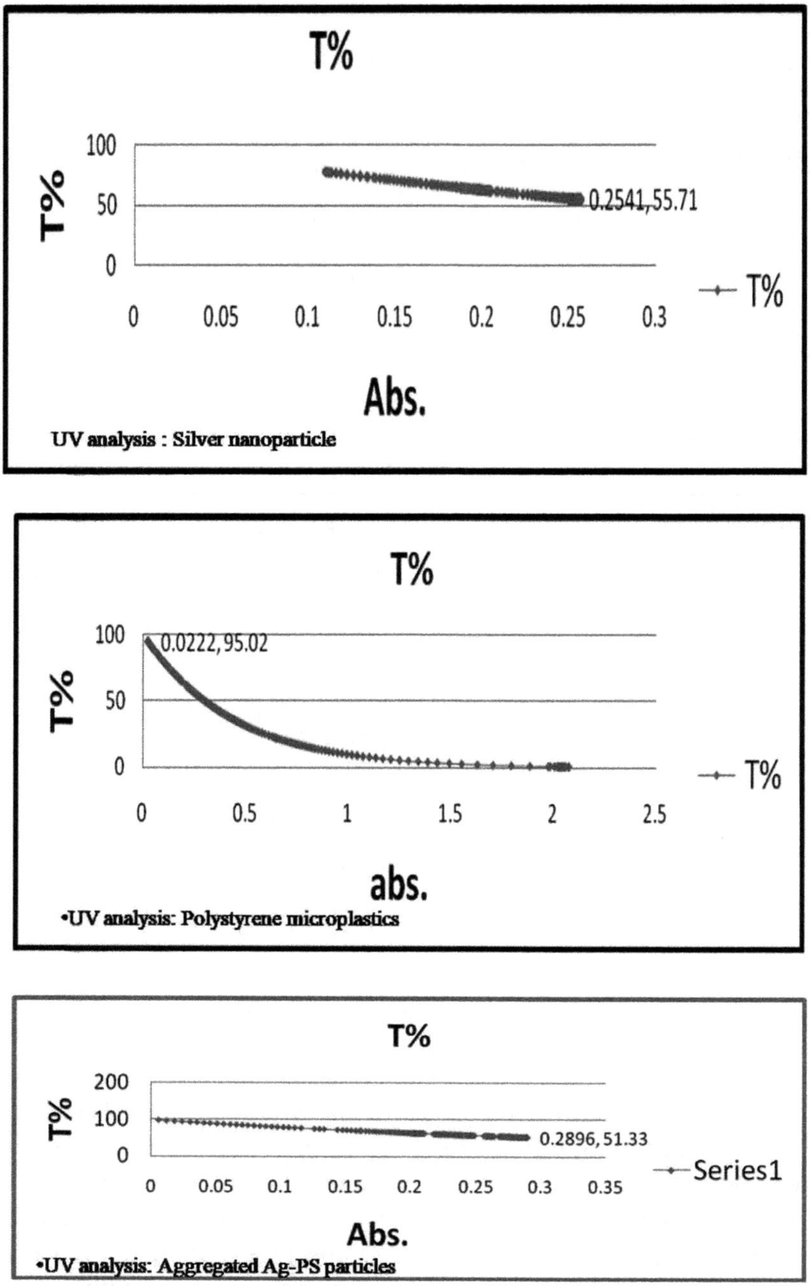

FIGURE 15.4 UV analysis. A. silver nanoparticles, B. polystyrene microplastic, C. aggregated Ag-PS particles.

15.7.2 BIOCHEMICAL ANALYSIS

Biochemical analysis was performed at lower concentrations of 1, 3, 9, 27 ppm. In control tank no mortality was recorded. Stress behavior was observed during 120 h of toxicity test of LC_{50}. Gills and liver were dissected from infected and control fishes for further analysis.

ROS assay: The ROS results revealed that oxidative stress releases the free radicals depending upon the reaction between the control and treated fishes at different concentrations, while increase in the concentration of PSNPs resulted in the enzymes reacting to release low percentage of radicals in the 100 and 50 nm size. The results are shown in Fig. 15.5(a) and (b).

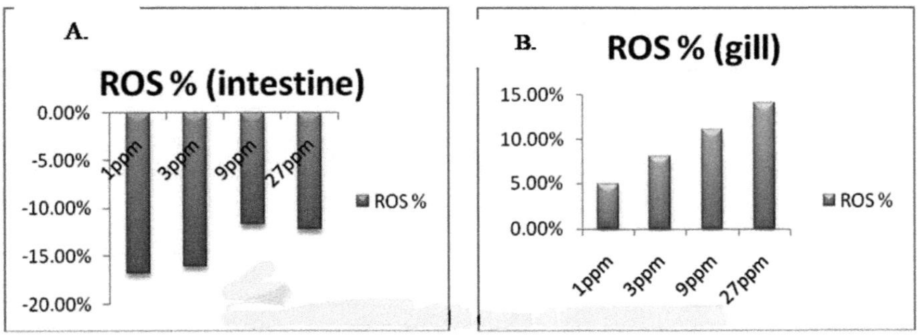

FIGURE 15.5 ROS in specific organs. A. Intestine. B. Gill.

CAT activity: The catalase activity was found to be lower when compared to control in 27 ppm. The results are shown in Fig. 15.6(a) and (b). Many studies have shown the role of oxidative stress due to nanoparticle as well as microplastics. To prevent stress condition, the enzymes SOD provide antioxidant defence, CAT.[19]

Protein activity: Protein activities were determined in specific organs (gill and intestine) by using standard BSA by the Bradford reagent method. Reduction in the value of protein compared to control and Ag-PS treated organs was observed respectively. The protein activity in the infected organs was lower when compared to control. The results are shown in Figure 15.7(a), (b), and (c)

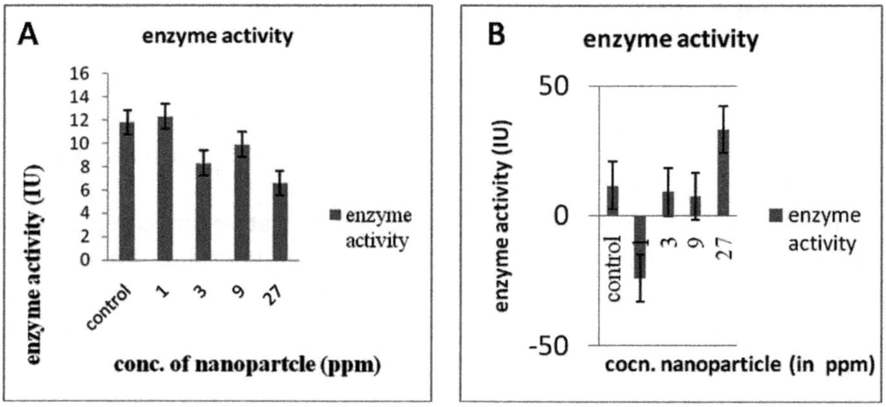

FIGURE 15.6 CAT activities. A. Gill, B. Intestine

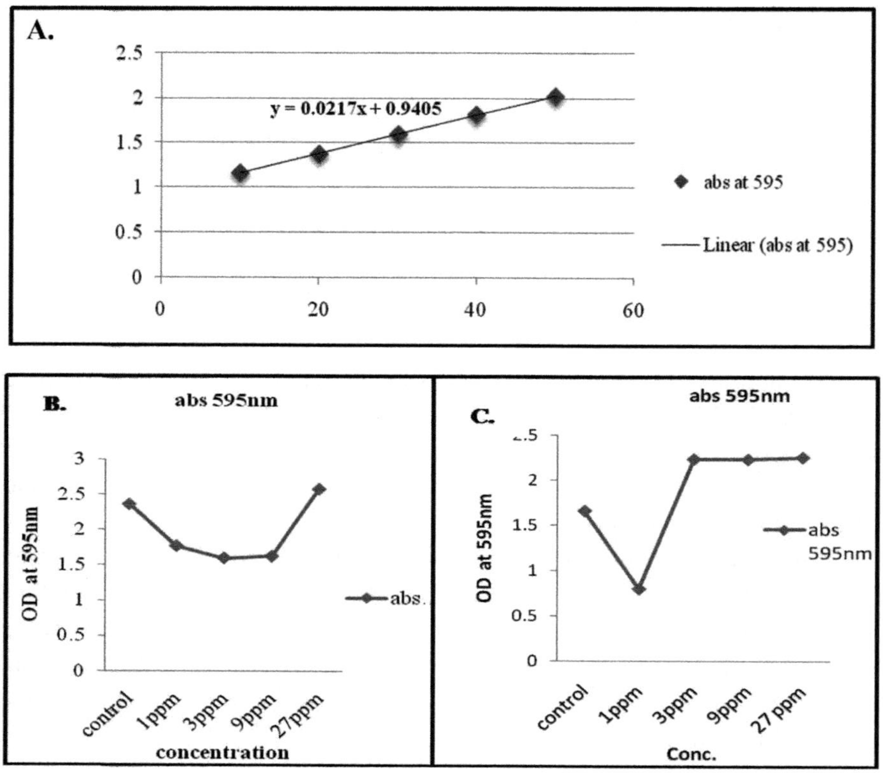

FIGURE 15.7 (A) Protein: Standard curve of protein analysis. (B) Protein content in gill, (C) Intestine

15.7.3 HISTOLOGICAL ASSESSMENT

The gills of the control fish showed a normal gill filament pattern under a light microscope at 10× (Figure 15.8). Primary lamella and secondary lamella were clearly visible, whereas in the Ag-PS (27 ppm) treated fish the gills were completely damaged. The control showed that the there was no damage in gills. The results are shown in Figure 15.8. Histopathology is a basic technique used in aquatic toxicity, which provides useful information for identifying target tissues and injuries to them as a result of their sensitivity to toxicological parameters such as mortality.[19] Among the toxicological endpoints in the toxicity test, the histological change in gill and intestine showed that they were more susceptible to Ag-PS exposure.

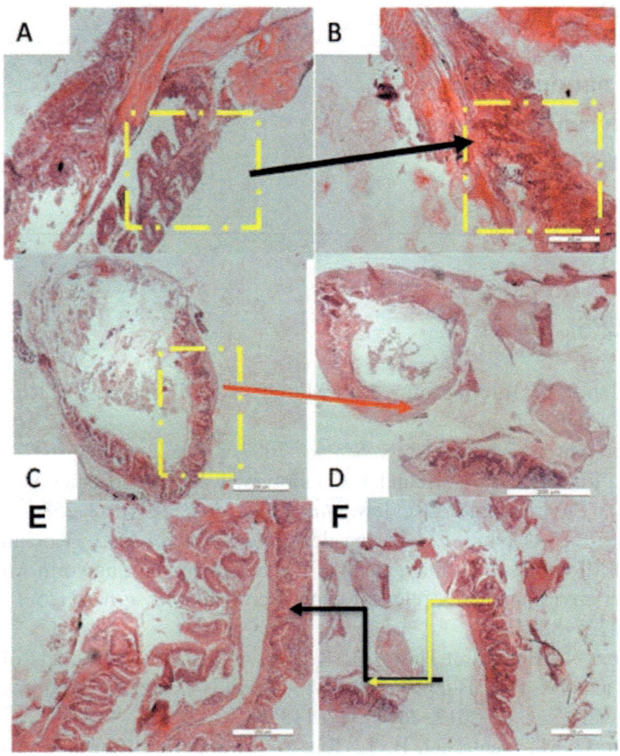

FIGURE 15.8 Histological section of zebra fish gill and intestine exposed to Ag-PS: view at 10X light microscope A. gill control B. gill infected (27 ppm); the blue dotted line path showed change in gill cellular structure after nanoparticle exposure. view at 10X light microscope C. and D. intestine control E. and F. intestine infected (27 ppm) red line between C and D and E-F black and yellow line showed a major change in similar cell after nanoparticle exposure

15.8 CONCLUSIONS

Silver-polystyrene aggregate (Ag-PS) particle induced great toxicity in zebra fish under chronic exposure condition and hence is toxic to aquatic organisms. It accumulates highly in gills and also affects intestine of zebra fish. The present study shows that toxic and stress effect of Ag-PS in gill is dose related dependent. The histological changes reveal a clear stress and apoptosis condition in gill and intestine tissues under microscope. This shows that polystyrene particles have the ability to bind silver nanoparticle and cause toxicity to living organisms. More studies need to be conducted to elucidate the mechanism of Ag-PS accumulation and its involvement in toxicity.

KEYWORDS

- **silver nanoparticles**
- **polystyrene microplastics**
- **toxicity**
- **histological changes**

REFERENCES

1. Maurer-Jones, M. A.; Gunsolus, J. I.; Murphy, C. J. L. Haynes, C. *Anal. Chem.* 2013, *85*, 3036–3049.
2. Nel, A. T.; Xia, L.; Madler; Li, N. *Science* 2007, *311*, 622–627.
3. Cole, M.; Lindeque, P.; Halsband, C.; Galloway, T. S. Microplastics as Contaminants in the Marine Environment: A Review. *Mar. Pollut. Bull.* 2011, *62*, 2588–2597.
4. Besseling, E.; Foekema, E. M.; van den Heuvel-Greve, M. J.; Koelmans, A. A. The Effect of Microplastic on the Uptake of Chemicals by the Lugworm Arenicola Marina (l.) Under environmentally relevant exposure conditions. *Environ. Sci. Technol.* 2017, *51*(15), 8795–8804.
5. Van Moos, N.; Burkhardt Holm, P.; Kohler, A. Uptake and Effects of Microplastics on Cells and Tissue of Blue Mussel Mytilus Edulis L after an Experimental Exposure. *Environ. Sci. Technol.* 2012, *46*, 11327–11335.
6. Rochman, C. M.; Hentschel, B. T.; Teh, S. J. Long Term Sorption of Metals Similar Among Plastic Types: Implications for Plastic Debris in Aquatic Environments. *PLoS One* 2014, *9*, e85433.
7. Giusti, L.; Hamilton Taylor, J.; Davison, W.; Newitt, C. N. Artifacts in Sorption Experiments with Trace Metal. *Sci. Total. Environ.* 1994, *152*, 227–238.
8. West, F. K.; West, P. W.; Iddings, F. A. Adsorption Characteristics of Traces of Silver on Selected Surfaces. *Anal. Chim. Acta.* 1967, *37*, 112–121.

9. Sekine, R.; Khurana, K.; Vasilev, K.; Lombi, E.; Donner, E. Quantifying the Adsorption of Ionic Silver and Functionalized Nanoparticles During Ecotoxicity Testing: Test Container Effects and Recommendations. *Nanotoxicology* 2015, *9* (8), 1005–1012.
10. Fako, V. E.; Furgeson, D. Y. Zebrafish as A Correlative and Predictive Model for Assessing Biomaterial Nanotoxicity. *Adv. Drug Deliv. Rev.* 2009, *61* (6), 478–486.
11. Kimmel, C. B.; Ballard, W. W.; Kimmel, S. R.; Ullmann, B.; Schilling, T. F. *Dev. Dyn. Off. Publs.* 1995, 203–310.
12. Lieschke, G. J.; Currie, P. D. Animal Models of Human Disease: Zebrafish Swim into View. *Nat. Rev. Genet.* 2007, *8*(5), 353–367.
13. Duan, J.; Yu, Y.; Li, Y.; Yu, Y.; Sun, Z. Cardiovascular Toxicity Evaluation of Silica Nanoparticles in Endothelial Cells and Zebrafish Model. *Biomaterials* 2013.
14. George, S.; Xia, T.; Rallo, R.; Zhao, Y.; Ji, Z.; Lin, S.; Wang, X.; Zhang, H.; France, B.; Schoenfeld, D.; Damoiseaux, R.; Liu, R.; Lin, S.; Bradley, K. A.; Cohen Y.; Nel, A. E. *ACS Nano*. 2011, *5*, 1805–1817.
15. Lee, W. S.; Cho, H. J.; Kim, E.; Huh, Y. H.; Kim, H. J.; Kim, B.; Jeong, J. Bioaccumulation of Polystyrene Nanoplastics and their Effect on the Toxicity of Au Ions in Zebrafish Embryos. *Nanoscale* 2019, *11*(7), 3173–3185.
16. Beutler, E.; Catalase. In: Beutler, E. ed. *Cell Metabolism, A Manual of Biochemical Methods*, New York: Grune and Statton Inc, 1982, 105–106.
17. Bradford, M. M. A. Rapid and Sensitive Method for the Qualification of Microgram Quantities of Protein Utilizing the Principle of Protein-Dye Binding. *Anal. Biochem.* 1976, *72*, 248–254.
18. Lin, P. C.; Lin, S.; Wang, P. C.; Sridhhar, R. Techniques for Physicochemical Characterization of Nanomaterials. *Boitechnol. Adv.* 2014, *32*, 711–726.
19. Wu, Y.; Zhou, Q. Silver Nanoparticles Cause Oxidative Damage and Histological Changes in Medaka (Oryzias Latipes) After 14 Days of Exposure. *Environ. Toxicol. Chem.* 2013, *32*(1), 165–173.

CHAPTER 16

Alginate-Based Wound-Healing Dressings

MEREENA LUKE PALLIKKUNNEL[1], TOMY MURINGAYIL JOSEPH[1], JÓZEF T. HAPONIUK[1], and SABU THOMAS[2]

[1]*Chemical Faculty, Department of Polymers Technology, Gdansk University of Technology, Gdansk, Poland*

[2]*International and Inter University Centre for Nanoscience and Nanotechnology, Mahatma Gandhi University, Kottayam, India*

ABSTRACT

The wounds can be treated by using dressings, antibiotics, and skin replacements and growth factor therapy. Wound-healing dressings effectively promote and help to provide a favorable milieu, which is conducive to the healing process. Many factors can obstruct one or more phases of wound healing, resulting in inadequate or impaired tissue recovery. Many natural and synthetic polymers are being used in the preparation of artificial dressing materials. The biopolymers such as alginate, collagen, chitosan, gelatin cellulose starch, and other polysaccharides based as well as protein, and lipid-based hydrogels. have been widely investigated as materials for wound dressings. Alginic acid is a polymer that derives from brown algae like *Laminaria* and *Ascophyllum*. Linear block copolymerization of D-mannuronic acid and L-guluronic acid produces alginic acid. Alginic acid and salts are used to treat wounds and burns because of their hemostatic properties. Due to their availability, superior biological capabilities, and low toxicity, they are biomaterials in the field of wound care management. When alginate dressings come into contact with a wound, they turn into a gel. The

Foundation and Growth of Macromolecular Science: Advances in Research for Sustainable Development. Meegle S. Mathew, PhD, Józef T. Haponiuk, PhD & Sabu Thomas, PhD, DSc (Eds.)

intrinsic features of alginic dressings depend on the ions in wound exudates. The absorbent properties of alginate hydrogels are determined by this cation interaction, which facilitates their usage as a dressing. The expansion of the hydrogel creates a moist environment healing the wound, while also preventing it from microbial invasion

16.1 INTRODUCTION

The skin, the outermost part and largest organ in the body, functions as a protective covering. Sweat and evaporation are two mechanisms by which the skin controls body temperature. The elimination of waste products is also aided by this mechanism. The acid mantle is a barrier that protects the skin from microorganisms. The sebaceous glands lubricate the epidermis, while fat cells insulate and protect the interior organs, maintaining water and electrolyte balance in the process. Nerve endings provide sensation to the body to protect it from pain, and also participate in vitamin D synthesis.[1] The skin retains essential substances and nutrients in the body, thus maintaining a barrier against toxic contaminants that penetrate the body, and provides a protection against the adverse effects of harmful radiations such as ultraviolet released from the sun. The epidermis, the skin's outermost layer, serves as a protective barrier and determines skin tone. Tough connective tissue, hair follicles, and sweat glands are all found under the epidermis in the dermis. Fat and connective tissue compose the deeper subcutaneous tissue (hypodermis) (Fig. 16.1).

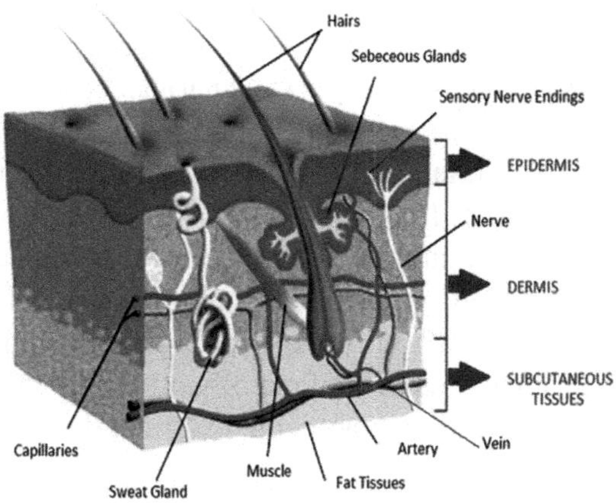

FIGURE 16.1 Structure of human skin.

Source: To Author: Please specify source since you have stated this is adapted

A wound is a confined or serious damage induced by an external factor that can disrupt living tissues and organs. Depending on the depth of the incision, damage occurs to the epidermis, local vascular, and perhaps the dermis and deeper tissues. This poses a risk to the organism's integrity by allowing infections and blood loss into the body. Wound healing is a complex process with multiple phases, such as (1) *hemostasis:* fibrin clot formation; (2) *inflammation*: neutrophils and monocytes rush into the injured skin when the immune system of the body has been triggered. This is the first step of wound healing, and it occurs in association with blood clotting; (3) *prolifération*: during the proliferative phase, the wound is repaired with new tissue composed of collagen and extracellular matrix; and (4) *remodeling*: remodeling is the final phase of the healing, when the granulation tissue grows into scar tissue and the tensile strength of the tissue is enhanced.[2]

The wounds can be treated by using dressings, antibiotics, and skin replacements and growth factor therapy. Wound-healing dressings actively stimulate healing and provide a microenvironment that is conducive to the healing process. Many factors can obstruct one or more phases of wound healing and lead to inefficient or delayed tissue repair. There are several varieties of wound products, all with distinctly different characteristics available. The ideal choice for wound care depends on the type, intensity, and color of the wound taken in accordance with the healing stage and what the primary care goals, such as debridement or protection. The dressing selection type of dressing depends on a number of parameters such as amount and nature of exudates, intensity of the wound etc.

Biodegradable polymers are gaining popularity as biomaterials, particularly for tissue engineering, gene therapy, wound healing, and drug delivery systems. The most significant benefit of biodegradable polymers is their biodegradability, which allows for the systematic and progressive elimination of foreign material implanted within the body.[3]

Hydrogels are substances with special properties that enable them to be used as an important type of wound-dressing material. Hydrogel wound dressings, alginate dressings, flat hydrogel dressings, amorphous hydrogel dressings, foam dressings, films, and composite products made of a fibrous substrate filled with polymer-forming hydrogel are examples of hydrocolloid dressings.[4]

This chapter considers the development and properties of various types of wound-healing materials, as well as recent advances in wound care management, particularly various techniques for preparing hydrogel wound dressings based on marine polysaccharides such as alginate. Additionally, this chapter focuses on the development of hydrogel-based wound care materials through polymerization of monomers and crosslinking of polymers. There are several crosslinking processes that may be utilized to create hydrogel dressings: Physical crosslinking; crosslinking by low molecular weight chemicals; ultraviolet, gamma, and electron beam irradiation; condensation processes between polymer functional groups; enzymatic crosslinking; and interpenetrating polymer network crosslinking.

16.2 WOUND-DRESSING MATERIALS

16.2.1 REQUIREMENTS FOR A WOUND-DRESSING MATERIAL

The optimal dressing for wounds is a combination of moist and dry conditions, with the latter being ideal for a very dry wound, and the excess exudate is absorbed in the case of an intensively exuding wound. It is also important to develop specialized dressings for the various phases of the wound. Wound-dressing materials are available in a variety of types. An ideal wound dressing is considered to be the one that ensures optimal healing by providing the patient with the ability to repair damage caused by an infection or injury. They should have the following properties:

Highly absorptive: To prevent maceration and minimize discomfort and swelling, exudates, bacteria, toxins, and dead cells are eliminated. Exudates can be managed by selecting the appropriate dressing, which will either absorb them or transfer them to a hydrophilic absorbent secondary dressing.

Nonadherent: The important point for patient comfort is a pleasant, conforming, flexible dressing that causes minimum pain when changed and that does not take an inordinate amount of time to replace. The dressing must be nonadhesive to the wound bed and shield the wound from further damage that will allow rapid and effective heal.

Gaseous exchange: Both hypoxia and normal quantities of oxygen are necessary at different stages of wound healing. In an anaerobic environment, the microcirculation is restored more quickly. For the formation of fibroblasts and collagen, high quantities of oxygen are required.

Antibacterial properties: A dressing should restrict germs from getting to the wound surface. A saturated or leaky dressing serves as a medium for germs to go in both directions.

Nonfiber shedding/nontoxic: Fibers in contact with the wound may cause an infection and discomfortness. Granulating tissue can develop into the open mesh of the dressing, permitting it to adhere to the wound. Iodine, for example, can induce topical pain or irritation. They should also fulfill characteristics including wear resistance, shelf life, and cost-effectiveness.[5]

16.2.2 CLASSIFICATION OF WOUND DRESSINGS

1. Cloth
2. Foam dressings
3. Transparent dressings
4. Hydrocolloid dressings
5. Hydrogel dressings
6. Alginate dressings, and
7. Collagen dressing.

1. Cloth: Cloth or gauze dressings are the most commonly used kind. They are made from natural fibers such as cotton and silk, as well as synthetic fibers such as linen, polyester, and rayon. These fibers may absorb fluids and are available in both woven and nonwoven varieties. They can be used to treat severe or open wounds, as well as minor burns cuts. These bandages come in a variety of shapes and sizes, ranging from small finger coverings to large bandage for treating wounds on broader areas of the body.[6]

2. Foam dressings: Foam dressings are another form of dressing; they comprise polyurethane solutions (foaming polymer solutions), and they contain tiny, open chambers that can absorb fluids. The thickness and composition of a material determine its absorption capacity. Foam dressings are available in pad, sheet, and hollow configurations, as well as with an adhesive surface and a transparent film covering that can function as a bacterial shield.

3. Transparent dressings: Transparent dressings are helpful when medical practitioners or nurses check the conditions of wound and progress of healing process, so these dressings protect the wound with a translucent film. This permits an easy way to detect possible risks such as infections, inflammations on the skin surface in each stage. Thus, this type of transparent dressing is also used in surgical incision areas, in burns and ulcers, and in IV procedures.

4. Hydrocolloid dressings: Hydrocolloid dressings are of a particular nature. They include gel-forming agents (gelatin or sodium carboxymethylcellulose) retained inside the adhesive compound that is laminated in place on foam or film, usually made of polyurethane. The final result is an innovative wound dressing that resembles an absorbent, durable wafer that is usually both waterproof and self-adhesive. Burns, mildly draining wounds, septic wounds, and pressure or venous ulcers are all popular uses for these dressings. They are one of the most long-lasting types of dressings, and their self-adhesive properties make them easy to use.[7]

5. Hydrogel dressings: Hydrogel wound dressings are usually used on a wide range of wounds. They are often used for wounds that are extremely traumatic or necrotic. Owing to the excess fluid in these dressings, this facilitates cell formation. Hydrogel dressings were intended to effectively manage fluid exchange on a wound surface. Hydrogels are capable of exchanging sodium and other therapeutic materials to the wound which help to heal the damaged skin layers, veins, and tissues.[8] The hydrogel dressings hydrate the region and remove dry, dead tissue, allowing it to heal faster. This method helps to improve the extent of patient comfort while minimizing pain and discomfort by dead tissues simultaneously. A cooling gel is used as an additional support in some hydrogel products. The hydrogel is suitable for second-degree burns and contaminated wounds, injuries from radiation exposure, severe wounds, partial- and full-thickness wounds, abrasions, and dry wounds.[9,10]

Different types of hydrogel dressings are as follows:

1. Sheet hydrogel: The hydrogel is enclosed within a thin net that can overlap with the skin without damaging it, which is a problem with several conventional wound dressings. However, if the wound does stick to the skin surface at all, saturating it with saline solution will usually easily loosen it. Although it is available in a number of sizes, the strips can typically be cut to fit the particular shape of the wound.

2. Impregnated hydrogel: This hydrogel is made by mixing the gel compound with cotton pads, sponge cordage, or clean fabric strips. This type of treatment usually necessitates a second dressing that covers the entire location to provide complete protection. The impregnated hydrogel can be put over the wound as well as packed into the deep or uneven wounds.

3. Amorphous hydrogel: This dressing is free-flowing, unlike the other two forms of hydrogels. It can ooze into the splits of puncture and other deep wounds due to the viscous nature, and are flexible, so it requires a secondary dressing to fix it on the wound surface.[11]

6. Alginate dressings: Alginate dressings exhibit highly absorbent capability and are utilized for wounds with excessive extrudate. The absorbency is up to 20 times higher, making it ideal for intense or deep wounds. In addition to absorption properties, alginate dressings often produce a gel that facilitates in the healing process. In several different scenarios, alginate dressings can be used in wound care regimens. They are used for treating diabetic and cavity wounds, pressure ulcers, venous wounds, and chronic wounds. In addition, they are used as combos of other dressing materials.[12]

*7. **Collagen dressing:*** These dressings are commonly used to treat chronic wounds that take a long time to heal. Pressure sores, transplant sites, surgical incisions, ulcers, burns, or bruises that cover a large area of the body are also treated with them. Collagen dressings serve as a temporary "second skin" that enables the growth and development of new cells. The recovery time will take even longer without it. As they tend to facilitate healing in a variety of different areas, collagen dressings are a better alternative to conventional bandaging. In addition to creating a scaffolding structure to grow new cells, collagen dressings help eliminate dead tissue, supporting the formation of new blood vessels, and help in the restoration of the wound's margins. Depending on the type of injury, collagen dressings are available as gels, granules, pastes, and even freeze-dried sheets that can be applied to the wound. In addition to these dressings, hydrogel dressings are one of the most adaptable and modern forms of wound dressing, particularly in the event of extraditing, wet wounds.[13]

16.2.3 HYDROGELS AND THEIR PROPERTIES

The concept of a hydrogel is very familiar, but usually the functional importance has developed. Natural polymer-based hydrogels are extensively employed in the biomedical industry and have made significant achievements as a modern substitute for synthetic or plastic materials due to their outstanding qualities. Hydrogels are three-dimensional cross-linked systems of macromolecular networks formed from synthetic and natural hydrophilic polymers that can swell, absorb, and hold a large amount of water or fluid and sustain their structural integrity.

Hydrophilicity of the hydrogels depends on the hydrophilic functionality of the system such as –OH, –CONH, –$CONH_2$, –SO_3H, –NH_2, and –COOH, and the degree of crosslinking the water uptake of the hydrogel is due to the hydrophilic functionality in the polymer and the resistance of dissolution is due to the crosslinking of polymer network. Water or fluids have an important role in shaping the distinct characteristics of hydrogels. The Flory–Huggins's theory explains how the volume of water absorbed by a hydrogel is affected by temperature and the specific interaction between water molecules and polymer chains.[14]

As seen in Figure 16.2, the solid part of the hydrogel is a network of cross-linked polymer chains, referred to as a network. The spaces are filled with a substance, normally vapor. These networks help to transport the fluid and provide an elastic force that may be achieved by the expansion and contraction of hydrogel, hence increasing the strength. By adopting different techniques and preparation techniques, the mechanical properties, biocompatibility, and porosity of the hydrogels can be enhanced, as the findings are available for various forms of hydrogels such as engineered tri-block copolymers, grafted hydrogels, and hybrid hydrogels with high stability, rapid reaction properties, and other enhanced properties. Because of its excellent properties, such as reversible swelling and deswelling behavior, high environmental sensitivity, high ionic conductivity, high permeability, surface properties, novel mechanical properties, and sorption capability, the smart hydrogel provides a medium for a variety of applications. Hydrogels can be categorized into different categories based on their physicochemical features.[6]

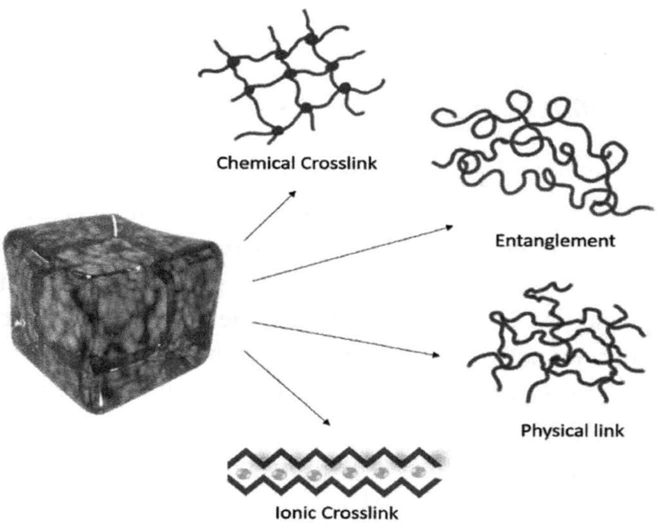

FIGURE 16.2 Possible structures of a hydrogel.

16.2.3.1 CLASSIFICATION OF HYDROGELS

Hydrogels can be cationic, anionic, or inert, depending on the charges on the attached groups. Another criterion for hydrogel classification is the type of crosslinking agent used, which determines whether the hydrogel is physical or chemical in origin. Hydrogels can also be classified according to their structure: amorphous, semicrystalline, crystalline, and hydrocolloid aggregates. Table 16.1 represents the classification of hydrogels based on their sources and properties, nature of swelling, method of preparation, origin, ionic charges, sources, and degree of biodegradation and nature of crosslinking.[15]

TABLE 16.1 Classification of Hydrogel.

1	Source	Natural
		Synthetic
		Hybrid
2	Crosslinking	Physically crosslinked
		Chemically crosslinked
3	Ionic charge	Cationic
		Anionic
		Neutral/nonionic
4	Synthesis	Homo polymerization
		Copolymerization
		Interpenetrating polymer networks
5	Type	Smart hydrogels
		Conventional hydrogels
6	Response	Chemically responsive
		Biochemically responsive
		Physically responsive
7	Degradability	Biodegradable
		Nonbiodegradable

Hydrogel classification is based on a number of variables, such as chemical nature monomer, composition, configuration, and interaction of polymer network. Besides these classifications the hydrogels can be categorized based on their origin and the method of synthesis. The chemical method of hydrogel synthesis is carried out by different techniques, some of which require simple single-step methods, including polymerization and multifunctional monomer crosslinking at the same time including the polymerization and simultaneous crosslinking of multifunctional monomers. Some processes require complicated and multiphase synthesis, such as the synthesis of polymer molecules using reactive chemicals, followed by reticulation with specialized crosslinking agents. The methods are variable based on desired attributes such as structure, matrix density, mechanical strength, biocompatibility, and biodegradability, which differ depending on the applications.

The types of gelation for hydrogel synthesis are generally of two categories.

(1) *Chemical hydrogels:* Chemical hydrogels are classified according to the presence of covalent bonds in the matrix that was used to make them. Because of their polymer–water connection and the degree of reticulation occurring in the matrix, these linkages govern the degree of swelling of the hydrogel. Because of changes in configuration, chemical hydrogels are permanent and irreversible.

(2) *Physical (temporary) hydrogels*: Ionic bonding, hydrogen bonding, and physical interactions all contribute to the formation of physical hydrogels. This is often done in physical gels by physical processes such as hydrophobic contact, chain aggregation, crystallization, polymer chain complex-ion, and hydrogen bonding. Physical hydrogels are reversible due to conformational changes.[16]

Based on the composition hydrogels can be classified as homopolymeric hydrogels, multipolymeric interpenetrating hydrogels, and copolymeric hydrogels. The polymer network is made up of homopolymeric hydrogels that are derived from a specific class of monomers. The type of monomer in a homopolymeric hydrogel, as well as the polymerization method used to make it, dictates whether the hydrogel backbone structure will be crosslinked.

Copolymeric hydrogels are composed of two or more different types of monomers, one of which is usually hydrophilic. On the backbone of the polymer network, this form of polymer configuration contains a chain randomly organized in block or irregular patterns.[17]

Multipolymeric hydrogels are otherwise known as hybrid hydrogels; they generally include both interpenetrating (IPNs) and semi-interpenetrating networks (Semi-IPNS), which involve the polymerization of two or more polymers. The formation of IPNs and semi-IPNS depends on the presence or absence of crosslinking initiator. Reactions in the presence of the cross-linking initiator result in the formation of complete IPNs, whereas the reaction without the crosslinking initiator leads to the formation of semi-IPNS.[18] Combining various polymer classes with required properties has been able to diminish limitations of the IPNs hydrogels, such as mechanical strength and absorbance potential. Due to an electrostatic interaction, this dual-network hydrogel was produced by combining physical and chemical cross-linked hydrogels. The hybrid IPNs exhibit some important advantages like improved controlled physical and mechanical properties; they have relatively dense hydrogel matrices, and thus they exhibit effective drug loading potential compared with other hydrogels. Because of their great liquid absorption capacity across a wide pH range and increased sensitivity, they can address the limitations of only using physical or chemical hydrogels. Hydrogels can be existing in different forms such as coatings, micro or nanoparticles, membranes, sheets, films, pressed powder matrices, liquids, solids, and gel.

16.2.4 NATURAL HYDROGELS

Natural polymers have explicit biomedical applications and properties, such as nontoxicity, biodegradability, and biocompatibility; hence, polymers having these characteristics are frequently used, but they have the limitations of poor mechanical strength and stability. These limitations can be rectified by hybrid hydrogels, interpenetrating polymer networks, or semi-inter penetrating polymer network. The most used polymers of natural origins include: hyaluronic acid, cellulose and its derivatives, dextran, alginate, chitosan, gelatin, and collagen (Table 16.2). Among these polymers, polysaccharides have been introduced and applied as potential candidates for numerous pharmaceutical and biotechnological applications, such as diagnostics, bioactive therapy, regulated drug delivery, and gene therapy by the benefit from their biocompatibility, biodegradability, low immunogenicity, and bioactive properties. Polysaccharides are considered to be important bio-macromolecules for all living organisms; they are found in various parts of plants, animals, fungi, microbes, and algae and play a key role in a number of physiological functions.[19]

16.2.5 ALGINATE HYDROGELS

I. SOURCES:

Alginate is an anionic polymer usually found on the cell walls of brown algae (Phaeophyceae). Alginate is derived from different sources of macroalgae (seaweeds), which is naturally available in many coastal areas of several countries. The cell wall of brown seaweed contains large amount of alginate (alginic acid) when compared with other red or green macroalgae and it is an

TABLE 16.2 Different Types of Natural Hydrogels.

Charge	Polymer	Origin/source	Type	Nature of hydrogel formation	Applications	Reference
Anionic	Hyaluronic acid	Extra cellular matrix of the skin, lenses, and cartilage of animals	Polysaccharides	Thermal or photo crosslinking	Skin and cartilage regeneration, cell differentiation	20
	Alginic acid	Sea weed [brown alga]		Ionic crosslinking [*By divalent cations*], thermal gelation chemical modification	wound healing, drug delivery, medical adhesives, tissue regeneration	21
	Pectin	Primary cell wall and intracellular layer of terrestrial plant cells [mostly in fruits]		Ionic crosslinking [*by divalent cations*]	Wound healing, medical adhesives, oral drug delivery, cosmetics	22
	Carrageenan	Marine alga		Heat reversible gelation, ionic crosslinking [*by divalent cations*], chemical modification		
Cationic	Chitosan	Crustaceans		Ionotropic gelation		
	Polylysine	Animal based	Amino acid	Enzymatic crosslinking	Drug delivery	
Amphipathic	Collagen	Extra cellularatrix	Protein	Chemical cross linking	Tissue regeneration	23–26
	Gelatin				Tissue engineering	
	Fibrin	Animal based				
Neutral	Dextran	Bacteria	Polysaccharides	Chemical crosslinking		
	Agarose, Pullulan	Sea weed				

insoluble salt. This salt can be extracted as water-soluble alginate to enhance the properties and for advanced applications. Through a series of processing steps, water-insoluble alginate is converted into a water-soluble salt. These soluble salts exhibit significant rheological characteristics, including gelling, viscosity, and dispersion stability.[27]

II. STRUCTURE:

Alginates are composed of *β*-D-mannuronic (M) and *α*-L-guluronic (G) acid units that form regions of M-blocks and G-blocks as well as blocks of alternating sequences (MG blocks). That is, the uronic acids can be arranged in a hetero polymeric manner that has both mannuronic and guluronic acid blocks with a near equal proportion of the monomers and homopolymeric (MM or GG acid blocks) block. The structural organization of the G and M units of the uronic acid depends on the source location and seasons of their harvesting (Fig. 16.3).

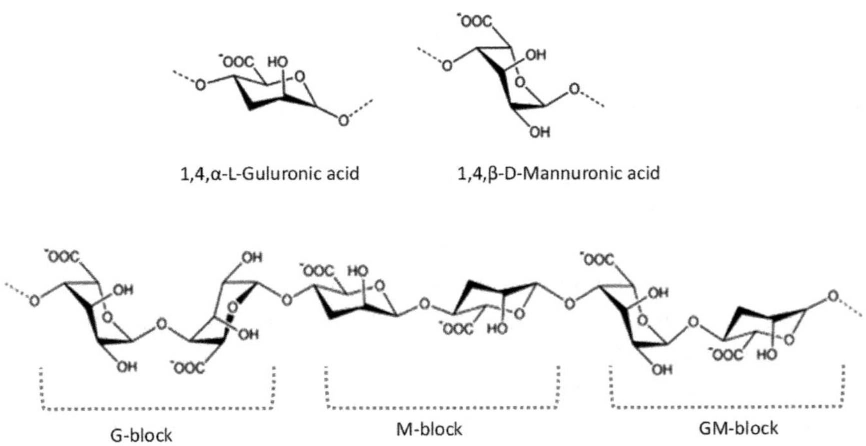

FIGURE 16.3 Structure of alginate.

The mannuronic blocks/guluronic acid blocks (M/G) ratio and their structure have a significant impact on the characteristics of alginate. Because of its ability to produce cytokine via human monocytes, alginate's high M block content is advantageous for chronic wound-healing applications.[28] The source of the algal species has the greatest influence on the quality of alginate.

Moreover, alginate is nontoxic, biocompatible, biodegradable, biostable and it is a hydrophilic biopolymer. These outstanding properties have aided in the development of several alginate-based products, enabling them to be used in a wide range of innovative clinical and biological applications, as well as in the food industry.

TABLE 16.3 Properties of Alginate.

Alginate	Biocompatible
	Biodegradable
	Inert nature
	Nontoxicity
	Regenerative
	Gel/film forming capacity
	Insoluble
	Swelling
	Low cost
	Natural sources

III. GENERAL PROPERTIES

Alginate is commonly utilized in biomedicine as a hydrogel for wound healing, drug administration, and tissue engineering. Hydrogels are three-dimensionally linked networks made up of highly water-soluble hydrophilic polymers.[29] The physicochemical properties of hydrogels are strongly dependent on the cross-linking type and density, as well as the molecular weight and chemical composition of the polymers.

Molecular weight and solubility: Sodium alginates with a molecular weight of 32,000–400,000 g/mol are commercially available. As the carboxylate groups in the alginate backbone become protonized and form hydrogen bonds, the viscosity of alginate solutions increases as the pH falls, reaching a median pH of 3-3.5.[30]

IV. METHODS OF GELLING

1. Ionic crosslinking

The most popular approach for preparing hydrogels from aqueous alginate solution is to mix ionic cross-linking agents, divalent cations (Ca^{2+}, Zn^{2+}).

Divalent cations are supposed to be attached exclusively to the guluronate blocks of the alginate chains; it permits the coordination of the divalent ions. The guluronate blocks of one polymer then form junctions with the guluronate blocks of the neighboring polymer chains in what is called the egg-box cross-linking model, resulting in a gel formation (Fig. 16.4). Calcium chloride is commonly employed as a crosslinking agent, but because of its high solubility in aqueous solutions, it exhibits rapid gelation but less control. This limitation can be effectively minimized by using buffer containing phosphate groups such as sodium hexametaphosphate, in which the phosphate groups in the buffer interact with the carboxylate groups of alginates and react with calcium ion, which will facilitate slow and controlled gelation. Because of their lower solubilities, calcium sulfate (CaSO4) and calcium carbonate (CaCO3) can both slow and accelerate the gelation process.

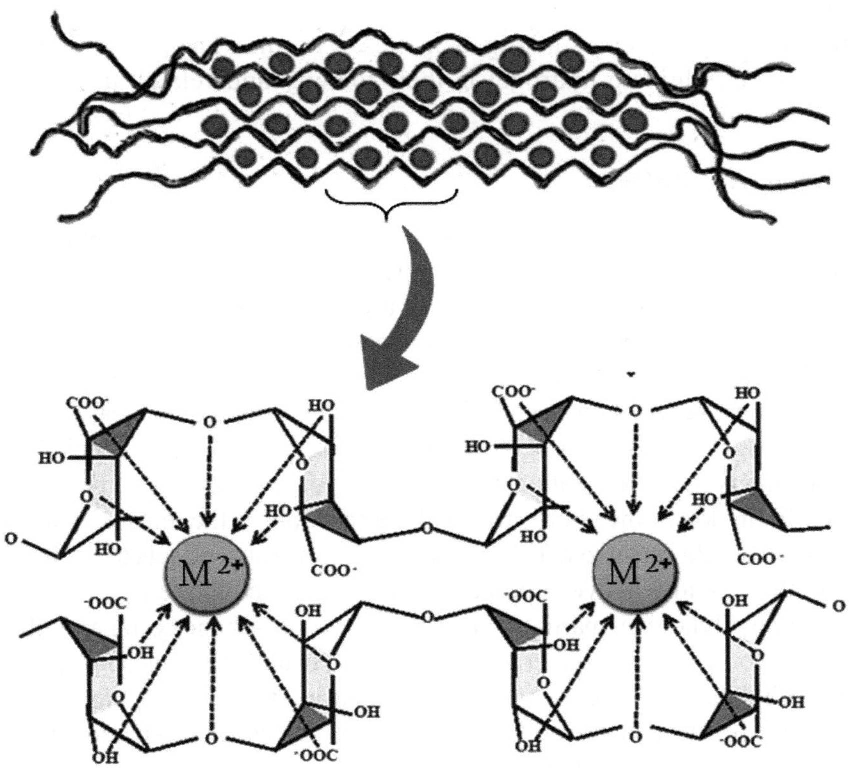

FIGURE 16.4 Ionic crosslinking of alginate hydrogel.

Slow gelling offers more homogeneous structures and superior mechanical integrity, which is important when utilizing divalent cations to regulate gel consistency and strength. The gelation temperature has an impact on the pace of gelation and the mechanical properties of the gels.[31]

2. Covalent crosslinking

Though water movement can induce stress relaxation in covalently cross-linked gels, the inability to dissociate and reform bonds causes extreme elastic deformation.[32] Multifunctional cross-linking molecules allow a larger range of degradation rates and control over the mechanical properties of hydrogels than bifunctional crosslinking molecules. Glutaraldehyde and polyethylene glycol (PEG) are the commonly used crosslinking agents. The usage of such a system improves the mechanical properties of alginate hydrogels, and the swelling of the hydrogels may be carefully controlled by utilizing different types of crosslinking molecules and adjusting the crosslinking densities. The drawback of chemically cross-linked alginate hydrogel is their toxicity and related undesirable side effects.

3. Photo cross-linking

It is a type of in situ gelation technique that makes use of covalent cross-linking. Photo crosslinking can be performed using the suitable chemical initiators under a mild environment or a controlled condition, even in the presence of close proximity to drugs and cells. For example, Alginate can be crosslinked with methacrylate by exposing it to a 514-nm argon ion laser for 30 s in the presence of eosin and triethanol amine, resulting in clear and flexible hydrogels. These hydrogels are used in ophthalmology; especially for the sealing corneal perforation also it facilitates the sutureless surgery.[33] The major limitation of the photo cross-linking reactions is the release of acid by the use of a light sensitizer. This may cause harmful side effects to the human body; thus, an alternative approach can be introduced to rectify this limitation—polyallylamine partially modified with α-phenoxycinnamyldiene acetyl chloride, which dimerizes on light exposure at about 330 nm; this approach did not evolve toxic by-products during the crosslinking reaction. Light irradiation considerably improved the mechanical characteristics of these hydrogels, and they were also easily permeable to cytochrome c and myoglobin.

4. Thermal gelation

Because of their versatile swelling properties in response to temperature changes, thermosensitive hydrogels have been extensively used in many drug delivery applications, providing for controlled drug release from the gels.[34] The most widely used thermosensitive gels are poly (N-isopropylacrylamide) (PNIPAAm) hydrogels, which undergo a reversible phase change in aqueous media near body temperature-lower critical solution temperature near 32°C. Copolymerization with hydrophilic monomers like acrylic acid and acrylamide will change the transition temperature.[35] Regarding the potential importance of thermo-sensitive hydrogels in biomedical applications, few alginate-based systems have been identified so far, as alginate is not naturally thermo-sensitive. Semi-interpenetrating polymer network (semi-IPN) structures were prepared via in situ copolymerization of N-isopropylacrylamide (NIPAAm) with poly (ethylene glycol)-*co*-poly (ε-caprolactone) (PEG-*co*-PCL) in the presence of sodium alginate by UV irradiation. At a specific temperature, the swelling ratio of the gels increased with the concentration of sodium alginate and decreased with an increase in temperature. The use of sodium alginate in semi-IPN structures enhanced the mechanical strength and total release of BSA from the gels, implying that it could be used in drug delivery.[36]

16.2.5.1 ALGINATE-BASED WOUND DRESSINGS

Wound healing is a dynamic and complex process. Wound healing includes regulated sequential phases including inflammation, migration, proliferation, and maturation.[37] This integrated series of cellular physiological and biochemical events can be complicated by tissue and skin layer destruction, which leads to infection spreading to nearby and interior organs, which also can intensify this process known as sepsis.[38] Wounds necessitate additional procedures and attention, as this process might be impeded by a variety of factors such as poor oxygenation, infections, and dressing selection that are inappropriate for the wound type.

The treatment of acute and chronic wounds is an urgent need. Synthetic polymers and biopolymers are commonly used as wound dressings. Compared to standard wound dressings, alginate-based wound dressings have numerous advantages. Alginate-based wound dressings are available as hydrogels, films, foams, nanofibers, membranes, and sponges. As sodium alginate is used as a wound dressing, it is commonly infused with calcium chloride to form sheets. The dressing fiber swells and is partly converted into a gel that moisturizes the wound bed and speeds up the healing process when Ca^{2+} from the dressing interact with Na^{+} from the wound fluid. Alginate dressing products are biocompatible, nontoxic, and nonallergenic, and they have various properties such as microbial resistance, air permeability, and mechanical resistance. These materials also enable the incorporation of a suitable drug. Alginate wound dressings absorb a lot of water and swell. Because of these features, it is commonly used in the treatment of intensely exuding wounds like surgical wounds, leg ulcers, and pressure sores.[39] They also absorb wound exudate, through a chemical reaction, and transform it to a hydrophilic gel. Alginate wound dressings can absorb liquid up to 15–20 times their weight.[19] These bandages are flexible, making them convenient to use on a regular basis. They are also permeable to water and air.

Wound dressings incorporating antimicrobial or curing agents have been formulated to protect against infectious infections, and as compared to consumer products, the product has higher epithelialization, granulation, and angiogenesis benefits. In line with emerging century's threats, natural materials have started to take the place of synthetic polymers, ceramics, and mesoporous materials.[27] Biomaterials are materials that are designed to interact with biological processes to evaluate, treat, improve, or substitute any tissue, organ, or function. Biomaterials must be sterile and noninteractive with the host organism to be used in the medical industry.

Generally, the development of the alginate wound dressing involves a series of steps. It involves the ethanol treatment for the prevention of alginate swelling and premature gelling on contact with water or other fluids. After that solution of alginate in distilled water is used for impregnation of woven or nonwoven material followed by drying and mechanical softening of these flexible dressing materials.

i. Hydrogel dressings

Sodium alginate was used to produce hydrogel dressings. The ability of hydrogel dressings to provide an optimal hydration state for healing makes them prominent. Because of its low interfacial tension, oxygen permeability, and mechanical and moisturizing properties, physiological soft tissue is reassembled. As a result, polysaccharides with hydrogel-forming properties are thought to be beneficial as a wound-dressing material.

Sodium alginate, deionized water, and H_2S are mixed for one hour to produce hydrogel dressings. $CaCl_2$ is added and gently mixed to start the ionic cross-linking of the alginate polymer chains.[40] Other cations divalent cations such as Ca^{2+}, Sr^{2+}, Cd^{2+}, Zn^{2+}, Ba^{2+}, Cu^{2+}, or trivalent cations Al^{3+}, Fe^{3+} can result in the development of ionically cross-linked alginate hydrogels. Alginate can be crosslinked with divalent or trivalent metal cations to produce highly porous materials (hydrogels) that are appropriate for bone tissue engineering. Injectable hydrogels are most widely used hydrogels they offer a simple way to encapsulate both soluble and insoluble pharmaceuticals.[41]

Chemically, alginate hydrogels can be altered and adjusted by changing the molecular weight (higher molecular weight alginates make stiffer gels, despite lower molecular weight alginates allowing for increased degradability and cell proliferation). Although hydrogel dressings based on alginate contain 70–90% water, they provide a moist environment around the wound bed. Furthermore, by dissolving necrotic tissue on the wound surface, this facilitates a self-cleaning wound process. Depending on the amount of extrudate, the hydrogel dressings can last up to three days. Minor burns, abrasions, infected wounds, partial- and full-thickness wounds, surgical wounds, pressure, diabetic, and venous ulcers are all suitable candidates for these dressings. For example, CarraSorb H® (Carrington Laboratories), Tegagel®™ (3 M Health Care) are used for less exudating wounds.[42] Hydrogels are used efficiently as regulated drug delivery and wound-healing systems because of their 3D cross-linked polymeric networks, high biocompatibility, and biodegradability. Hydrogels must have adhesiveness, elasticity, and stability, as well as be occlusive and bacteria-resistant. At 25°C, alginate can form hydrogels.[43]

ii. Foam Dressings

Recent years have seen foam dressings largely replace wound dressings. Since the foam dressing is heavier than the wound dressing, it has a greater supportive padding action over the wound. They not only avoid exudate accumulation, but also retain hydration, facilitating epithelialization and healing by enabling cell migration in the appropriate moisture environment.[44] Because of their exceptionally sorptive characteristics, which facilitate necrotic tissue autolysis, foam dressings are ideal for partial- and full-thickness wounds, surgical wounds, pressure ulcers, venous and diabetic ulcers, and highly exudation wounds. The duration of wear is up to 7 days, but most people only wear it for 3–5 days. Foam dressings are appropriate for infected wounds when changed on a daily basis, but they are not suggested for nonextruding wounds since they may cause the wound surface to dry up.[45]

Alginate foam dressings are made by homogenizing alginate with a sequestering agent, plasticizer, and surfactant, then adding a divalent or trivalent metal ion to create a water-insoluble hydrogel. The mixture then poured into the molds and freeze for developing the frozen hydrogels. The final step is to lyophilize the hydrogel foam to remove any remaining moisture. Silver, which is used as an antimicrobial agent, or other required wound-healing agents, can be added to enrich the alginate foam dressings; finally, the alginate dressing retains a moist environment around the wound, restores wound tissue to its original state, and reduces tissue granulation over time, eliminating the danger of tissue damage when the lesion is removed.[46]

iii. Alginate-Based Fibers Nanofibers and Hydrogel Fibers

Calcium alginate fiber is found in the commercially available bandages. They absorb considerable amount of exudate and form a gel-like protective layer over the wound surface, which keeps the wound moist and prevents the wound bed from drying out. They are appropriate for wounds with mild-to-heavy exudation, partial- and full-thickness wounds, surgical wounds, infected wounds, pressure ulcers, diabetic ulcers, and venous ulcers. The main benefit of these dressings is that they do not change regularly; depending on the amount of extrudate, they can be stable for 2–7 days.[47] Alginate fibers are not apposite for wounds with few or no exudate because they have the potential to dry out the wound bed and split or stick fibers to the wound bed. Furthermore, alginate fibers frequently necessitate a secondary dressing. AlgiDERM® (Bard), Algosteril® (Johnson & Johnson Medical), Kalginate (DeRoyal), KendallTM Calcium Alginate Dressing (Covidien), and Sorbsan® are some of the most popular alginate fiber dressings (Dow B. Hickam, Inc.).

Alginate bandage is also available as a hydrofiber dressing, which is a nonwoven tape or pad constructed of hydrocolloid fibers and sodium carboxymethylcellulose. Alginate hydrofiber dressings are highly absorbent and, when coupled with exudate, generate a gel that retains the wound wet and allows for painless removal without injuring sensitive granulation tissue. They are appropriate for acute or chronic wounds with a considerable volume of exudate, partial- and full-thickness wounds, donor sites, surgical wounds, pressure ulcers, and other similar conditions, for example ConvaTec AQUACEL Hydrofibre Wound Dressing.[48] Alginate nanofibers are potential materials for wound dressing. Nanofibers act as a substitute for the extracellular matrix, promoting epithelial cell proliferation and tissue formation. Their nanofibrous meshes promote hemostasis of wounded tissues, optimize fluid absorption, enhance drug delivery, cell respiration, and high-air permeation, and thus avoid bacterial infections.[49] Electro spinning techniques are generally used for the preparation of alginate nanofibers; it is very difficult to control the pore structures and size and these fibers are not uniform. The addition of synthetic polymers such as polyethylene oxide and polyvinyl alcohol allowed the alginate solution to be spun into nanofibers. When therapeutic nanoparticles were added to the nanofibers, they showed good antibacterial activity.[50]

iv. Wafers

Topical drugs that are applied to the skin, such as creams, gels, and lotions, etc have a slight effect; it relies on passive diffusion through the skin. They lose their controlled release properties. Thus, it is essential to develop a novel topical dosage formulation that can deliver drugs to the wound site in a continuous and localized manner. Wafer dressings, for example, deliver active medicines to accelerate wound healing. A wafer or sponge is a dispersion of gas (usually air) in a solid matrix of interlinked pores; wafers are made using the lyophilization method. Wafers are made using a freeze–drying technique, which is designed to remove any liquids or water from frozen material. This freeze–drying method is critical for preserving the texture, bioactivities, and other product characteristics. The wafers produced by this freeze–drying technique have low moisture content and can be kept for a prolonged period of time.[51] The qualities of wafers are heavily influenced by the type of polymer utilized as well as the concentration of polymer and medicines in the final formulation. Boateng et al. prepared lyophilized wafers containing sodium alginate and gelatine in various concentrations, along with 0.1% w/w silver sulfadiazine.

The use of sodium alginate in the formulation resulted in the polymeric network having uniform pores. The presence of sodium alginate has an effect on the water absorption capability of wafers. Water absorption was highest in sodium alginate-containing wafers (2299.79 ± 151.29%). Compositions containing a higher quantity of sodium alginate release the loaded drugs more quickly than other compositions throughout a 7-hour period. Because of their high swelling capacity, mechanical features, and releasing properties, wafers made of sodium alginate can be used as bandages for severely leaking wounds.[52]

v. Hydrocolloids

Hydrocolloids are wafer dressings that combine sticky elastomers with absorbent colloidal components. They produce gel in conjunction with a little to moderate quantity of exudate, which can offer a moisture environment to keep the wound bed from drying. Hydrocolloids are simple to apply and remove, and they do not necessitate the use of a second wound dressing and are useful for seven days. They are water resistant, which means that they are resistant to bacterial infection. These dressings are oxygen impermeable, which can result in an unpleasant odor after the removal, and they are therefore not suitable for infected wounds. The most commonly used hydrocolloids on granulating and epithelizing wounds, partial- and full-thickness wounds, surgical wounds, necrotic wounds, pressure ulcers, diabetic and venous ulcers are alginate-based hydrocolloids. They are not appropriate for severe exudation wounds because the significant amount of exudate may lead to more frequent dressing changes. The most well-known alginate-based hydrocolloids are Comfeel PlusTM (Coloplast AS), Restore® Calcium Alginate (Hollister Woundcare), TegasorbTM (3 M Health Care), and UrgoSorbTM (URGO Medical).[53]

vi. Films

Film is a semipermeable membrane layer that is impervious to liquid and bacterial contamination from the outside, while permitting the passage of water vapor, carbon dioxide, and oxygen.[54] It provides a moist environment with a little quantity of exudate, allowing necrotic tissue to be removed and cellular migration to facilitate wound healing. Alginate-based films provide a number of advantages, including the ability to avoid or minimize friction, as well as the ability to manage wound-healing process through the use of a transparent film. Reducing exudation wounds, superficial wounds, blisters, wounds on heels, elbows, and flat surfaces can all benefit from a dressing like TegadermTM (3 M Health Care).[55] For wounds with a lot of exudation, these bandages are not suggested.

vii. Antimicrobial Dressings

Antibacterial dressings made of alginate with silver nanoparticles, charcoal, or other active medicinal components have antibacterial, antifungal, and antiviral properties. They can absorb a lot of exudate and minimize smells, but they are only good for a short time and require a secondary dressing. The moist wound environment provided by these alginate dressings aids in the autolysis process and speeds wound healing.[56] Dressings like Algicell® Ag (Derma Sciences), AQUACEL Ag Hydrofibre Dressing with Silver (Conva

Tec), Suprasorb A (+ Ag) (Lohmann & Rauscher GmbH and Co. KG), and UrgoSorbTM Ag/Silver (URGO Medical) are suitable for chronic nonhealing wounds, wounds with moderate-to-high exudate, surgical wounds, pressure ulcers, and diabetic foot and leg ulcers. It is not advised for silver-sensitive patients or for use in combination with an oil-based medicine.[57]

Medihoney® antimicrobial honey dressings developed by Derma Sciences, Princeton, NJ, USA are new and unique dressings. The advantage of using y Active Manuka honey (*Leptospermum scoparcium*) can diminish the toxicity of ionic silver but also exhibit a long-term antibacterial activity similar to ionic silver. This dressing is also used to increase the rate of healing and reduce methicillin-resistant *Staphylococcus aureus* (MRSA) infections, as well as effectively aid in debridement. It also supports in moisture retention and odor prevention, which are typical issues with contaminated or heavily colonized wounds.[58] For example, medihoney gel sheet and apinate dressings (comvita) are the honey impregnated on mechanically bonded calcium alginate fiber and algivon dressing by Advancis acts as a barrier against antibiotic resistant strains and other wound pathogens reducing the risk of infection. Outflow of exudates induced by osmotic action removes the wound bacteria, endotoxins debris, and slough. The optimal healing environment enhances granulation and epithelialization. This is suitable in the treatment of leg ulcers, burns, donor graft sites, and infected wounds.

viii. Alginate Bio-Aerogels

Bio-aerogels are a new family of materials with unique qualities such as low density, high porosity, heat resistance, lightweight weight, and low dielectric constant that make them ideal candidates for a wide range of applications, including medical and pharmaceutical ones. Polysaccharide-based aerogels made from alginate, agar, starch, cellulose, chitosan, chitin, or pectin are biomedical and pharmaceutical materials that can replace silica aerogels in applications such as tissue engineering, drug delivery, and biosensing.[59] Because of their outstanding biocompatibility, high surface/volume ratios, and huge interior surface areas, these aerogels can imitate extracellular matrices in the body and allow for substantial drug loading.[60]

Various strategies for loading pharmacological molecules into aerogels before gelation, during the aging stage, or during the adsorption have been devised to take use of these features. The structure, composition, and hydrophobic nature of biopolymer aerogels all influence the dispersion of the drug in the aerogel matrix, the crystalline/amorphous phases of the drug, and the accessibility of the loaded drug to the solvent during the release process. A combination of these features influences the drug release profile of aerogels. The size and surface area of the aerogel are the primary factors governing drug release. To utilize this ability, 3D multimembrane alginate-based aerogels with layered (onion-like) topologies have been developed. These special features produce open spaces ideal for cell or drug integration, increasing drug loading and prolonging drug release.[61]

ix. Combination Dressings

Wound-healing dressings can be used as a combined form to attain both properties of individual dressing. They can be fabricated by combining two or more various wound-dressing materials such as foam, hydrocolloid, and charcoal. PolyMem® (Ferris Mfg.), SeaSorb® (Coloplast), Tromboguard® (tricomed), etc. are the examples of combination dressings based on alginate hydrogel. They can be used to treat low to moderate fluid exudating wounds, partial- and full-thickness wounds, surgical wounds, pressure ulcers, venous, and diabetic ulcers. Combination dressings are commonly available as waterproof, which protects against bacterial and other types of contaminations. These dressings are not recommended for wounds with a lot of exudates because their use length is usually determined by the amount of exudate.[62]

16.2.6 HEMOSTATIC PROPERTIES OF CALCIUM ALGINATE

Wound dressings made of alginate are well known and widely used in wound care. Alginate-based dressings are used to treat bleeding wounds because calcium alginates are a natural hemostat. By providing a moist environment, it promotes rapid granulation and re-epithelialization. Alginate is a biocompatible substance that has excellent absorbent, gel-forming, and hemostatic properties. As a result, alginate is an excellent choice for treating burns and wounds. Its hemostatic properties, however, are widely established. Calcium alginate promoted fibroblast growth while inhibiting microvascular endothelial cell and keratinocyte proliferation. It slowed down the movement of fibroblasts but had no effect on the motility of keratinocytes. When calcium alginates come into contact with blood, they release calcium ions in exchange for sodium ions. Platelet aggregation is prompted by migrating calcium ions, which act as a cofactor (clotting factor IV) in the coagulation cascade, resulting in instant blood clotting.[63–65] Calcium alginate may be used as a hemostatic agent in a splenic injury, according to rat reports. Alginate and chitosan (Alchite) have been shown to form fibers that can be used in wound treatment. Its antibacterial properties suggest that it may be helpful in the production of wound dressings. This may also facilitate in the development of a hemostatic wound dressing.[66]

16.3 CONCLUSIONS

Alginates are natural compounds generated from marine brown algae that have traditionally been widely employed as food additives and ingredients. Alginates are an excellent candidate for the development of wound dressings due to their features such as biocompatibility, nontoxicity, biodegradability, and functional adaptability with diverse matrices and substrates. Because of its hemostatic and antibacterial qualities, both in vitro and in vivo data emphasize the benefits of utilizing alginate in the treatment of wounds. Dressings available in a variety of forms, including wound dressings, hydrogels, foams, and aerogels, depend on the use. Wound dressings are used for the care of fluid exudating wounds such as surgical wounds, leg ulcers, and pressure sores. They are biocompatible, biodegradable, and nontoxic, having healing characteristics. Foam dressings protect the wound by its unique cushioning effect and maintain it hydrated, in this suitable condition allowing epithelialization and healing. Hydrogel dressings are used to administrate the controlled drugs or heal wounds, and their qualities can be enhanced by adding natural or synthetic medications.

Alginate-based products could be used in a variety of applications, including insulation, energy storage devices, conservational cleaning, biomedical, food, and agricultural industries. Because they are completely natural, they are absorbed much more rapidly by the human body, lowering the risk of negative effects.

KEYWORDS

- **polysaccharides**
- **alginate**
- **hydrogel**
- **wound dressing**
- **ionic gelation**

REFERENCES

1. Wooi, Ng. K.; Lau, W. M. *Skin Deep: The Basics of Human Skin Structure and Drug Penetration Skin Deep: The Basics of Human Skin Structure and Drug Penetration*; Springer: Berlin, Heidelberg, 2015; pp 1–12.
2. Nishino, K. *The Appearance of Human Skin The Appearance of Human Skin*; Hanover, MA, USA: Now Publisher Inc., 2014; pp 1, 1572–2740.
3. Smith, A. M.; Moxon, S.; Morris, G. A. Biopolymers as Wound Healing Materials. In *Wound Healing Biomaterials*. Elsevier: Sawston, United Kingdom, 2016; p. 261–87.
4. Zhou, Y. M.; Mizuno, T.; Takehisa, T.; Haraguchi, K.; Taenaka, Y. In *Nanocomposite Hydrogels: A Novel Wound Dressings*. Proceedings - 2010 3rd International Conference on Biomedical Engineering and Informatics, BMEI 2010, 2010; pp 432–434.
5. Harding, K. *Wound Dressings*; 2014.
6. Uzun, M. A Review of Wound Management Materials. *J. Textile Eng. Fashion Technol.* **2018,** *4* (1), 53–59.
7. Tavakoli, S.; Klar, A. S. Advanced Hydrogels as Wound Dressings. *Biomolecules* **2020,** *10*, 1169.
8. *https://www.cardinalhealth.com/en/patient-education/home-care/wound-management/types-of-wounddressings.html*
9. *https://www.clhgroup.co.uk/news-article/2017/09/12/7-types-of-wound-dressings-when-to-use-each/258*
10. Zarrintaj, P.; Mozafari, M. Can Regenerative Medicine and Nanotechnology Combine to Heal Wounds ? The Search for the Ideal Wound Dressing. *Nanomedicine* **2017,** *12* (19), 2403–2422.
11. Paper, C.; Onar, N. Usage of Biopolymers In Medical Applications Usage of Biopolymers In Medical Applications **2014,** 2011–2013.
12. Detsch, R.; Sarker, B.; Zehnder, T.; Frank, G.; Boccaccini, AR. Advanced Alginate-Based Hydrogels. *Mater. Today* **2015,** *18* (10), 590–1.
13. Anumolu, S. S.; Menjoge, A. R.; Deshmukh, M.; Gerecke, D.; Stein, S.; Laskin, J.; Sinko, P. J. Doxycycline Hydrogels with Reversible Disulfide Crosslinks for Dermal Wound Healing of Mustard Injuries. *Biomaterials* **2011,** *32* (4), 1204–1217.
14. Ganji, F.; Vasheghani-Farahani, S.; Vasheghani-Farahani, E. Theoretical Description of Hydrogel Swelling: A Review. *Iran. Polym. J.* **2010,** *19*, 375–398.
15. Qiu, Y.; Park, K. Environment-Sensitive Hydrogels for Drug Delivery. *Adv. Drug Deliv. Rev.* **2001,** *53*, 321–339.
16. Ullah, F.; Bisyrul, M.; Javed, F.; Akil, H. Classification, Processing and Application of Hydrogels: A Review. *Mater. Sci. Eng. C Mater. Biol. Appl.* **2015,** *57*, 414–33.
17. Takashi, L.; Hatsumi, T.; Makoto, M.; Takashi, I.; Takehiko, G.; Shuji, S. Synthesis of Porous Poly (N-isopropylacrylamide) Gel Beads by Sedimentation Polymerization and their Morphology. *J. Appl. Polym. Sci.* **2007,** *104* (2), 842.
18. Haque, M. D. A.; Kurokawa, T.; Gong, J. P.; Super Tough Double Network Hydrogels and their Application as Biomaterials. *Polymers* **2012,** *53* (9), 1805–1822.
19. Babo, P. S.; Reis, R. L.; Gomes, M. E. Engineering Applications. *Materials* **2020,** 1–29.
20. Saranraj, P. Hyaluronic Acid Production and its Applications - A Review. *Int. J. Pharm. Biol. Sci. Arch.* **2017,** 4.
21. Peteiro, C. *Alginate Production from Marine Macroalgae, with Emphasis on Kelp Farming*. Springer: Singapore, **2017**; pp 27–66.

22. Smith, A. M.; Moxon, S.; Morris, G. A. Biopolymers as Wound Healing Materials. In *Wound Healing Biomaterials*, 2016; pp 261–287.
23. Applications, B.; Ruiz-caro, R.; Veiga, M. Carrageenan: Drug Delivery Systems and Other Biomedical Applications. *Mar. Drugs* **2020,** *18* (11), 583.
24. Dinescu, S.; Kaya, M. A.; Chitoiu, L.; Ignat, S. Collagen-Based Hydrogels and Their Applications for Tissue Engineering and Collagen-Based Hydrogels and Their Applications for Tissue Engineering and Regenerative Medicine. In *Cellulose-Based Superabsorbent Hydrogels*; Springer: Cham, 2018; pp 1–21.
25. Schneider-Barthold, C.; Baganz, S.; Wilhelmi, M.; Scheper, T. Hydrogels Based on Collagen and Fibrin – Frontiers and Applications. *Bio. Nano. Materials* **2016,** *17*, 3–12.
26. Samoila, I.; Dinescu, S.; Pircalabioru, G. G. Pullulan/Poly (Vinyl Alcohol) Composite Hydrogels. *Materials* **2019,** 12, 3220
27. Lee, K. Y.; Mooney, D. Alginate: Properties and Biomedical Applications. *Prog. Polym. Sci.* **2012,** *37*, 106–126.
28. Lee, K. Y.; Mooney, D. J. Hydrogels for Tissue Engineering. *Chem. Rev.* **2001,** *101*, 1869–1879.
29. Kong, H. J.; Lee, K. Y.; Mooney, D. J. Decoupling the Dependence of Rheological/ Mechanical Properties of Hydrogels from Solids Concentration. *Polymer* **2002,** *43*, 6239–6246.
30. Draget, K. I.; Skjåk Bræk, G.; Smidsrød, O. Alginic Acid Gels: The Effect of Alginate Chemical Composition and Molecular Weight. *Carbohydr. Polym.* **1994,** *25* (1), 31–8.
31. Kuo, C. K.; Ma, P. X. Ionically Crosslinked Alginate Hydrogels as Scaffolds for Tissue Engineering: Part 1. Structure, Gelation Rate and Mechanical Properties. *Biomaterials* **2001,** *22*, 511–521.
32. Zhao, X. H.; Huebsch, N.; Mooney, D. J.; Suo, Z. G. Stress-Relaxation Behavior in Gels with Ionic and Covalent Crosslinks. *J. Appl. Phys.* **2010,** *107*, 063509/1–5.
33. Smeds, K. A.; Grinstaff, M. W. Photocrosslinkable Polysaccharides for in situ Hydrogel Formation. *J. Biomed. Mater. Res.* **2001,** *54*,115–121.
34. Roy, D.; Cambre, J. N.; Sumerlin, B. S. Future Perspectives and Recent Advances in Stimuliresponsive Materials. *Progr. Polym. Sci.* **2010,** *35*, 278–301.
35. Rzaev, Z. M. O.; Dincer, S.; Piskin, E. Functional Copolymers of N-Isopropylacrylamide for Bioengineering Applications. *Progr. Polym. Sci.* **2007,** *32*, 534–595.
36. Zhao, S. P.; Cao, M. J.; Li, H.; Li, L. Y.; Xu, W. L. Synthesis and Characterization of Thermo-Sensitive Semi-IPN Hydrogels based on Poly (ethylene glycol)-co-Poly (epsilon-caprolactone) Macromer, N-Isopropylacrylamide, and Sodium Alginate. *Carbohydr. Res.* **2010,** *345*, 425–431.
37. Kataria, K.; Gupta, A.; Rath, G.; Mathur, R. B.; Dhakate, S. R. In vivo Wound Healing Performance of Drug Loaded Electrospin Composite Nanofibers Transdermal Patch. *Int. J. Pharm.* **2014,** *469*, 102–110.
38. Bindu, T.; Vidyavathi, M.; Kavitha, K.; Sastry, T.; Suresh Kumar, R. Preparation and Evaluation of Chitosan-Gelatin Composite Films for Wound Healing Activity. *Trends. Biomater. Artif. Organs.* **2010,** *24*, 123–130.
39. Qin, Y. The Gel Swelling Properties of Alginate Fibers and their Applications in Wound Management. *Adv. Polym. Technol.* **2018,** *19*, 6–14.
40. Sun, J.; Tan, H. Alginate-Based Biomaterials for Regenerative Medicine Applications. *Materials* **2013,** *6*, 1285–1309.

41. Roxana Gheorghita Puscaselu.Alginate: From Food Industry to Biomedical Applications and Management of Metabolic Disorders. *Polymers* **2020,** *12*, 2417. doi:10.3390/polym12102417
42. Aderibigbe, B. A.; Buyana, B. Alginate in Wound Dressings. *Pharmaceutics* **2018,** *10* (2), 42, DOI: 10.3390/pharmaceutics10020042
43. Kurczewska, J.; Pecyna, P.; Ratajczak, M.; Gaj ̨ecka, M.; Schroeder, G. Halloysite Nanotubes as Carriers of Vancomycin in Alginate-based Wound Dressing. *Saudi. Pharm. J.* **2017,** *25*, 911–920.
44. Vowden, K.; Vowden, P. Wound Dressings: Principles and Practice. *Surgery* **2014,** *32*, 462–467.
45. Sood, A.; Granick, M. S.; Tomaselli, N. L. Wound Dressings and Comparative Effectiveness Data. *Adv. Wound Care* **2014,** *3* (8), 511–529. DOI: 10.1089/wound.2012.0401
46. Namviriyachote, N.; Lipipun, V.; Akkhawattanangkul, Y.; Charoonrut, P.; Ritthidej, G. Development of Polyurethane Foam Dressing Containing Silver and Asiaticoside for Healing of Dermal Wound. *Asian. J. Pharm. Sci.* **2019,** *14*, 63–77.
47. Rajendran, S.; Anand, S. C. Hi-tech Textiles for Interactive Wound Therapies. In *Handbook of Medical Textiles*; Woodhead Publishing, 2011, 38–79.
48. Zahedi, P.; Rezaeian. I.; Ranaei-Siadat, S. O.; Jafari. S. H.; Supaphol, P. A Review on Wound Dressings with an Emphasis on Electrospun Nanofibrous Polymeric Bandages. *Polym. Adv. Technol.* **2009,** *21*, 77–95. DOI: 10.1002/pat.1625.
49. Abrigo, M.; McArthur, S. L.; Kingshott, P. Electrospun Nanofibers as Dressings for Chronic Wound Care: Advances, Challenges, and Future Prospects. *Macromol. Biosci.* **2014,** *14*, 772–792. DOI: 10.1002/mabi.201300561
50. Dahlin, R. L.; Kasper, F. K.; Mikos, A. G. Polymeric Nanofibers in Tissue Engineering. *Tissue. Eng. Part B. Rev.* **2011,** *17*, 349–364. DOI: 10.1089/ten.teb.2011.0238.
51. Boateng, J. S.; Matthews, K. H.; Stevens, H. N.; Eccleston, G. M. Wound Healing Dressings and Drug Delivery Systems: A Review. *J. Pharm. Sci.* **2008,** *97*, 2892–2923.
52. Boateng, J.; Burgos-Amador, R.; Okeke, O.; Pawar, H. Composite Alginate and Gelatin Based Bio-Polymeric Wafers Containing Silver Sulfadiazine for Wound Healing. *Int. J. Biol. Macromol.* **2015,** *79*, 63–71. DOI: 10.1016/j.ijbiomac.2015.04.048
53. Fredric, S.; Gowda, D. V.; Yashashwini, M. Review Article Wafers for Wound Healing. *J. Chem. Pharm. Res.* **2015,** *7* (9), 450–468.
54. Albuquerque, G. S.; Silva, L. C. N.; Lima-Ribeiro, M. H. M.; et al., Healing Activity Evaluation of the Galactomannan Film Obtained from Cassia Grandis Seeds with Immobilized C ratylia Mollis Seed Lectin. *Int. J. Biol. Macromol.* **2017,** *102*, 749–757. DOI: 10.1016/j.ijbiomac.2017.04.064
55. Paul, W. P.; Sharma, C, *Advances in Wound Healing Materials: Science and Skin Engineering. Smithers Rapra Technology*, Smithers Rapra: Shrewsburry, UK, England. 2015.
56. Zahedi, P.; Rezaeian, I.; Ranaei-Siadat, S. O.; Jafari, S. H.; Supaphol, P. A Review on Wound Dressings with an Emphasis on Electrospun Nanofibrous Polymeric Bandages. *Polym. Adv. Technol.* **2009,** *21*, 77–95. DOI: 10.1002/pat.1625
57. Aderibigbe, B. A.; Buyana, B. Alginate in Wound Dressings. *Pharmaceutics* **2018,** *10* (2), 42. DOI: 10.3390/pharmaceutics10020042
58. Molan, P. Medical dressings comprising gelled honey. European patent EP1,237,561, 2009

59. Mikkonen, K.; Parikka, K.; Ghafar, A.; Tenkanen, M. Prospects of Polysaccharide Aerogels as Modern Advanced Food Materials. *Trends. Food. Sci. Technol.* **2013,** *34*, 124–136.
60. Macan, J. Definitions of Terms Relating to the Structure and Processing of Sols, Gels, Networks, and Inorganic-Organic Hybrid Materials. *Kem. Ind.* **2011,** *60*, 135–153.
61. Ruban, S. Biobased Packaging-Application in Meat Industry. *Vet. World* **2009,** *2*, 79–82.
62. Paul, W.; P. Sharma, C. *Advances in Wound Healing Materials: Science and Skin Engineering*. Smithers Rapra Technology: Shrewsburry, UK, England, 2015; p 208.
63. Qin, Y. M. Absorption Characteristics of Alginate Wound Dressings. *J. Appl. Polym. Sci.* **2004,** *91* (2), 953–957.
64. McNicol, A.; Israels, S. J. Platelets and Anti-platelet Therapy. *J. Pharmacol. Sci.* **2003,** *93* (4), 381–396.
65. Segal, H. C.; Hunt, B. J.; Gilding, K. The Effects of Alginate and Non-Alginate Wound Dressings on Blood Coagulation and Platelet Activation. *J. Biomater. Appl.* **1998,** *12* (3), 249–257.
66. Taskin, A. K.; Yasar, M.; Ozaydin, I.; Kaya, B.; Bat, O.; Ankarali, S.; Yildirim, U.; Aydin, M. The Hemostatic Effect of Calcium Alginate in Experimental Splenic Injury Model. *Ulus. Travma. Acil. Cer.* **2013,** *19* (3), 195–199.

CHAPTER 17

Pharmaceutical 3D Printing: Insights into Polymeric Armamentarium and Saga of Product Fabrication

PRACHI KHAMKAR[1] and DEBARSHI KAR MAHAPATRA[2]

[1]*Pharmaceutical Manufacturing Operations, CiREE Edutech, Pune, Maharashtra, India*

[2]*Department of Pharmaceutical Chemistry, Dadasaheb Balpande College of Pharmacy, Nagpur, Maharashtra, India*

ABSTRACT

Polymers are the most often used class of excipients in current pharmaceutical science, contributing significantly to the advancement of medication dosage formulations. Three-dimensional (3D) printing has recently been recognized in the pharmaceutical sector as a revolutionary manufacturing method that allows the rapid development of pharmaceutical formulations feasible with complex geometries. Furthermore, it also has paved the way for customized medication, since it allows pharmacists to produce pharmaceutical products incorporating the precise amount of active pharmaceutical ingredients (APIs) that successfully serve each patient's specific needs. A great range of polymers for 3DP-based drug delivery systems have been used to create highly advanced medication formulations for patient-customized medication. Polymers are used to produce 3D-filled dosage formulations, as they are capable of modulating the release rate to provide physical consistency to the drug that is in the dosage form. Polymers such as polyvinyl alcohol (PVA), hydroxypropyl cellulose (HPC), hydroxypropyl methylcellulose (HPMC), polyurethane (PU), and polycaprolactone (PCL) play an important

Foundation and Growth of Macromolecular Science: Advances in Research for Sustainable Development. Meegle S. Mathew, PhD, Józef T. Haponiuk, PhD & Sabu Thomas, PhD, DSc (Eds.)

role in designing the dosage form. The use of polymers as a backbone does play a key role in the final properties of the dosage type. It is a project with an emphasis on developing devices with an interconnected collection of fundamental components and technologies required for a 3DP medication delivery system with self-customized customizable dosage types. To support the current chapter, we have attempted to concentrate on 3DP, accompanied by 3DP technology used in pharmaceuticals and innovative pharmaceutical polymers used in 3DP technology to customize the dosage shape.

17.1 INTRODUCTION

We are now living in a technological age, which is referred to as the "next technological transition," and in which cloud computing, contactless digital, artificial intelligence, and 3D printing (3DP) have now become a fact that was once just a fantasy. The pharmaceutical industry must consider and harness the power of imagination to aid in the production of tailor-made medicines and 3DP.[1] The 3DP methods have also proven their beneficial ability in the printing of prostheses and different implants as well as in tissue techniques where they are introduced to the medical industry, and the entire functioning organs have been 3D printed.[2] 3DP allows complex structures to be built into materials with chemical, physical, and mechanical properties that are essential for the manufacturing medical applications such as tailored prostheses, implants, and equipment.[3] A modern methodology enables dosage forms to be formulated in a single component that formerly is an inexpensive multistage procedure. Besides, it has unlocked the door to the personalization of medication by supplying pharmacy specialists with the right quantity of active pharmaceutical ingredients (APIs) to fulfill particular patient's requirements.[4] 3DP offers a number of advantages, including cost efficiency, design flexibility, and automation freedom. Thus, 3DP technology has become an economical process, with the increase of manufacturing products at an industrial start. While examining the many functions that 3DP can play in medical sciences, one intriguing idea is customized medicine, in which a treatment is adapted to the unique needs of each patient. Due to the fact that the majority of these limitations are less severe in this sector, 3DP can provide advantages in three primary areas such as the sophistication of drug products, personalization, and on-demand fabrication.[2] The 3DP methods are mostly used for small-batch processing (or also for one-of-a-kind products). For drug delivery devices (DDD), this may be useful in turn, such as incorporating a variety of APIs, which have different release kinetics, into one DDD, or tailoring dose to a patient's individual mass and metabolic requirements.[5–7] Another advantage of small-scale processing in the area of drug distribution is the possibility of producing DDD at the point of treatment. This can be particularly beneficial where the medications being used have a short shelf life or are being used in an emergency situation. Furthermore, 3DP can provide a secure and cost-effective process for producing rare-disease dosage forms, obviating the need for mass-production equipment for smaller-scale tablet production. Now though, science is based on the extrusion and fusion techniques. This would bring customized and affordable medication to fruition in the immediate future. However, there is continuous proof of other approaches being studied that may continue throughout the next decades. By contrast, conventional methods of manufacture are top-down techniques. They carve through a block of material to shape the object. They are capable of extracting large amounts of unnecessary waste. This discarded content is either trashed or saved for future use. Although some are recyclable, some take additional time, effort, and resources. By contrast, 3DP is a bottom-up approach. It makes adequate use of the material used to manufacture the component. Additionally, these items are mostly recyclable and reusable. As a result, almost no waste is generated. Additionally, 3DP is a comparatively energy-efficient method. This decline in electricity demand and waste production contributes to a drop in environmental effects, thus increasing the degree of sustainability.

3DP is a promising approach for the production of a diverse range of differentiated personalized products, that is, for mass adaptation at a low cost.[8] To comprehend, suitable polymers for 3DP in medicine should be biocompatible, among other characteristics of the drug. To round out the list, higher permeability, strong mechanical properties, tolerance to chemicals, and sterility are essential.[9] Polymers are the most widely used materials for 3DP because of their flexibility and a wide variety of mechanical and chemical properties. 3DP polymers include thermal plastics, elastomers, polymers, biopolymers, and polymers combined with organic materials.[10] The designing and polymer selection would result in improved material functions, mechanical properties, porosity, and stability. Various 3DP-based drug delivery systems, such as oral drug released systems, microchips, implants, pills, tablets for immediate release (IR), and multiple release dosage forms, have, for example, been developed.[11] Polymers are a foundation stone in the manufacture of 3DP dosages since they can modulate the release rate and give the integrated APIs physical stability.

In this chapter, we have attempted to concentrate on 3DP, accompanied by 3DP technology used in pharmaceuticals and innovative pharmaceutical polymers used in 3DP technology to customize the dosage shape and implants.

17.2 TYPES OF 3D-PRINTING TECHNOLOGY

3DP technology allows 3D solid structures to be created directly from a computer-aided design (CAD) model. The CAD model is first converted into a printer-friendly format before being sliced into thin, well-defined layers with slicer tools. The layer model is then translated into G-code, which instructs the computer numerical control (CNC) printer to deposit the layers in a sequential manner. If the model is built, export it as an .STL file, which is the popular 3DP file format known as standard tessellation language (Fig. 17.1). Depending on the dimensions of the 3D printer, your template may easily be cut into several layers.

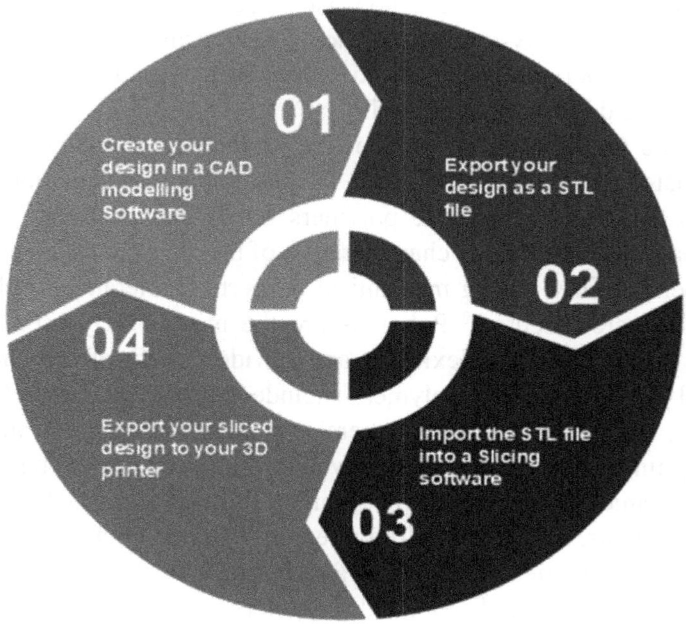

FIGURE 17.1 3D-printing process.

17.2.1 STEREOLITHOGRAPHY

Due to the usage of photopolymer resin, stereolithography is member of the Vat Photopolymerization family. The starting material is light-sensitive polymers (also called photopolymers).[4] The laser beam is guided along a fixed path by a pair of mirrors called Galvos that are located in the X-axis and Y-axis. The galvos contributes to the laser beam's projection onto the material. The tank's bottom is transparent with a silicon coating that enables the laser beam to move through and treat the resin, which uses photopolymerization a treatment of photosensitive materials. The construction platform shifts down after a sheet is imaged on the resin surface and a recovery bar travels around the platform to apply the next resin layer. The method is performed layer by layer for making an object. When the first layer is cured, the bed or substrate goes up in the Z-axis. The wiper assists in documenting the surface and prepares it for the subsequent new layer. For the second sheet, the bed is lowered down. Smartest laser ally (SLA) gives us the polished surface finishing capability, the smart materials are susceptible to UV light and we can alter their mechanical properties by too much exposure to the sunshine.[12] There is now a greater interest in SLA printing by companies. As one can imagine, some of these innovations involve the automobile sector, healthcare, aviation, sports, and general consumer goods.[13]

SLA has the significant benefit of high definition. Furthermore, since SLA is a nozzle-free 3DP process, it allows for continuous 3DP by removing nozzle clogging and allows for 3DP of viscous resins, which are often incompatible with nozzle-based 3DP. The main disadvantages of SLA include expensive facilities, the need for postprinting care to ensure interlayer assembly and proper solidification, shrinkage-induced warpage, and a trade-off between high throughput and high resolution. Since SLA would only use photocurable resins, this 3DP approach requires the inclusion of photoinitiators in SLA resins. Traces of monomers and photoinitiators may be cytotoxic if not fully healed.[14] To improve SLA 3DP performance and reduce printed component defects, a thorough understanding of photopolymer properties such as curing kinetics and printing parameters, light sensitivity, and penetration depth is essential.[15]

17.2.2 BINDER JETTING

Raw material in the form of powder can be used in powder-based 3DP. Powder has been bonded together using liquid binder chemicals. Unlike

SLS, metal powder is not heated. In this process, the powder is mounted on the creation platform and binder droplets are jetted selectively onto a portion of the powder. The binder agent acts like an adhesive, allowing the powder particles to adhere to one another and create a solid until the primary layer is full. The substrate is lowered and a fresh layer of powder and binder is deposited. This procedure is replicated until the 3D-printed component is finished. Binder jetting recovers 100% of the powder content. Postprinting treatment may be needed in certain cases to obtain the material's optimum physical and chemical properties.[16] Pharmaceutical researchers have also demonstrated an interest in the use of 3DP technology for delayed-release manufacture, complex release profile tablets such as immediate, dual-pulsate, and zero-order release profiles.[4] This is due to the fact that this method has the capacity to design complex formulations, even for those with loose powders in the interior. In 2015, the binder jet method was adapted as a mass manufacturing technique to allow the manufacture of Spritam® (the first FDA-approved 3D-printed tablet). Over the next decade, binder jet printing is projected to continue to have a significant effect on formulation manufacturing. Binder jet printing, in particular, enables the production of oral dosage types with a variety of specific release properties, varying from rapid dissolution to controlled-release platforms.[17]

17.2.3 FUSED DEPOSITION MODELING

Fused deposition modeling (FDM) was invented by Stratasys in the 1990s and became commercially available in the 1980s. FDM has improved its reliability, accuracy, and material choice, making it ideal for a range of production applications.[18] The product is created with the help of a thread or filament type of material that is unwound from a spiral coil and fed into the extrusion nozzle. The coil melts and extrudes the filaments onto a base, which is often called a building base or table. The nozzle and the base are all operated by a device that transforms the object measurements into X, Y, and Z coordinates for the nozzle and the printing base. The extrusion nozzle travels horizontally and vertically over the build platform and "print an object" onto the platform a cross-section of an entity.

To mitigate thermal shrinkage impact, which may impair internal layer adhesion and mechanical properties, FDM machines allow to regulate cooling rate by adjusting the temperature of the building level.[19]

During the filament fabrication phase, the polymers are normally subjected to hot-melt extrusion (HME), hot blending, and polymer composites. This phase must be designed for each new feedstock content and can be difficult at times due to the requirement that materials subjected to HME have the necessary thermal, mechanical, and rheological properties. Filaments must have an accurate diameter size and a low melt viscosity to allow for a smooth flow in FDM.[20]

FDM's operating flexibility and low costs render it common and widely available, including small businesses and private users. However, due to printing inaccuracies, filament nonuniformity, surface roughness, layered composition, and interlayer adhesion, FDM-fabricated objects are vulnerable to structural flaws and decreased mechanical power.[21]

17.2.4 SELECTIVE LASER SINTERING

The mechanism depicted in selective laser sintering is a laser beam projected over a layer of thermoplastic powder material that is closely compact. A roller helps to spread the powder on the platform. To fit the layer of the powder piston moves down according to the layer thickness. In each layer, the excess powder will assist the component through the construction. A focused infrared beam is generated by the laser under sensor control of the scanner machine. The laser emits an infrared powerful heating beam. No section of the machine is allowed to go beyond the melting temperature of the powder. Laser heat helps to coarsen and speed up the operation.

The scan patterns for these segments are predetermined based on the properties of the final object. By laser sintering/melting between particles, the substance is heated to an appropriate temperature (below the melting point) to initiate fusion. The scanner moves the laser in a two-dimensional path, and the powder bed's height is set to concentrate the laser beam on the freshly created surface. On the construction base, coarse powder particles provide help throughout the operation. The powder bed surface is reduced by one sheet width, and the laser deposits and fuses another layer of powder. This procedure is replicated until the whole object is created. Inside the printer, the finished product is permitted to cool.[18] The substance is extracted manually or by sieving from the loose powder.

17.3 POLYMERS: BACKBONE OF 3D-PRINTING DOSAGE FORMS

Polymers serve as the foundation for the manufacture of 3DP dosage types so they can be used to monitor the rate of release and have physical stabilization for integrated APIs wide range. An ever-increasing number of polymers derived from both natural and synthetic sources are used in pharmaceutical 3DP. Natural polymers such as gelatin, collagen, alginate, and chitosan are commonly used, but they often need cytotoxic cross-linkers.[22–24] As a result, synthetic polymers have recently gained attention for 3DP due to their ability to circumvent the disadvantages associated with natural polymers. Table 17.1 represents the polymers used in 3DP-based drug delivery systems, respectively.

TABLE 17.1 List of Polymers in 3D-Printing-Based Drug Delivery Systems.

Polymers	Drug delivery
Hydroxypropyl methylcellulose	Matrix tablet
Poly(lactic-co-glycolic acid)	Microsphere, capsules, tablets, nanosphere
Poly(methacrylates) (Eudragit)	Nano capsules, tablets
Poly(ethylene glycol) diacrylate	Hydrogels
Polyvinyl alcohol	Tablets, capsules
Polylactic acid	Nanofiber
Polyvinylpyrrolidone (Kollidon®)	Tablets
Poly(ε-caprolactone)	Tablets, carbon nanotubes
Polyurethane	Tablets, hydrogels
Pluronic	Hydrogels
Poly(*N*-(2-hydroxypropyl) methacrylamide-mono/dilactate)-polyethylene glycol triblock copolymer (M15P10)	Hydrogels
Ethyl cellulose	Tablets
Ethylene vinyl acetate	T-shaped intrauterine system and subcutaneous rods
Poly(methyl methacrylate)	Tablets
Methacrylic/cellulosic polymers	Tablets

17.3.1 POLYVINYL ALCOHOL (PVA)

Polyvinyl alcohol (PVA) is manufactured by the polymerization of vinyl acetate (VA) followed by partial or full hydrolysis. The degree of hydrolysis is determined by the stop time of the saponification reaction. The water solubility of PVA depends on the degree of polymerization and the degree of hydrolysis. The highest solubility can be achieved with the hydrolysis degree between 87 and 89%. The printability of PVA is excellent with good thermoplasticity, a low melt viscosity, and a suitable solidification rate. Therefore, this polymer is extensively used in FDM 3DP as a supporting material, which helps to fabricate complicated structures and can be easily removed by dissolving.[25]

Goyanes et al. used FDM to make tablets containing drug molecules paracetamol and caffeine. The filaments used in the printing process were created by dissolving each drug in PVA using HME. Filaments with two separate drug amounts were created for each compound, yielding paracetamol drug loads of 4.3 and 8.2% and caffeine drug loads of 4.7 and 9.5%, respectively. To achieve a range of medication release profiles, 3D-printed tablets with two distinct designs were used. As shown in Figure 17.2, the first version used 1-mm alternating layers of caffeine and paracetamol. The second version, dubbed DuoCaplets (9.0 mm in length × 3.34 mm in diameter), featured a capsule-shaped core containing one medication within and an outer layer containing the opposite drug. To model various parts of the gastrointestinal tract (GIT), release studies were conducted in a bicarbonate buffer at various pH values. The results indicated that the experiment with alternate layers released both drugs simultaneously, while the DuoCaplets model released the drug located on the outer layer first, accompanied by the drug located on the inner side of the capsule. Additionally, it was shown that the rate of drug release increased as the drug content inside the filaments extruded increased.[26]

Using an FDM 3D printer with dual nozzles, Tagami et al. created composite tablets with a drug-loaded PVA part and a PVA or PLA filler component. The elevated extrusion and printing temperatures used in the printing phase, which restricts its use in 3DP drugs, are the main drawbacks of FDM 3DP in pharmaceutical applications.[27] Since pharmaceutical excipients and active drugs can degrade at high temperatures during the extrusion and printing processes, it is not suitable for producing thermolabile drugs; many investigators have tried to solve this issue.

17.3.2 HYDROXYPROPYL CELLULOSE

Cellulose is the most abundant organic polymer that is renewable and biodegradable. It is obtained from plants and plant products. It is a natural expert who has shown significant opportunities for usage in the automotive, plastics, and healthcare industries. Hydroxypropyl cellulose (HPC) is fully soluble in cold water, but not in chemical solvents like methanol, ethanol, isopropyl alcohol, or acetone.[28] HPC has a temperature-dependent solubility profile, since it is insoluble in hot water and precipitates as a strongly swollen froth at temperatures between 40 and 45°C. HPC is commercially manufactured in a variety of grades of different melting points and viscosities.

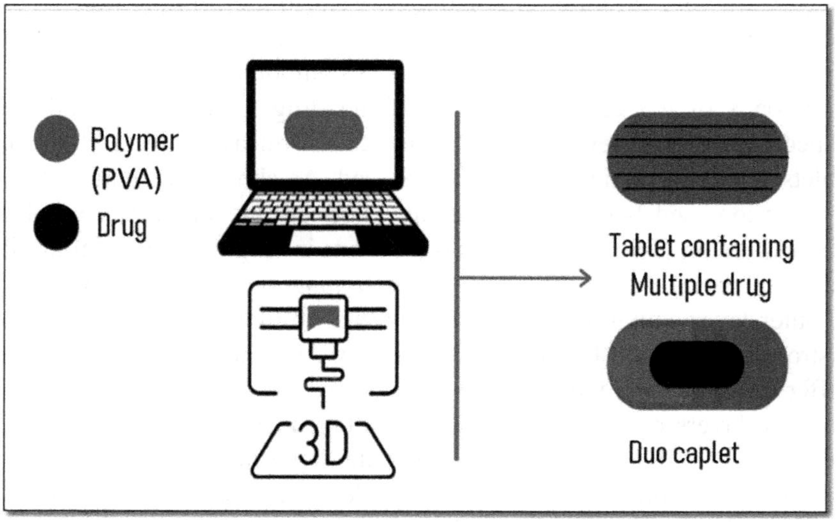

FIGURE 17.2 3D-printed tablet containing multiple drugs.

Arafat et al.,[29] for example, used a creative design approach to fabricate immediate-release tablets with special built-in gaps, called "Gaplets," printed using HME-FDM 3DP. To print capsule-shaped devices with interconnected blocks, filaments of SSL-grade HPC polymer were filled with theophylline. Thermal analysis showed that thermal processing had almost no effect on the crystallinity of the drug. The unusual geometry of the printlets, as well as the incorporation of the HPC polymer, facilitated rapid drug release (>80% in 30 min) from the devices.

A capsule-shaped system for the oral pulsatile release of acetaminophen was successfully prepared by Melocchi et al. Depending on the concentration of HPC in the mixture, in-house filament extrusion was performed at 50–165°C. The hollow capsular system 3D-printed from the HPC filaments exhibited a typical pulsatile–release profile, with a lag period of approximately 70 min followed by total drug release within 10 min. As a consequence, the obtained results were compatible with those obtained from capsule shells formed with the same formulation using injection molding (IM) technology.[30]

17.3.3 HYDROXYPROPYL METHYLCELLULOSE

Hydroxypropyl methylcellulose (HPMC) is a perfect polymer for drug-infused printable filament. Applications: Many solid dispersion products based on HPMC are commercially available such as Sporanox®, Prograf®, Intelence®, Zortress®, Onmel®, and other products, and Afinitor®. HPMC is available in a range of molecular grades hydroxyl and methyl group weights and replacement ratios.[25] HPMC-based filaments manufactured are used alone or in combination with other polymers for the preparation of 3D-printed tablets.

According to Hossam et al., the creation of a filament using a single pharmaceutical polymer, without additives, may be multipurpose and manipulated by means of a computational design for the preparation of tablets with the desired release and absorption patterns. HPMC and diltiazem, a model API, for the preparation of both drug-free and drug-infused filaments.[31] Thus, even the internal geometry of the polymer is extremely important for modifying the release pattern of the drug that can be achieved by 3DP, as HPMC was loaded in differential geometry in the tablet, and the innovative computational design enables tablets to be produced with the ability to create distinct absorption patterns and structured hollow tablets.[32]

17.3.4 POLYURETHANE

Polyurethane (PU) is a thermoplastic polymer possessing excellent biocompatibility and the ability to mimic tissue mechanics. PU swells and helps to lengthen the diffusion pathway of the dosage form. According to Salimi et al., in the preparation of solid implants by moderate temperature 3D extrusion printing, PU has been used because it is thermally reversible. Polyethylene glycol (PEG) and paracetamol were fused with polyurethene.[33] According to Juan et al., for the treatment of pelvic organ prolapse surgical implant as vaginal mesh was used. But multiple controversies were associated with traditionally manufactured polypropylene mesh; this substance was not suitable because it leads to severe pain and risk of infection. Later, it was suggested to use thermoplastic polyurethane (TPU) filled with antibiotic conjugation by using FDM 3D-printer and was more secured than the traditional one. For this reason, TPU filaments various concentrations of levofloxacin (LFX). Incorporated into the TPU matrix, LFX can be used to manufacture anti-infective vaginal meshes with improved mechanical properties in contrast to existing and polypropylene vaginal meshes.[34]

17.3.5 POLYCAPROLACTONE

Polycaprolactone (PCL) is an FDA-approved biodegradable polymer for surgical, continuing drug release, and tissue engineering therapeutic applications. 385°C has been recorded a degradation temperature for PCL and thus has a big processing window. PCL is hydrophobic relative to PLA and thus degrades slowly leading to drug release. Saraha et al. planned to design an implant that is biodegradable. Therefore, there is no need to remove the body implant until the aim is met. Drug and polymer were loaded into the nozzle and the implant was 3D printed. The implant was later wrapped with PCL using the solution casting method to monitor drug release from the implant.[35] The purpose of this study was to create medicated T-shaped models of intrauterine system (IUS) with the drug deposited inside the entire framework of the medical device using the 3DP technique, based on fused deposition modeling (FDM™). Indomethacin was chosen as a drug delivery model because of its broad therapeutic range. The key aim of this work was to fabricate an implantable IUS utilizing PCL-based high molecular polymers using a 3D printer.[36]

Combining scaffold materials may be helpful in building a tracheal construction because one material can give the structure with mechanical integrity and the other material can be used to create biomimetic design as a biocompatible and designed especially environment for the cell. As a 3D-printing material, PCL was extensively implemented as FDA authorized, ready for impression, and biodegradable (Li et al., 2020). PCL also possesses the mechanical characteristics to be a tracheal replacement and a slow biodegradation rate in vivo that can minimize difficulties with fast deterioration, such as postoperational tracheal relaxation, failure, and restenosis (Ghorbani et al., 2017).

17.3.6 ETHYL CELLULOSE

Ethyl cellulose (EC) has been widely used in the pharmaceutical industry, where it is found in a number of oral and topical pharmaceutical formulations. Due to its hydrophobic nature and swelling capacity, it has the power to modulate and improve the physiological efficacy of drug dosage styles. The primary purpose of EC usage is to optimize prescription dosage formulations of modified-release (MR), as EC guarantees full breakdown of the product in the GIT, retaining a stable drug concentration, and removing the need for multiple doses per day, thereby improving pharmacotherapeutic effectiveness.[37] According to Yang et al., the tablet with internal scaffold geometry was designed using 3DP to adjust the release pattern of the drug. Ibuprofen and ethylcellulose were combined together and the filament was extruded using HME. Later on, the filaments were loaded into the nozzle of the fused deposition machine (Fig. 17.3). Drug material, release modifications, and printing conditions affect the release behavior of dosage form. Research indicated that 3DP is a super comfortable and computer-controlled process that can be used to combine various polymers and API with customized release profile.[38]

According to Shiv et al., to develop the medication with matrix tablets for achieving the zero-order release, the research examined the possibility of coupling HME and 3D printing. The impact on mechanical and printability characteristics of extruded filaments was studied of the combination ratio of EC and HPC, carbamazepine (CBZ), model and triethylcitrate (TCE). This formulation of CBZ, EC, and HPC (3, 64.7, and 32.3% w/w) and 20% TEC (by weight on the dry powder) printed into tablets (370 mg, 13 mm diameter, 3.5 mm thickness, cylindrical) at 187°C exhibited the best mechanical and printability characteristics. The printed tablets exhibited excellent drug uniformity and adequate mechanical characteristics. The optimal filament exhibited the first-order pattern of medication release, whereas the printed 3D tablets displayed zero-order release of medicines and a lower rate of release than the ideal filament. Therefore, the combination of HME and 3D printing systems with the capacity to reduce dosage frequency and side effects may be achieved from a 2:1 ratio of EC and HSC with zero-order release tablets.[39]

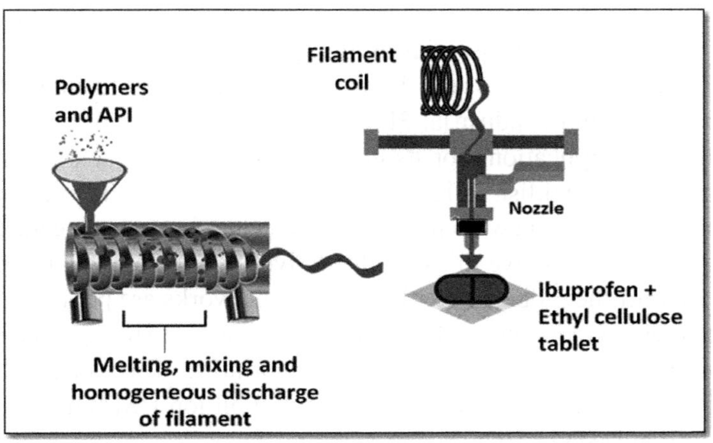

FIGURE 17.3 3D-printed tablets using ethyl cellulose and ibuprofen for sustained release.

17.4 CONCLUSIONS

The various uses of pharmaceutical 3DP are evident. There are tremendous changes made by 3DP in various industries due to the incredible opportunities. While there are still obstacles to address such as widening the variety of raw materials that can be used, such advances can contribute to more advantages for businesses. In terms of the many benefits of 3DP in the supply of drugs, certain challenges such as low mechanical properties and anisotropic behavior are still needed to overcome, which are impediments to technology scale-up. Polymers, as a construction substance, undoubtedly contribute significantly to the final property of the dosage process. The technology could revolutionize the development of formulations by shifting away from mass production to produce on-demand highly versatile, personalized dosage forms. The first approved tablet by the United States Food and Drug Administration (USFDA) is Spritam® for human use in the treatment of epilepsy in July 2015. However, there is still potential for improving manufacturing speed and minimizing total development costs with the aid of advancements in 3D system design. Additional care should be taken when choosing the right polymer, as their properties have proven to be very significant influence on drug content, release, and surface morphology. 3D-bioprinted organs such as the liver and ear will simplify the preclinical assessment of new medicines. Need to study more polymer from current polymers from a 3D point of view. New polymers (biocompatible) must also be produced in view of 3DP technologies. 3DP can have personalized dosing for virtually every disease situation with the aid of polymers. The creation of potential materials in 3DP would eventually be the use of certain "green natural polymers." Although fast-growing sectors like 3D printing get a lot of media attention, it is essential to look past the hype and stay current on their technical capabilities and limits. This enables us to understand not just the true worth of industrial 3D printing, but also to concentrate on the most practical applications for its use. The issue necessitates foresight and cautious work to establish solid regulatory standards that will eventually lead to the success of this new technology. One of the important features of 3D printing for the future, we think, is the delivery of customized medicines to distant areas of the globe where drug supply networks are poorly structured. Among other developments, 3DP is now often known to be a powerful tool to advance medicine and pharmacology by bioprinting. The worldwide size of the 3DP industry in 2018 was estimated at 9.9 billion USD and is estimated at 34.8 billion USD in 2024.[40]

CONFLICT OF INTEREST

The authors have no conflict regarding the publication of this article.

FUNDING INFORMATION

No funding agency provided any financial support.

KEYWORDS

- **3D printing technology**
- **polymers**
- **dosage forms**
- **implants**
- **instrumentation**
- **techniques**

REFERENCES

1. Trenfield, S. J.; Madla, C. M.; Basit, A. W.; Gaisford, S. The Shape of Things to Come: Emerging Applications of 3D Printing in Healthcare. In *3D Printing of Pharmaceuticals*; Basit, A., Gaisford, S., Eds.; AAPS Advances in the Pharmaceutical Sciences Series 31; Springer: Cham, 2018.
2. Uziel, A.; Shpigel, T.; Goldin, N.; Lewitus, D. Y. Three-Dimensional Printing for Drug Delivery Devices: A State-of-the-Art Survey. *J. 3D Print. Med.* **2019,** *3* (2), 95–109. DOI: 10.2217/3dp-2018-0023
3. Poologasundarampillai, G.; Nommeots-Nomm, A. Materials for 3D Printing in Medicine: Metals, Polymers, Ceramics, Hydrogels. In *3D Printing in Medicine*; Kalaskar, D., Ed.; Woodhead Publishing, 2017; pp 43–71.
4. Gioumouxouzis, C. I.; Eleftheriadis, G. K.; Fatouros, D. G. Emerging 3D Printing Technologies to Develop Novel Pharmaceutical Formulations. In *3D and 4D Printing in Biomedical Applications: Process Engineering and Additive Manufacturing*; 2019; pp 153–184.
5. Norman, J.; Madurawe, R.; Moore, C.; Khan, M.; Khairuzzaman, A. A New Chapter in Pharmaceutical Manufacturing: 3D-Printed Drug Products. *Adv. Drug Deliv. Rev.* **2017,** *108*, 39–50. DOI: 10.1016/j.addr.2016.03.001
6. Zema, L.; Melocchi, A.; Maroni, A.; Gazzaniga, A. Three-Dimensional Printing of Medicinal Products and the Challenge of Personalized Therapy. *J. Pharm. Sci.* **2017,** *106* (7), 1697–1705. DOI: 10.1016/j.xphs.2017.03.021

7. ZipDose Technology | Spritam | Aprecia. 2021. https://www.aprecia.com/technology/zipdose (accessed May 18, 2021).
8. Jain, A.; Bansal, K.; Tiwari, A.; Rosling, A.; Rosenholm, J. Role of Polymers in 3D Printing Technology for Drug Delivery - An Overview. *Curr. Pharm. Des.* **2019,** *24* (42), 4979–4990.
9. Home - Kumovis [Internet]. Kumovis. 2020. https://kumovis.com/ (accessed Nov 17, 2020).
10. Park, B. J.; Choi, H. J.; Moon, S. J.; Kim, S. J.; Bajracharya, R.; Min, J. Y.; Han, H. K. Pharmaceutical Applications of 3D Printing Technology: Current Understanding and Future Perspectives. *J. Pharm. Investig.* **2019,** *49* (6), 575–585.
11. Beg, S.; Almalki, W.; Malik, A.; Farhan, M.; Aatif, M.; Rahman, Z.; et al. 3D Printing for Drug Delivery and Biomedical Applications. *Drug Discov. Today* **2020,** *25* (9), 1668–1681.
12. 3D Printing with Polymers: All You Need to Know - AMFG. 2021. https://amfg.ai/2019/01/17/3d-printing-with-polymers-all-you-need-to-know/#:~:text=Using%20plastic%20instead%20of%20metal&text=The%20ability% (accessed Jan 19, 2021).
13. Types of 3D Printers: Complete Guide - SLA, DLP, FDM, SLS, SLM, EBM, LOM, BJ, MJ Printing. 2021. https://3dinsider.com/3d-printer-types/ (accessed Jan 19, 2021).
14. Wang, X.; Jiang, M.; Zhou, Z.; Gou, J.; Hui, D. 3D Printing of Polymer Matrix Composites: A Review and Prospective. *Compos. B. Eng.* **2017,** *110*, 442–458. DOI: 10.1016/j.compositesb.2016.11.034
15. Raeymaekers, B. 3D Printing: How Materials Are Printed Layer-By-Layer Using Stereolithography. *Sci. Trends* **2017**. DOI: 10.31988/scitrends.4951
16. Flynt, J. What is Powder-Based 3D Printing? 2020. https://3dinsider.com/powder-based-3d-printing/ (accessed Apr 14, 2021).
17. Trenfield, S. J.; Madla, C. M.; Basit, A. W.; Gaisford, S. Binder Jet Printing in Pharmaceutical Manufacturing. In *3D Printing of Pharmaceuticals*; Basit, A., Gaisford, S., Eds.; AAPS Advances in the Pharmaceutical Sciences Series 31; Springer: Cham, 2018; pp 41–54.
18. What is FDM?: Fused Deposition Modeling Technology for 3D Printing | Stratasys. 2021. https://www.stratasys.com/fdm-technology (accessed Jan 19, 2021).
19. Yang, Y.; Chen, Y.; Wei, Y.; Li, Y. 3D Printing of Shape Memory Polymer for Functional Part Fabrication. *Int. J. Adv. Manuf. Technol.* **2015,** *84* (9–12), 2079–2095. DOI: 10.1007/s00170-015-7843-2
20. Postiglione, G.; Natale, G.; Griffini, G.; Levi, M.; Turri, S. Conductive 3D Microstructures by Direct 3D Printing of Polymer/Carbon Nanotube Nanocomposites via Liquid Deposition Modeling. *Compos. A Appl. Sci. Manuf.* **2015,** *76*, 110–114. DOI: 10.1016/j.compositesa.2015.05.014
21. Nikzad, M.; Masood, S.; Sbarski, I. Thermo-Mechanical Properties of a Highly Filled Polymeric Composites for Fused Deposition Modeling. *Mater. Des.* **2011,** *32* (6), 3448–3456. DOI: 10.1016/j.matdes.2011.01.056
22. Stansbury, J.; Idacavage, M. 3D Printing with Polymers: Challenges Among Expanding Options and Opportunities. *Dent. Mater.* **2016,** *32* (1), 54–64. DOI: 10.1016/j.dental.2015.09.018
23. Pekkanen, A.; Mondschein, R.; Williams, C.; Long, T. 3D Printing Polymers with Supramolecular Functionality for Biological Applications. *Biomacromolecules* **2017,** *18* (9), 2669–2687. DOI: 10.1021/acs.biomac.7b00671

24. Nadgorny, M.; Ameli, A. Functional Polymers and Nanocomposites for 3D Printing of Smart Structures and Devices. *ACS Appl. Mater. Interfaces* **2018,** *10* (21), 17489–17507. DOI: 10.1021/acsami.8b01786
25. Lu, M. Novel Excipients and Materials Used in FDM 3D Printing of Pharmaceutical Dosage Forms. In *3D and 4D Printing in Biomedical Applications: Process Engineering and Additive Manufacturing*; 2019; pp 211–237.
26. Acosta-Velez, G. F.; Wu, B. M. 3D Pharming: Direct Printing of Personalized Pharmaceutical Tablets. *Polym. Sci.* **2016,** *2* (1). DOI: 10.4172/2471-9935.100011
27. Xu, X.; Zhao, J.; Wang, M.; Wang, L.; Yang, J. 3D Printed Polyvinyl Alcohol Tablets with Multiple Release Profiles. *Sci. Rep.* **2019,** *9* (1), 1–8. DOI: 10.1038/s41598-019-48921-8
28. Giri, B.; Poudel, S.; Kim, D. Cellulose and Its Derivatives for Application in 3D Printing of Pharmaceuticals. *J. Pharm. Investig.* **2020,** *51* (1), 1–22. DOI: 10.1007/s40005-020-00498-5
29. Arafat, B.; Wojsz, M.; Isreb, A.; Forbes, R.; Isreb, M.; Ahmed, W.; et al. Tablet Fragmentation Without a Disintegrant: A Novel Design Approach for Accelerating Disintegration and Drug Release from 3D Printed Cellulosic Tablets. *Eur. J. Pharm. Sci.* **2018,** *118*, 191–199. DOI: 10.1016/j.ejps.2018.03.019
30. Melocchi, A.; Parietti, F.; Loreti, G.; Maroni, A.; Gazzaniga, A.; Zema, L. 3D Printing by Fused Deposition Modeling (FDM) of a Swellable/Erodible Capsular Device for Oral Pulsatile Release of Drugs. *J. Drug Deliv. Sci. Technol.* **2015,** *30*, 360–367. DOI: 10.1016/j.jddst.2015.07.016
31. Kadry, H.; Al-Hilal, T. A.; Keshavarz, A.; Alam, F.; Xu, C.; Joy, A.; et al. Multi-Purposable Filaments of HPMC for 3D Printing of Medications with Tailored Drug Release and Timed-Absorption. *Int. J. Pharm.* **2018,** *544* (1), 285–296.
32. Chai, X.; Chai, H.; Wang, X.; Yang, J.; Li, J.; Zhao, Y.; et al. Fused Deposition Modeling (FDM) 3D Printed Tablets for Intragastric Floating Delivery of Domperidone. *Sci. Rep.* **2017,** *7* (1), 2829.
33. Salimi, S.; Wu, Y.; Barreiros, M.; Natfji, A.; Khaled, S.; Wildman, R.; et al. A 3D Printed Drug Delivery Implant Formed from a Dynamic Supramolecular Polyurethane Formulation. *Polym. Chem.* **2020,** *11* (20), 3453–3464.
34. Domínguez-Robles, J.; Mancinelli, C.; Mancuso, E.; García-Romero, I.; Gilmore, B. F.; Casettari, L.; et al. 3D Printing of Drug-Loaded Thermoplastic Polyurethane Meshes: A Potential Material for Soft Tissue Reinforcement in Vaginal Surgery. *Pharmaceutics* **2020,** *12* (1), 63.
35. Stewart, S.; Domínguez-Robles, J.; McIlorum, V.; Gonzalez, Z.; Utomo, E.; Mancuso, E.; et al. Poly(caprolactone)-Based Coatings on 3D-Printed Biodegradable Implants: A Novel Strategy to Prolong Delivery of Hydrophilic Drugs. *Mol. Pharm.* **2020,** *17* (9), 3487–3500.
36. Holländer, J.; Genina, N.; Jukarainen, H.; Khajeheian, M.; Rosling, A.; Mäkilä, E.; et al. Three-Dimensional Printed PCL-Based Implantable Prototypes of Medical Devices for Controlled Drug Delivery. *J. Pharm. Sci.* **2016,** *105* (9), 2665–2676.
37. Wasilewska, K.; Winnicka, K. Ethylcellulose-A Pharmaceutical Excipient with Multidirectional Application in Drug Dosage Forms Development. *Materials (Basel, Switzerland).* **2019,** *12* (20), 3386. DOI: 10.3390/ma12203386
38. Yang, Y.; Wang, H.; Li, H.; Ou, Z.; Yang, G. 3D Printed Tablets with Internal Scaffold Structure Using Ethyl Cellulose to Achieve Sustained Ibuprofen Release. *Eur. J. Pharm. Sci.* **2018,** *115*, 11–18. DOI: 10.1016/j.ejps.2018.01.005

39. She, Y.; Fan, Z.; Wang, L.; Li, Y.; Sun, W.; Tang, H.; et al. 3D Printed Biomimetic PCL Scaffold as Framework Interspersed with Collagen for Long Segment Tracheal Replacement. *Front. Cell Dev. Biol.* **2021,** *9*. DOI: 10.3389/fcell.2021.629796
40. The Overall 3D Printing Market Is Expected to Grow from USD 9.9 Billion in 2018 to USD 34.8 Billion by 2024 at a CAGR of 23.25%. 2021. https://www.prnewswire.com/news-releases/the-overall-3d-printing-market-is-expected-to-grow-from-usd-9-9-billion-in-2018-to-usd-34-8-billion-by-2024-at-a-cagr-of-23-25-300806896.html#:~:text=2018%20to%202024-,The%20overall%203D%20printing%20 (accessed Jan 19, 2021).

Index

D

E

S

T

U

9781774913673